Nanoscience and Nanoengineering: Advances and Applications

Nanoscience and Nanoengineering: Advances and Applications

Editor: Mindy Adams

NY RESEARCH
P R E S S

New York

Published by NY Research Press
118-35 Queens Blvd., Suite 400,
Forest Hills, NY 11375, USA
www.nyresearchpress.com

Nanoscience and Nanoengineering: Advances and Applications
Edited by Mindy Adams

Cataloging-in-Publication Data

Nanoscience and nanoengineering : advances and applications / edited by Mindy Adams.
     p. cm.
Includes bibliographical references and index.
ISBN 978-1-63238-557-4
1. Nanoscience. 2. Nanotechnology. 3. Nanostructured materials. 4. Nanoparticles.
I. Adams, Mindy.
T174.7 .N36 2017
620.5--dc23

Printed in the United States of America.

Contents

Preface

Path-breaking studies in the field of nanoscience and nanoengineering are presented in this book. Nanoengineering and nanoscience are related fields which focus on the engineering methods practiced on a nanoscopic scale. Development in this field benefits various industries like automobile, chemicals, energy, etc. Analyzing material at nanoscale levels can give new insights into the formation and behavior of the materials that immensely benefits material scientists. This book traces the advances in this field. It also discusses the applications that help the reader to develop a comprehensive understanding of this field. A number of latest researches have been included to keep the readers up-to-date with the global concepts in this discipline.

The information shared in this book is based on empirical researches made by veterans in this field of study. The elaborative information provided in this book will help the readers further their scope of knowledge leading to advancements in this field.

Finally, I would like to thank my fellow researchers who gave constructive feedback and my family members who supported me at every step of my research.

Editor

1

A comparison between Raman scattering from GaN nanowires and polyhedrons

D. Sathish Chander[a,b], J. Ramkumar[b] and S. Dhamodaran[a]*

[a]Department of Physics, Indian Institute of Technology Kanpur, Kanpur 208016, India; [b]Department of Mechanical Engineering, Indian Institute of Technology Kanpur, Kanpur 208016, India

GaN polyhedrons and nanowires have been synthesised on various substrates using chemical vapour deposition with gold catalysis. We observed hexagonal polyhedrons on silicon and pyramid-like nanostructures on GaN and AlN substrates with a size of about $2\,\mu$ m in all the cases. The nanostructures were subjected to field emission scanning electron microscopy, X-ray diffraction (XRD) and Raman spectroscopy studies for analysing the morphological and structural properties. The XRD clearly indicates the hexagonal wurtzite structure with intense (0 0 2) peak on silicon substrate. However, the pyramid-like structures on GaN and AlN substrates show intense (1 0 1) diffraction peaks. Raman scattering shows intense peaks at 565, 730, 530, 555 cm^{-1} assigned to E$_2$(high), A$_1$(LO), A$_1$(TO) and E$_1$(LO) modes, respectively. Additional peaks at 420 cm^{-1} and broad peaks at 650 and 690 cm^{-1} were attributed to acoustical overtone and surface optical phonons, respectively. A small red-shift was observed for the phonon modes with respect to the literature values, which may be due strain. Raman spectra of nanowires show increased red-shift in comparison to the polyhedra, indicating nano size-effects.

Keywords: GaN; polyhedra; nanowires; Raman scattering

1. Introduction

GaN nanostructures have attracted a lot of research interest due to their direct bandgap nature. They are useful for nano-photonic and nano-electronic applications in the UV region [1,2]. The strong polarisation along the c-axis limits certain device applications. The possibility of non-polar GaN along m- and a-axes has driven a lot of research on the utilisation of the anisotropic properties of GaN for various applications [3,4]. Most of the reports indicate that the chemical vapour deposition of nanowires is governed by the celebrated vapour–liquid–solid process for which a catalytic particle is necessity [5,6]. The laterally overgrown structures reported in literature are chiefly with the use of masks [7]. But it is also known that the structures grown with masks have properties depending on the mask itself. Major problems due to the mask are the impurity incorporation and distribution of strain [8,9]. Here, we report the synthesis of controlled morphologies such as laterally over grown polyhedral and pyramid-like structures on different substrates which are self-assembled without any mask. From the morphological analysis using scanning electron microscopy (SEM), the polarity and orientation of the nanostructure are investigated, as they can be related as shown in ref. [10]. As mentioned above, the shape/morphology is of great relevance for understanding the directional dependent properties useful for device applications. We correlate the morphology and the polarity

*Corresponding author. Email: kdams2003@gmail.com

investigations with the X-ray diffraction (XRD) and Raman scattering studies. XRD studies show almost (0 0 2) orientation of the polyhedrons and the pyramid-like structures are almost (1 0 1) oriented, supporting the SEM observations. Raman studies clearly indicate the scattering from polar and non-polar surfaces of the nanostructures as we observed all possible Raman modes of GaN. A detailed peak fit analysis shows that there is residual strain on structures grown in silicon and AlN substrates, which may be due to lattice and thermal mismatch. Surface effects were also noticed by the appearance of surface optical phonon (SOP) modes.

2. Experimental

GaN nanostructures were grown on silicon, GaN and AlN substrates with 2 nm thermally evaporated gold as catalysis. The substrates are (0 0 1) oriented silicon of about 350 μm in thickness, Ga-face terminated epitaxial p-GaN film on *c*-plane sapphire about 5 μm in thickness and Al-face terminated epitaxial AlN film on *c*-plane sapphire about 1 μm in thickness. Highpurity Ga-metal and ammonia (NH_3) gas were used as source and the growth temperature was 1323 K, while the ammonia flow rate was fixed to be 25 standard cubic centimetres per minute (SCCM). Before the start of the growth process, the flow rate of 500 SCCM was maintained for an hour continuously to remove residual oxygen. The ammonia flow rate is then slowly reduced to desired rate and the temperature was increased to the growth temperature at a rate of 30 K/min. The growth was carried out for 3 h and then the furnace allowed to cool to room temperature while maintaining the ammonia flow. The nanostructures were characterised by field emission scanning electron microscopy, energy-dispersive x-ray analysis, XRD and Raman spectroscopy. All SEM images and the EDAX were recorded using 10 keV electron beam while the X-ray used a Cu-Kα of wavelength (1.5Å). The Raman spectra were excited using 488 nm of Ar-ion laser in the backscattering geometry. The incident power was 5 mW through a 40*x* objective and the collection time of 300 s. Prior to the measurements, the monochromator was calibrated using 520 and 1332 cm^{-1} peaks of Si and diamond, respectively.

3. Results and discussion

Figure 1(a) and (b) represents the GaN nanostructures grown with silicon substrate and the respective EDAX spectra. The polyhedrons have clear hexagonal facets with size ranging to about 2 μm and the composition of gallium to nitrogen ratio almost 1. Figure 1(c) and (d) represents the GaN nanostructures grown on GaN and AlN substrates, respectively. The pyramid-like structures are clearly seen without the bottom facet compared to the polyhedrons grown on silicon. The important difference observed with the polyhedrons grown on silicon and the nitride substrates is the polarity. The shape of the nanostructures grown on silicon is hexagonal polyhedrons with flat surface at the top as observed in Figure 1(a). The facets identified as *m*-plane {10−10} on the sides, *r*-plane {10−11} at the angled top surface and *c*-plane {0 0 0 1} as the flat top. It is to be noted that the *m*-plane is non-polar and the *r*- and *c*-planes are polar. Typically, the *c*-plane with Ga-face is observable only at high temperatures as reported during epitaxial overgrowth of GaN by metal organic chemical vapour deposition and the size of the facet would indicate the approximate diffusion length of gallium [11]. On nitride substrates, pyramid-like structures with sharp tip or stacked pyramids at the tip were observed, see Figure 1(c) and (d). The *m*-plane was not observed but *r*-plane was observed corresponding to N-face and consistent with earlier report on bulk GaN crystallite polarity [12]. This

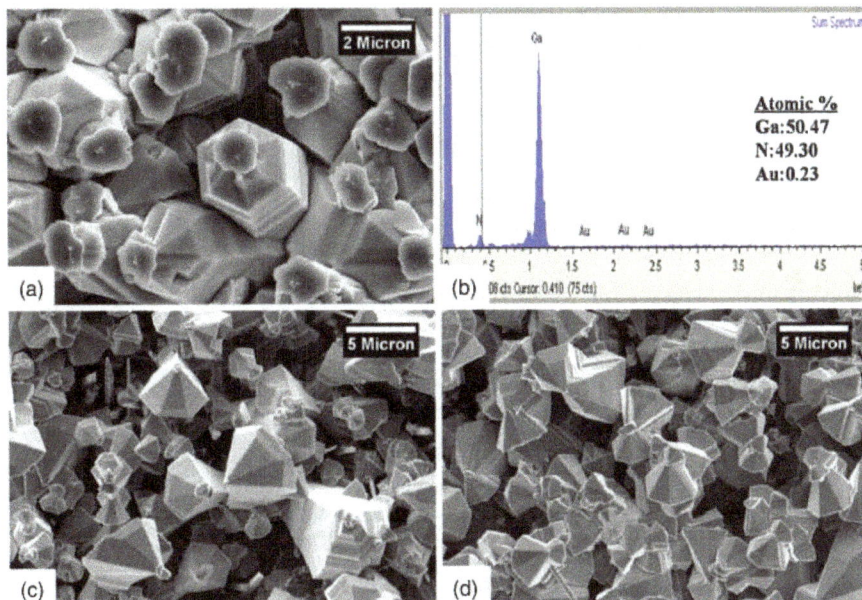

Figure 1.   (a) SEM images of polyhedra grown on silicon substrate and (b) the corresponding EDAX result, (c) and (d) polyhedra grown on GaN and AlN substrates, respectively.

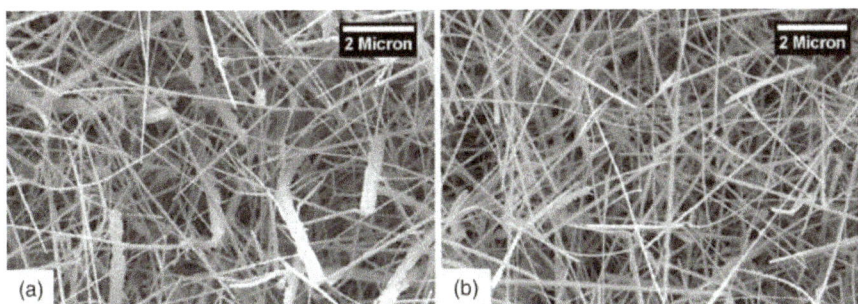

Figure 2.   (a) SEM images of nanowires grown on GaN and (b) AlN substrates both at 1273K and 25 SCCM of ammonia flow rate.

indicates that the nanostructures grown on silicon substrate do have a non-polar surface and the structures grown on nitride substrate lack a non-polar surface. Figure 2(a) and (b) shows the SEM images of GaN nanowires grown at 1273 K on GaN and AlN substrates, respectively. The nanowire diameters have a wide distribution between 50 and 300 nm on both the substrates.

The XRD studies show three major peaks between 30 and 40° for nanostructures grown on Si, GaN and AlN substrates (Figure 3). Typically, the peaks beyond 40° were negligible in intensity and hence not shown. The three major peaks were identified as diffractions from (1 0 0), (0 0 2) and (1 0 1) planes, confirming hexagonal wurtzite structure. In the case of the silicon substrate, the (0 0 2) peak is predominant but for the structures grown on nitride substrates, (1 0 1) peak is predominant. This is consistent with the orientation of

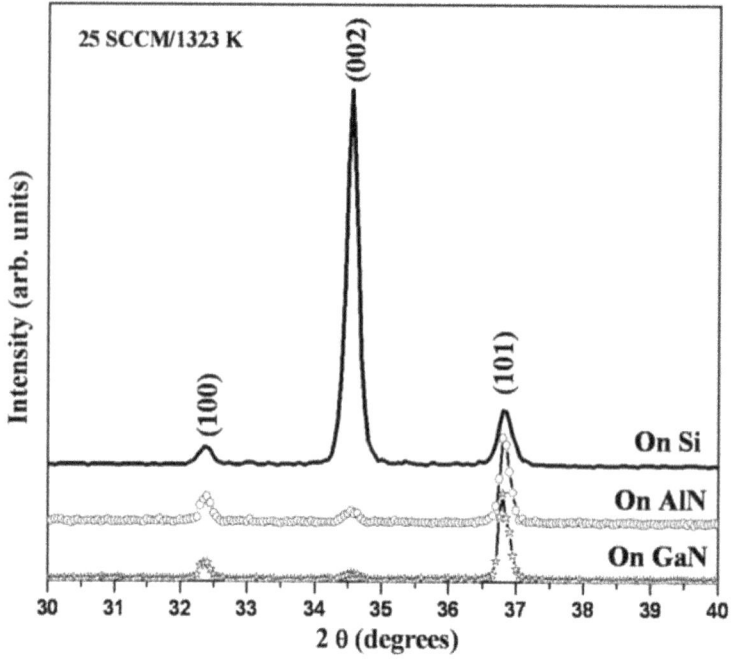

Figure 3.   XRD from the GaN polyhedra grown on different substrates (the corresponding SEM images are shown in Figure 1).

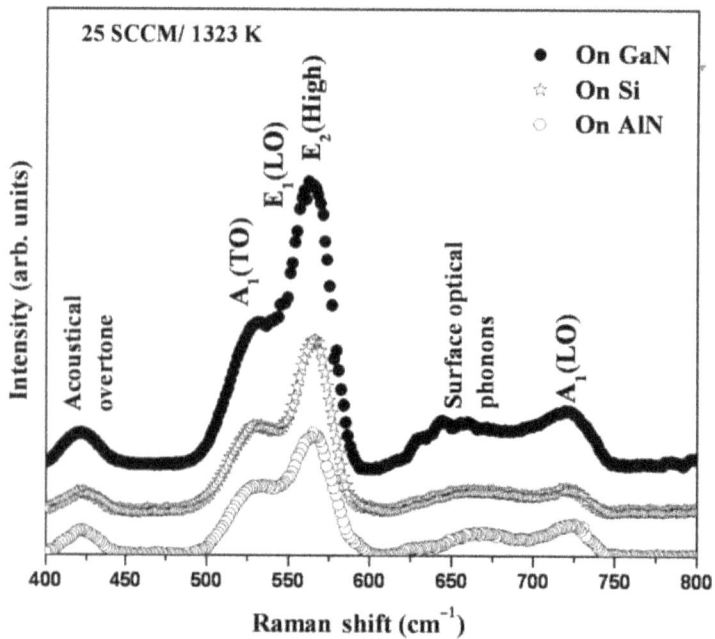

Figure 4.   Raman spectra of GaN nanostructures grown on Si, GaN and AlN substrates at 1323K and 25 SCCM of ammonia flow rate.

Table 1. Summary of the Raman results for the GaN polyhedra.

S. no	Si Peak (cm^{-1})	Si Width (cm^{-1})	GaN Peak (cm^{-1})	GaN Width (cm^{-1})	AlN Peak (cm^{-1})	AlN Width (cm^{-1})	After ref. [15] Assignment	After ref. [15] Frequency (cm^{-1})
1.	420	–	420	–	420	–	Acoustical	420
2.	531.59	39.41	532.39	40.74	525.48	30.44	A_1(TO)	533
3.	–	–	557.16	18.70	554.67	44.80	E_1 (TO)	560
4.	566.21	24.22	569.32	21.21	567.24	23.95	E_2 (high)	569
5.	–	–	631.49	16.83	626.31	14.99	A_1 (overtone)	640
6.	654.65	51.82	655.03	19.24	654.74	35.01	Surface OP[a]	652[a]
7.	688.33	37.47	673.46	32.50	677.37	44.04	Surface OP[a]	692[a]
8.	712.71	24.51	710.08	42.14	713.14	27.64	TPSB[b]	710[b]
9.	724.45	20.14	724.57	24.06	726.34	19.71	A_1 (LO)	735

Note: [a]Ref. [16]; [b]TPSB, two-phonon summation band as explained in the text after ref. [17].

the polyhedrons seen in the SEM (Figure 1). The growth shape, which may depend on several parameters among all the substrates, is very important. The adatom diffusion depends on the substrate and even on different planes on nitride substrates, as reported in the literature [13,14]. Different diffusion rate on different planes and substrates can lead to clear facets and shapes of nanostructures. In this study, it is understood that silicon substrate leads to *c*-axis growth but nitride substrates lead to *m*-axis growth on *c*-axis substrates. This indicates that with silicon substrate, it is almost epitaxial and with nitride substrates, it is not.

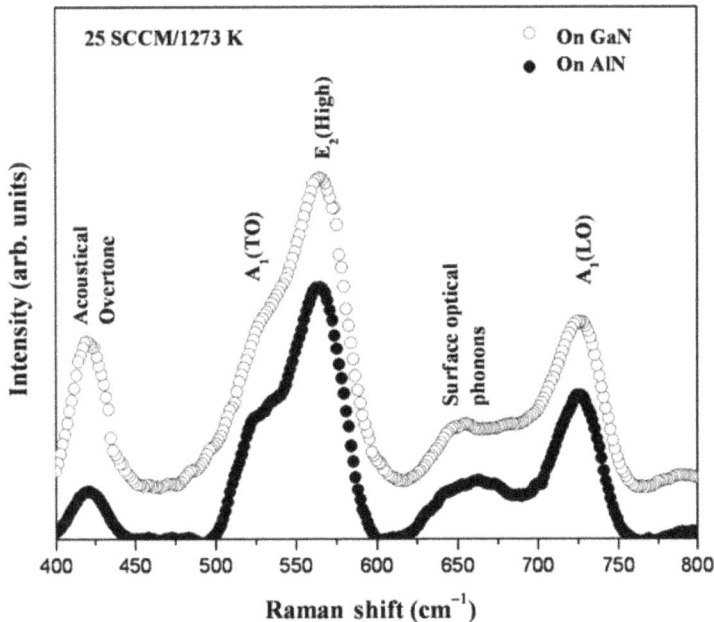

Figure 5. Raman spectra of GaN nanostructures grown on GaN and AlN substrates at 1273K and 25 SCCM of ammonia flow rate.

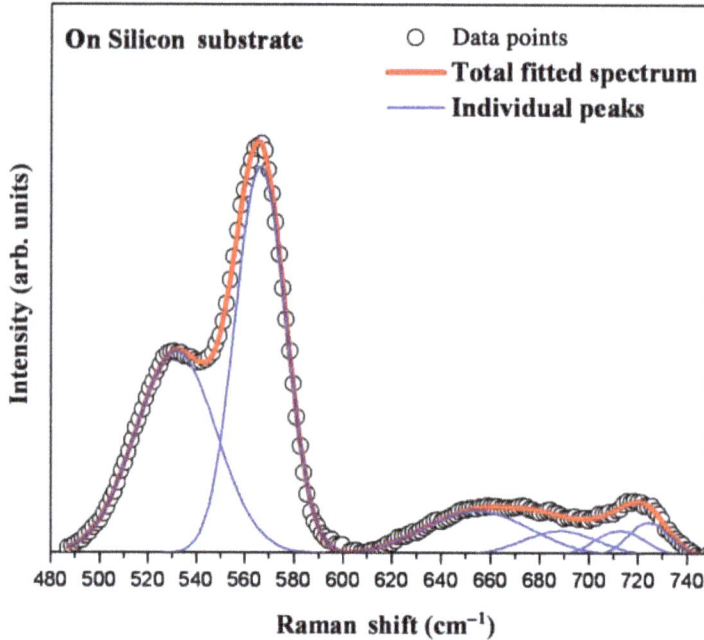

Figure 6.   A detailed Lorentian fit of the Raman spectra grown on silicon substrate showing total fitted spectrum and the individual peaks.

Raman spectra of the three samples are given in Figure 4 for the range between 400 and $800\,cm^{-1}$. All the observed modes are identified and assigned to the respective modes, as listed in Table 1. The E_2(high) was predominant for the nanostructures grown on all the three substrates. It is interesting to note that all the possible modes of GaN are present while recording the spectra in backscattering geometry, indicating scattering from all the exposed planes of GaN. For example, the presence of intense A_1(TO) mode confirms the phonon scattering from all the facets of the polyhedral structures. A slight red-shift with respect to the literature values in most cases indicates that there is residual strain in the nanostructures. Yet, a minimum strain or no strain for the nanostructures grown on GaN substrate is a clear indication that the lattice and thermal mismatch is probably the reason of the residual strain. The Raman spectra of the nanowires grown on GaN and AlN substrates at 1273 K and 25 SCCM of ammonia flow rate are shown in Figure 5. In the case of nanowires also, most of the modes observed for the polyhedra are present. The E_2(high) mode shows increased red-shift (up to $4\,cm^{-1}$) for the nanowires in comparison to the polyhedra. However, no considerable shift was observed for the A_1(LO) modes, but the acoustical phonon mode was found to be more intense for the nanowires grown on GaN substrate. The increased red-shift and broadening are typically attributed to nano size-effects and in our case also the nanowires show such effects in comparison to the polyhedra samples. A detailed peak fit analysis of all the spectra shows that apart from the expected Raman active modes, there are peaks at 420, 650, 690 and $712\,cm^{-1}$. A representative fit analysis is shown in Figure 6 for GaN nanostructures grown on silicon substrate. The peak at $420\,cm^{-1}$ is normally assigned to sapphire substrate; in our case, this particular peak is observed from structures grown on silicon substrate also and hence it could be an acoustical overtone [15]. The modes observed at 650 and $690\,cm^{-1}$ are observed in nanowires as reported recently and attributed to SOP modes [16]. In the present case, we observed these modes; though weak, the peak at $650\,cm^{-1}$ is quite intense. The peak around $712\,cm^{-1}$, according to Bulbul et al., corresponds to a two-phonon summation band which can originate at L-point of the zone with a combination of longitudinal acoustical phonon and transverse optical at 144 and $569 \, \text{cm}^{-1}$ [17].

4. Conclusions

In summary, the shape and polarity of the overgrown polyhedral nanostructures have been demonstrated using different substrates. The structures grown on silicon substrate shows a non-polar surface; however, on nitride substrates, all the surfaces were polar. The size of the structures was typically $2 \, \mu$ in all the cases with high density. The SEM micrographs also corroborate the XRD results on the morphology to structural and polarity correlation. The Raman results show a slight residual strain due to lattice and thermal mismatch. The intense $A_1(\text{TO})$ mode is observed and evidenced for scattering from different facets of the polyhedral structures. The observation of SOP modes is a proof of surface effects from these nanostructures. The results were compared with the results of nanowires grown at slightly lower temperatures. The Raman results indicate size-effects in nanowires in comparison to the polyhedra.

Acknowledgements

The authors are grateful to Department of Science and Technology, India and Indian Institute of Technology, Kanpur for funding.

References

[1] S. Nakamura, M. Shenoh, S. Nagahama, N. Iwasa, T. Yamada, T. Matsushita, Y. Sugimoto, and H. Kiyoku, *Room-temperature continuous-wave operation of InGaN multi-quantum-well structure laser diodes*, Appl. Phys. Lett. 69 (1996), pp. 4056–4058.

[2] O.B. Shchekin, J.E. Epler, T.A. Trottier, T. Margalith, D.A. Steigerwald, M.O. Holcomb, P.S. Martin, and M.R. Krames, *High performance thin-film flip-chip InGaN–GaN light-emitting diodes*, Appl. Phys. Lett. 89 (2006), pp. 71109-1–71109-3.

[3] Y.L. Wang, F. Ren, U. Zhang, Q. Sun, C.D. Yerino, T.S. Ko, Y.S. Cho, I.H. Lee, J. Han, and S.J. Pearton, *Improved hydrogen detection sensitivity in N-polar GaN Schottky diodes*, Appl. Phys. Lett. 94 (2009), pp. 212108-1–212108-3.

[4] R. Sharma, P.M. Pattison, R.M. Farrell, T.J. Baker, B.A. Haskel, F. Wu, S.P. DenBaars, J.S. Speck, and S. Nakamura, *Demonstration of a semipolar (10-1-3) InGaN/GaN green light emitting diode*, Appl. Phys. Lett. 87 (2005), pp. 231110-1–231110-3.

[5] C.C. Chen, C.C. Yeh, C.H. Chen, M.Y. Yu, H.L. Liu, J.J. Wu, K.H. Chen, L.C. Chen, J.Y. Peng, and Y.F. Chen, *Catalytic growth and characterization of gallium nitride nanowires*, J. Am. Chem. Soc. 123 (2001), pp. 2791–2798.

[6] N. Wang, Y. Cia, and R.Q. Zang, *Growth of nanowires*, Mat. Sci. Eng. R 60 (2008), pp. 1–51.

[7] B. Beaumont, P. Gibart, M. Vaille, S. Haffouz, G. Nataf, and A. Bouille, *Lateral overgrowth of GaN on patterned GaN/sapphire substrate via selective metal organic vapour phase epitaxy: A route to produce self supported GaN substrates*, J. Cryst. Growth 189–190 (1998), pp. 97–102.

[8] K.C. Zeng, J.Y. Lin, and H.X. Jiang, *Optical properties of GaN pyramids*, Appl. Phys. Lett. 74 (1999), pp. 1227–1229.

[9] K. Hiramatsu, H. Matsushima, T. Shibata, Y. Kawagachi, and N. Sawaki, *Selective area growth and epitaxial lateral overgrowth of GaN by metalorganic vapor phase epitaxy and hydride vapor phase epitaxy*, Mat. Sci. Eng. B 59 (1999), pp. 104–111.

[10] T. Kuykendall, P.J. Pauzauskie, Y. Zhang, J. Goldberger, D. Sirbuly, J. Denlinger, and P. Yang, *Crystallographic alignment of high-density gallium nitride nanowire arrays*, Nat. Mater. 3 (2004), pp. 524–528.

[11] S. Kitamura, K. Hiramatsu, and N. Sawaki, *Fabrication of GaN Hexagonal pyramids on dot-patterned GaN-sapphire substrates via selective MOVPE*, Jpn J. Appl. Phys. 34 (1995), pp. L1184–L1186.

[12] M. Aoki, H. Yamane, M. Shimada, T. Kajiwara, S. Sarayama, and F.J. DiSalvo, *Morphology and polarity of GaN single crystals synthesized by the Na flux method*, Cryst. Growth Des. 2 (2002), pp. 55–58.

[13] L. Lymperakis and J. Neugebauer, *Large anisotropic adatom kinetics on nonpolar GaN surfaces: Consequences for surface morphologies and nanowire growth*, Phys. Rev. B. 79 (2009), pp. 241308-1–241308-4.

[14] C.Y. Nam, D. Tham, and J.E. Fischer, *Effect of the polar surface on GaN nanostructure morphology and growth orientation*, Appl. Phys. Lett. 85 (2004), pp. 5676–5679.

[15] H. Siegle, G. Kaczmarczyk, L. Filippidis, A.P. Litvinchuk, A. Hoffmann, and C. Thomsen, *Zone-boundary phonons in hexagonal and cubic GaN*, Phys. Rev. B 55 (1997), pp. 7000–7004.

[16] P. Sahoo, S. Dhara, S. Dash, A.K. Tyagi, B. Raj, C.R. Das, P. Chandramohan, and M.P. Srinivasan, *Surface optical phonon mode in GaN nanowire*, Int. J. Nanotech. 7 (2010), pp. 823–832.

[17] M.M. Bulbul, S.R.P. Smith, B. Obradovic, T.S. Cheng, and C.T. Foxon, *Raman spectroscopy of optical phonons as a probe of GaN epitaxial layer structural quality*, Eur. Phys. J. B 14 (2000), pp. 423–429.

2

Analysis of mixtures of C_{60} and C_{70} by Raman spectrometry

J. Brett Kimbrell[a], Chritopher M. Crittenden[b], Walter J. Steward[a], Farooq A. Khan[a], Anne C. Gaquere-Parker[a] and Douglas A. Stuart[a]*

[a]*Department of Chemistry, University of West Georgia, Carrollton, GA 30118, USA;* [b]*Department of Chemistry, University of Texas at Austin, Austin, TX 78712, USA*

A non-destructive method to determine the percent composition and relative purity of bulk solid phase mixed fullerene samples using Raman spectroscopy is described. This ratiometric method has a significant advantage in ease of use over other methods, in that the samples are analyzed in a short time, as solids in suspension, with minimal sample pre-preparation, using easy to operate instrumentation. The ratio of unique vibrational bands vs. percent composition is used to construct a calibration curve, and a detection limit estimated in the presence of chemical noise.

Keywords: Raman spectroscopy; quantitative analysis; fullerenes; carbon nanomaterials; ratiometric calibration

1. Introduction

The discovery of C_{60}, that resulted in the 1996 Nobel prize in Chemistry for Smalley, Kroto and Curl, was first made in 1985 in a time-of-flight mass spectrometer.[1] Subsequent mass production of fullerenes by Huffman, Kratschmer and co-workers [2] made possible what is currently a vast and exciting area of research, as evidenced by a large number of publications devoted to fullerenes. The reactivity of fullerenes is an area of immense interest for synthetic organic chemists.[3] In addition, fullerenes find a wide variety of applications that include material science [4] and pharmacology.[5]

The ability to determine the purity of materials is of perhaps greater critical importance to the burgeoning field of nanotechnology than conventional bulk scale chemistry. The mass production of fullerenes by the heating graphite under helium results in a mixture of C_{60}, C_{70} and large fullerenes, the precise relative concentrations of which are largely dependent on experimental conditions.[6] The product mixture contains soot as well as fullerenes, and is typically analyzed via separation by high-performance liquid chromatography, followed by mass spectrometry.[7] It is highly desirable to analyze the composition of such product mixtures *in situ*, or with minimal sample preparation. Such methods are few and far between, an example [8–12] being near-IR spectrophotometric determination of the composition of a mixture of fullerenes C_{60} and C_{70}.

*Corresponding author. Email: dstuart@westga.edu

Raman spectroscopy provides a method whereby one can rapidly determine the composition of a mixed fullerene sample.

We report a simple method for analyzing a mixture of C_{60} and C_{70} via Raman spectrometry. This relies on the distinct Raman spectra of these species, which can be acquired in a facile fashion using an inexpensive, portable Raman spectrometer. The advantage of our method is that it can be applied to a solid sample *in situ*, and would be of considerable interest as an analytical tool to large-scale producers of fullerenes.

2. Experimental

Samples of [5,6]-Fullerene-C_{70} (98%) and [5,6]-Fullerene-C_{60} (Sublimed, 99.9%) were obtained from Sigma-Aldrich (St. Louis, MO) and used without further purification. Ultrapure water was prepared by a reverse osmosis Myron L Company: Series 750 HYDRO water purification system. All samples were massed on an analytical balance (Denver Instrument Company – A-160), placed into capped 1 mL glass vials (VWR International, West Chester, PA), 0.200 mL of pure water added, and mixed by a Vortex Genie-2 (Scientific Industries). Spectra were collected on an Advantage NIR 785 nm Raman Spectrometer (Delta Nu, Laramie, WY). The spectrometer's operating parameters were as follows: laser power, low (10 mW); resolution, low; baseline, on; integration time, 1 second and averages, 16. The samples were manually rotated in the sample holder during spectral acquisition.

For the construction of the standard curve, the initial sample contained the stock Fullerene-C_{60} (99.9%) and 0.200 mL of DI water. Masses of C_{70} were added in increments of $\sim 5\%$ by mass

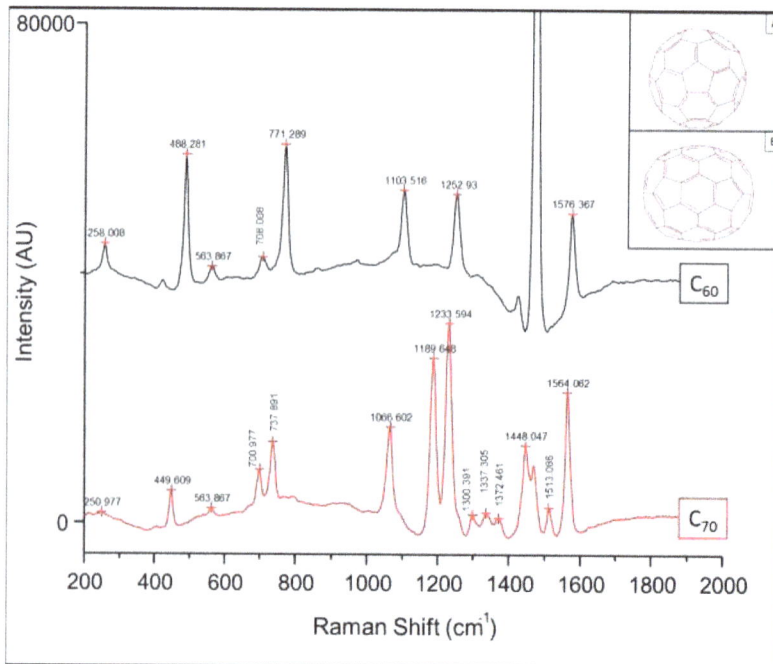

Figure 1. Raman spectra and structure (inset) of C_{60} and C_{70} fullerenes. The Raman spectra of the two fullerenes are expected to be highly similar, given the extensive structural similarities between the two molecules. There exist a sufficiency of unique spectral features specific to the vibrational modes to readily distinguish between C_{60} and C_{70}.

to the C_{60} sample until the mass percent of 50% C_{60}–C_{70} was obtained. Spectra were obtained after each addition. A second vial was prepared by the additions of 0.200 mL DI water and stock Fullerene-C_{70} (98%), but with mass additions of $\sim$5% by mass of C_{60}, until the mass by percent reached 50% C_{60}–C_{70}. Spectra were obtained after each addition.

To test the robustness of the method, a sample of mixed C_{60}–C_{70} was prepared as above, and its spectrum recorded. Masses of graphitic carbon were added to the sample to $\sim$11% and $\sim$20% graphitic carbon, and spectra taken as previously described. To test the robustness of the method in the presence of a constant contaminate, samples with a constant 2% by mass of graphite were prepared, and the relative mass percentage of C_{60} and C_{70} were varied, and spectra acquired as above.

3. Results and discussion

3.1. *Comparison of spectra*

The Raman spectra of C_{60} and C_{70} show many similar features, but are unique in several vibrational bands, as shown in Figure 1. This is expected due to the structural similarly between the two molecules, see in Figure 1, inset. In effect, C_{70} is similar to C_{60}, save for the addition of a band of hexagonal faces about its equator. This is reflected in the Raman spectrum, where the band at 1462 cm^{-1} for C_{60} is absent for C_{70}, and a "new" band at $\sim$1448 cm^{-1} appears, and is likely a shift of the pentagonal "pinch" mode. Several of the peaks in the data are slightly shifted from the solid-state values. This is likely due to interactions between the nanotubes and the aqueous environment, similar to shifts in peaks observed in solvated nanotubes.[9]

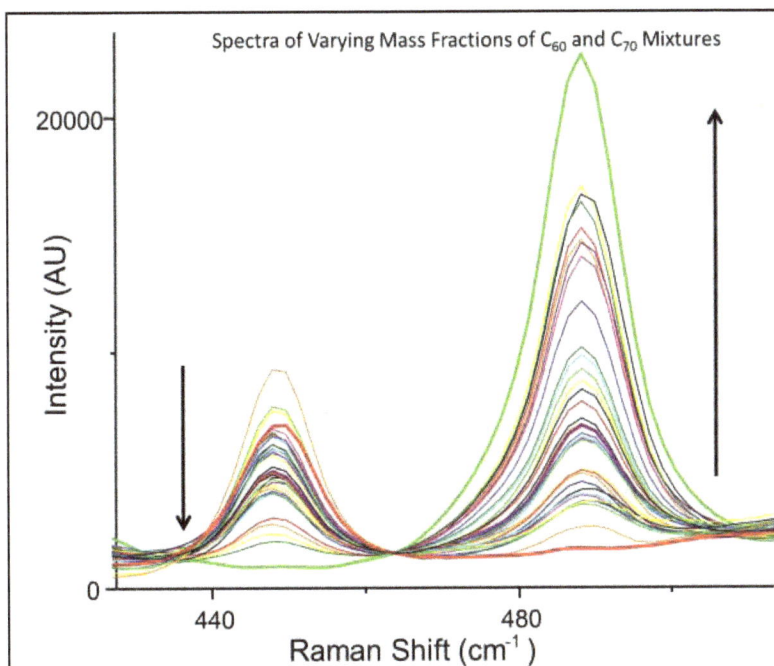

Figure 2.   Raman spectra of varying mass fractions of C_{60} and C_{70} mixtures. Spectra are offset to be centered at 463.7 cm^{-1}. The arrows indicate the direction of the change in the peaks at 447.8 and 488.3 cm^{-1} as the mass fraction of C_{70} increases from 0% to 100%. The 447.8 cm^{-1} band is unique to C_{60}, and decreases as the relative mass fraction of C_{70} increases. The 488.3 cm^{-1} peak is specific to C_{70}, and increases as the mass percent of C_{70} increases.

3.2. Sampling methodology

A common nuisance in Raman spectroscopy is the presence of a broad background continuum, or "hump", often of indeterminate origin. This is partially controlled by the application of baseline subtraction methods, here implemented by the spectrometer control software. This spectral noise can come from fluorescence (e.g. of the glass container, sample matrix or sample itself) or, as more probably the case here, from photothermal emission due to excitation from the laser. Indeed, broad band photon emission originating at laser focal point is visible to the naked eye from dry fullerene samples when acquiring Raman spectra from this instrument. Significant gains in the signal-to-noise ratio were realized in this experiment by the addition of a small volume of water to the sample vial. It is likely that the high specific heat of water provides an effective heat sink, lowering the effective temperature below which photothermal emission occurs. Modest improvements in signal quality were also achieved by manually rotating the sample vial in the holder during the course of spectral acquisition. Similar methods have been used in Raman spectroscopy, but often involved mechanical spinners and high spin rates.

3.3. Construction of the calibration curve

Raman spectroscopy is an incredibly useful analytical tool because it is able to provide both quantitative and qualitative information about a sample. Raman spectroscopy is a vibrational technique, in many way complementary to IR absorbance, but with several distinct advantages. Raman spectroscopy is a scattering technique, and can readily be used on opaque and solid samples, unlike IR absorbance spectroscopy. The exact intensity of Raman scattering from a particular vibrational mode is a complex matter, but, generally, will increase as the number of

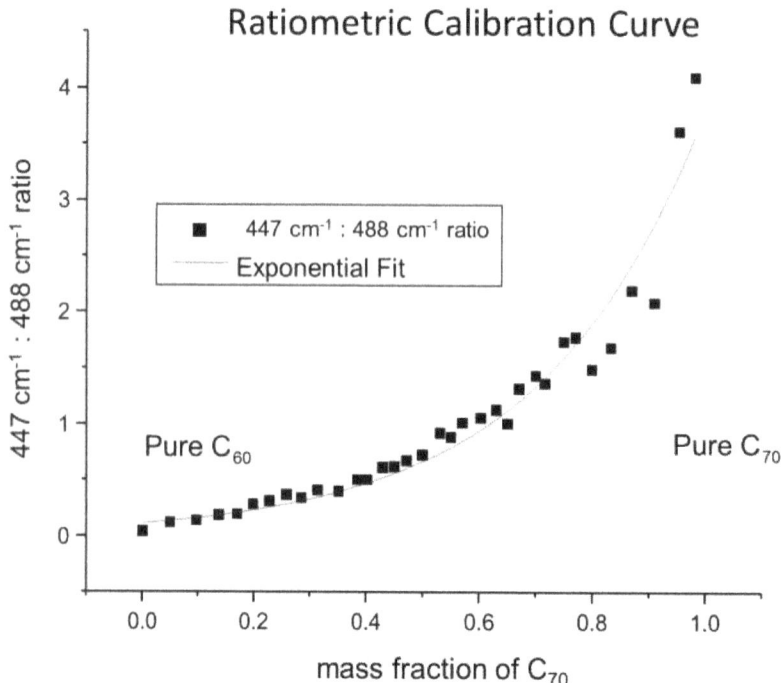

Figure 3.   Ratiometric calibration curve. The ratio of the relative (to intensity 463.7 cm^{-1}) peak intensity between the 447.8 and 488.3 cm^{-1} is plotted as a function of mass fraction of C_{70}. An exponential function was fit to the data with an equation $y = 0.1265e^{3.3989x}$, with an R^2 value of 0.98.

scatters increases. The vibrational modes for every molecule are characteristic of that molecule, and can act as spectroscopic fingerprint for identification. Although all fullerenes are structurally similar, there are several Raman bands for C_{60} and C_{70} that do not significantly overlap and are thus free of spectral congestion. For purposes of convenient illustration, two peaks that lie close together were chosen for analysis, although any spectroscopically distinct peaks could be used for construction of the calibration curve. Figure 2 shows the peak at 488.3 cm^{-1} characteristic of C_{70} increasing as the mass fraction of C_{70} is increased. Likewise, the peak at 447.8 cm^{-1}, unique to C_{60}, decreases as the mass fraction of C_{70} increases. The curves were offset so that the valley (463.7 cm^{-1}) between the peaks at 447.8 and 488.3 cm^{-1} was the same for all spectra. A calibration curve can thus be created by plotting the ratio of selected scattering peaks unique to C_{60} and C_{70} vs. mass percent or other measures of concentration. Figure 3 shows that a nonlinear curve is generated. However, an exponential curve was readily fit to the data using a graphical analysis software package (Origin 6.1, OriginLab Corp., Northampton, MA). Analysis of the spectroscopic data results in a best-fit line with the equation $y = 0.1265e^{3.3989x}$ and a very reasonable fit ($R^2 = 0.98$).

This ratiometric method is preferred because Raman spectra will frequently be superimposed on a broad continuum baseline. The spectrometer's software will correct for this broad baseline "hump", however the exact function used to subtract the baseline is not always immediately accessible. Therefore, one expects variation in the absolute peak intensities even between sample runs, however, the *ratio* of one peak to another should depend systemically only on the relative number of each sort of scatterers in the laser's focal spot. These ratios were stable ($< 1\%$ variation) between spectra taken up to weeks apart.

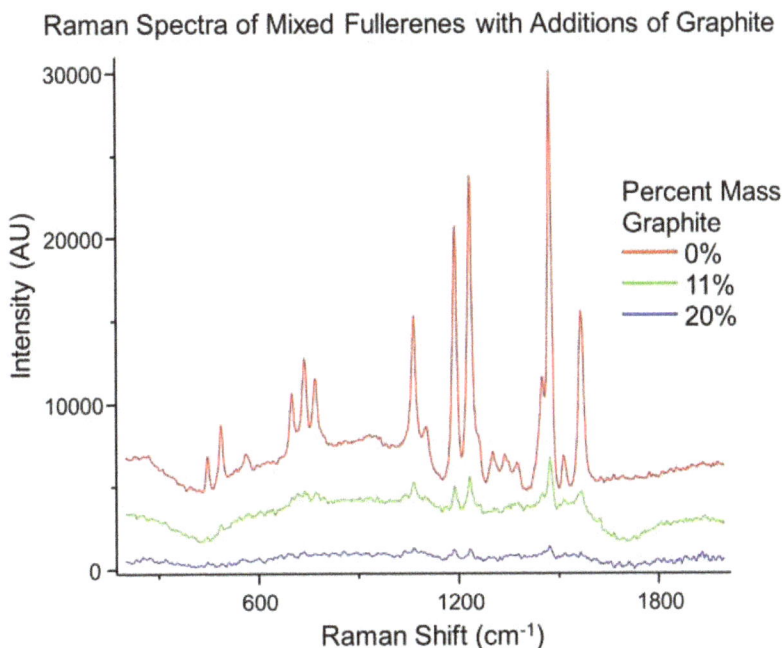

Figure 4.   Spectra of fullerene mixture in the presence of graphitic carbon. Spectra were collected of a neat fullerene mixture, and after the addition of graphite to the 11% and 20% mass to simulate a carbon contaminate. The 3σ limit of detection is reached when the sample contains 20% by mass of the graphitic carbon. The peak ratios for the neat sample (0.780) agree favorably with that of the 11% by mass contaminant sample (0.730), indicating a mass fraction of 53% and 51% C_{70}, respectively.

Various Fullerene Ratios with Fixed Mass Percent Graphite

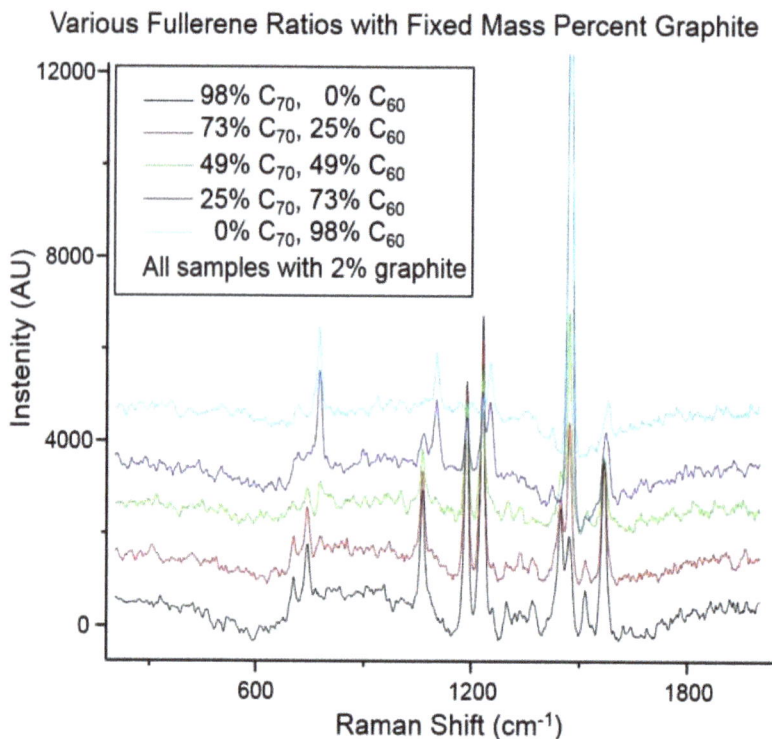

Figure 5. Spectra of various fullerene ratios in the presence of a fixed percent mass of graphitic carbon. Mass ratios of 98%, 73%, 49%, 25% and 0% C_{70} in the presence of 2% by mass graphite and the 0%, 25%, 49%, 73% and 98% C_{60} were prepared to show how the peak ratios vary with mass percent of fullerenes, even in the constant presence of a contaminant. Spectra have been offset for clarity. Peaks associated with C_{70} diminish and those unique to C_{60} increase as the mass ratios change from 98% C_{70} to 98% C_{60}.

3.4. *Application in the presence of a contaminant*

Spectra were obtained of a mixture of fullerenes in the presence of graphitic carbon to test the robustness of this method, and to estimate a limit of detection. Figure 4 demonstrates very useable spectra at ∼ 11% mass of contaminant. The limit of detection is reached at ∼ 20% mass contaminant, estimated as 3σ signal-to-noise ratio. Using the calibration curve generated above, the neat mixture is 53% C_{70}. The application of the calibration curve to the spectra of the mixture with 11% mass of contaminant shows that a 51% C_{70}, an error of 2% C_{70}, in good agreement with the ratios determined for the neat sample. Similarly, spectra were obtained of varying mass ratios of fullerenes in the presence of a constant mass percent of the contaminant, as shown in Figure 5. Here, a steady 2% by mass of graphite was used with decreasing mass percentages (∼ 100%, 75%, 50%, 25% and 0%) of C_{70} with a corresponding increase in the mass percent of C_{60}. Peaks associated with C_{70} diminish and those unique to C_{60} increase as the mass ratios change from 98% C_{70} to 98% C_{60}.

4. Conclusion

In conclusion, we have demonstrated a simple method for the determination of the relative composition of fullerene mixtures using Raman spectroscopic analysis. We anticipate that this method can be extended to other fullerene systems, for example, higher order fullerenes, multicomponent systems and nanotubes.

Acknowledgements

This work was supported by the National Science Foundation [TUES 1043847]; the University System of Georgia Board of Regents [Stem 2 Initiative] and the University of West Georgia, Department of Chemistry.

References

[1] Kroto HW, Heath JR, O'Brien SC, Curl RF, Smalley RE. C_{60}: buckminsterfullerene. Nature. 1985;318:162–163.

[2] Krätschmer W, Lamb LD, Fostiropoulos K, Huffman DR. Solid C_{60}: a new form of carbon. Nature. 1990;347:354–358.

[3] Nierengarten J, Holler M, Deschenaux R. Fullerene containing dendrimers: synthesis and properties. RSC Nanosci Nanotechnol. 2012;20:162–191.

[4] Stara IG, Stary I. Synthesis of extended polyaromatic hydrocarbons: graphite, fullerene, and carbon nanotube substructures. Sci Synth. 2010;45b:1115–1146.

[5] Martin N, Solladie N, Nierengarten J. Advances in molecular and supramolecular fullerene chemistry. Electrochem Soc Interface. 2006;15:29–33.

[6] Tzirakis MD, Orfanopoulos M. Radical reactions of fullerenes: from synthetic organic chemistry to materials science and biology. Chem Rev. 2013;113:5262–5321.

[7] Lin CM, Lu TY. C_{60} fullerene derivatized nanoparticles and their application to therapeutics. Recent Pat Nanotechnol. 2012;6:105–113.

[8] Montañez MI, Ruiz-Sanchez AJ, Perez-Inestrosa E. A perspective of nanotechnology in hypersensitivity reactions including drug allergy. Curr Opin Allergy Clin Immunol. 2010;10:297–202.

[9] Talyzin A, Jansson U. C_{60} and C_{70} solvates studied by Raman spectroscopy. J Phys Chem B. 2000;104:5064–5071.

[10] Aoyama S, Mieno T. Effects of gravity and magnetic field in production of C_{60} by a DC arc discharge. Jpn J Appl Phys Part 2. 1999;38:L267–L269.

[11] Jinno K, Sato Y, Nagashima H, Itoh K. Separation and identification of higher fullerenes by high-performance liquid chromatography coupled with electrospray ionization mass spectrometry. J Microcolumn Sep. 1998;10:79–88.

[12] Tran CD, Grishko VI, Challa S. Near-infrared spectrophotometric determination of compositions of fullerene samples. Spectrochim Acta A Mol Biomol Spectrosc. 2005;62:38–41.

3

Effect of coating and plasma treatments on the induced coupled plasma-reactive ionic etching of boron-doped diamond for microelectromechanical systems (MEMS) applications

Thanh Son Le[a]*, Marc Cretin[b], Patrice Huguet[b], Philippe Sistat[b] and Frederic Pichot[c]

[a]*Institute of Environmental Technology, Vietnam Academy of Science and Technology, 18 Hoang Quoc Viet Road, Cau Giay District, Hanoi, Vietnam;* [b]*Institut Européen des Membranes, ENSCM-CNRS-Université Montpellier 2, CC047, Place Eugène Bataillon, 34095 Montpellier, France;* [c]*Centrale de Technologies en Micro et Nanoélectronique, Université Montpellier 2, Place Eugène Bataillon, 34095 Montpellier, France*

Induced coupled plasma etching was performed on poly-crystalline boron-doped diamond (BDD) films synthesized by microwave assisted chemical vapor deposition (MWCVD). Effects of the input power and of the gas composition and pressure on etching performances have been investigated. A basic aluminium layer and a new hybrid material have been studied as protecting film for the patterning. Best conditions to get high etch rates were a gas ratio Ar/O_2 of 27% (vol.) under 10 mTorr at 200 W (input power). Aluminium (Al) masking induced un-intentional whiskers formation which was removable using CHF_3 as a reactive gas with the drawback to lower the process selectivity. In order to get both a high etching rate and a high selectivity, a hybrid material based on an organic/mineral resin was then used as the masking material. This performing photoresist film allowed etching a 3.6-μm-thick BDD film at 200 nm min^{-1}.

Keywords: BDD film; Al coating; hybrid photoresist coating; whiskers density; diamond etching

1. Introduction

Microfabrication technologies have drawn considerable interest in the field of fabrication of electrodes for electrochemical sensors especially because of the facility of low-cost mass production with a well-defined patterning. In the aim to build a micro total analysis system devoted to the detection of heavy metals ions, all components (i.e. detection and microfluidic) have to be integrated on a lab-on-a-chip device. Very sensitive electrochemical microsensors with mercury (Hg) [1–3] or bismuth [4] working electrodes have been developed for the detection of heavy metals with very low detection limit. However, the toxicity of Hg and the stability of bismuth are needed to develop new materials for electrochemical sensing. A boron-doped diamond (BDD) has been recently proposed for this purpose because of its great electrochemical properties toward especially heavy metals reduction.[5] In this way, Madore et al. [6] have synthesized BDD films by microwave plasma-assisted chemical vapor deposition (CVD) and patterned the material by photolithography: an Si_3N_4 layer was then deposited and etched to produce a hexagonal array of 5 μm diameter BDD microdisks spaced to 100 μm. Tsunozaki et al. [7] have used a photoresist film to prepare BDD tips. The first stage was an isotropic etching of a Si surface followed by a

---

*Corresponding author. Email: thanhson96.le@gmail.com

deposition of BDD, a spin-coating and hence a mechanical etching of a polyimide film to prepare the tips array (tips size 25–30 μm in diameter). The as-prepared diamond microelectrodes array showed a great interest in electroanalysis. More recently Pagels et al. [8] have prepared an original microsystem constituted of a hexagonal BDD microdisks array on an insulated diamond matrix. The main property being to obtain a coplanar system instead of an array made of BDD wells or tips.

Despite the numerous works devoted to BDD for sensing applications, from the best of our knowledge, BDD arrays devoted to electrochemical measurements have never been directly integrated on a unique substrate constituting the working electrode, the counter electrode and the reference electrode. For this purpose, an efficient method of BDD etching has to be developed to get a well-patterned film.

For a deep and clean removal and patterning of a diamond-like material not only a sufficient etching rate but also an adequate selectivity of the applied process relative to the masking layer is required. Therefore, different masks were tested here and several etching conditions were studied referring to their applicability to the removal of BDD.

Reactive ionic etching (RIE) comprises two related mechanisms: chemical reactive etching and physical sputtering etching. The former leads to a high etch rate but is isotropic whereas the latter leads to a lower etch rate but is anisotropic. In the gas mixture Ar/O_2, Ar (as the inert gas) induces the physical sputtering etching and oxygen as the reactive gas the chemical ion-enhanced etching. The way to realize high-quality patterning (i.e. high aspect ratio and/or high etching rate) is to control the proportion of these two factors. Main parameters are the Ar/O_2 ratio, the radiofrequency (RF) power, the gas pressure, the presence of a reactive gas as CHF_3 and the choice of the mask.

In the present work, we have studied the effect of the physical and chemical conditions on the etching rate and on the surface aspect of BDD films.

2. Experimental

2.1. BDD film synthesis

The poly-crystalline BBD films were grown by plasma-assisted CVD in a bell-jar reactor (PLASSYS BJS 150) previously described in [9]. An <100> oriented silicon wafer (2″ in diameter and 1 mm thick) was used as a substrate. The wafer was treated in an ultrasonic bath with a diamond grit to promote a surface seeding and then the diamond nucleation. High-purity methane and diborane diluted in hydrogen were used as the gas mixture. The (B/C) ratio in the gas phase and the methane concentration were set at 1000 ppm and 4%, respectively. The total pressure was set at 180 mbar, the input microwave power at 3 kW for 2 h and the substrate temperature at 850°C. These conditions are suitable for high-quality diamond growth with relatively high growth rates. The as-prepared boron diamond film had a thickness of 3.6 μm. The conductivity and the charge carrier density (determined by the four probes method associated with the Hall effect (donner le nom de l'appareil)) were 563 ± 73 S cm^{-1} and 1.5×10^{21} cm^{-3}, respectively. The charge carrier density is superior to the metallic transition (3×10^{20} cm^{-3}) which proved the high doping of the structure.

2.2. Sample preparation for etching experiments

2.2.1. Al mask

About 25 mm^2 BDD samples were first washed with acetone and ultrapure water and dried under nitrogen. A part of the area was then coated by an organic resin (Picéine®). The Al mask was

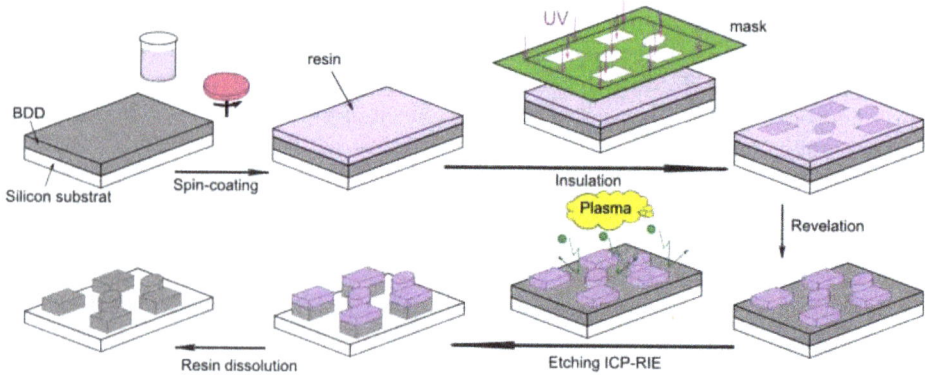

Figure 1.   Description of the etching process of the BDD film based on the use of a hybrid organic/mineral photoresist mask.

deposited on the whole surface by thermal evaporation at 500°C and 10^{-6} mbar using a current of 100 A. The Al deposit rate was 15 Å s^{-1} and the time deposit was 3 min (i.e. mask thickness was 270 nm). The next step was the dissolution of the Picéine® resin by immersion in trichloroethylene for 30 s to get samples constituting of BDD and Al separated by a 270 nm high step (Al layer thickness). Before etching, samples were washed in ultrapure water and acetone (twice). Finally to recover a full BDD surface after etching, the Al mask was dissolved for 10 min in the following acid mixture $H_3PO_4/H_2O/CH_3COOH/HNO_3 = 16/2/1/1$ (vol/vol), washed in ultrapure water and dried under nitrogen.

2.2.2. *Hybrid resin-based mask*

The stages of the deposit of the hybrid mask and the etching are depicted in Figure 1. The first step was the spin-coating (1200 rpm for 30 s, spin-coater DELTA 10BM) of the negative resin (K-CL® type, KLOE France). The resin was a methacrylate/siloxane hybrid material prepared by sol–gel and used as bought. After spin-coating, samples were dried at 80°C for 10 min to get an optimal adhesion on the BDD surface. Samples were then insulated at 365 nm for 90 s and immersed in isobutanol (90 s) for the liftoff. After etching, the mineral mask was removed by dissolution in acetone and methanol.

2.3. *Etching method and condition*

The RIE experiments were performed using an induced coupled plasma (ICP) etcher Corial 200IL (using a 800 W and 2 MHz ICP source) with RF excitation at 13.56 MHz adjustable from 50 to 300 W. The cathode was 22 cm in diameter and was continuously cooled with a helium flow. The pressure in the chamber was maintained by a turbomolecular magnetic pump (ATH 500 Alcatel). Prior to each experiment, the electronic grade gases of either Ar/O_2 or CHF_3/O_2 were introduced into the reaction chamber through a gas box by means of gas controllers. The compositions examined were a ratio Ar/O_2 from 5/40 to 30/40 and a ratio CHF_3/O_2 of 5/40 (sccm/sccm).

The etch depth of the BDD film was measured using a DEKTAK 3 stylus profilometer with an average of three scans performed after each experiments. In addition, a selection of features on the surfaces was examined by scanning electron microscopy (SEM) (Cambridge 200) and whiskers density was determined by image analysis using the software IMAGE J.

3. Results and discussions

3.1. *BDD etching patterned by an Al mask*

3.1.1. *Ar/O$_2$ ratio effect*

In this part the RF power at 13.56 MHz was 100 W, etching time 10 min and the working pressure 10 mTorr. Figure 2 shows the surface morphology of the etched BDD at a ratio Ar/O$_2$ = 5/40 (sccm/sccm). The deposits (called whiskers) identified at the grain boundaries of the etched diamond surface are due to Al sputtering. Indeed during the etching process, Al atoms are ionized in Al(III) ions which reacted with the oxygen plasma to form Al$_2$O$_3$ crystals, identified by energy dispersive X-ray.

Figure 3 shows the surface of the BDD samples etched at different argon contents. The whiskers density was determined by image analysis (Image J® software). The micromask densities were 12%, 9% and 7% of the analysed area for samples etched with 11%, 27% and 43% of Ar in the mixture, respectively, which proved that Al sputtering and Al$_2$O$_3$ deposition were mainly due to oxygen chemical etching.

Moreover, the etching rates of BDD by Ar/O$_2$ plasma are shown in Figure 4. The highest etching rate (120 nm min^{-1}) is obtained for an Ar content in the Ar/(O$_2$ + Ar) plasma of 27%. RIE of the diamond in Ar-based mixtures was studied by Leech and Reeves.[10] Authors determined high (30 nm min^{-1}) and low-rate (7–10 nm min^{-1}) regimes for ratios O$_2$/Ar > 25% and O$_2$/Ar < 25%, respectively. Both regimes were identified as ion-enhanced chemical etching and physical sputtering, respectively. The high etching rates observed in our cases for BDD were then linked to the use of high oxygen contents in the mixture (i.e. >57% of O$_2$) in accordance with Leech's results. In pure oxygen plasmas, the etching has been attributed to a dissociation of oxygen into O radicals [11] accompanied by the formation of volatile etch products of CO and CO$_2$. For diamond etching, pure O$_2$ plasmas were much more reactive than pure Ar plasmas (28 nm min^{-1} and 8 nm min^{-1}, respectively [10]). As observed from Figure 4 for a value of argon superior to 27% in the mixture, an increase in the Ar content led to a decrease in the etching rate. We assume that for Ar values lower than 27%, physical sputtering was negligible compared to chemical etching and that the high oxygen content led to a high micromask density (Figure 3) which acted as a protected layer for the BDD surface and then induced lower etch rates. The best conditions

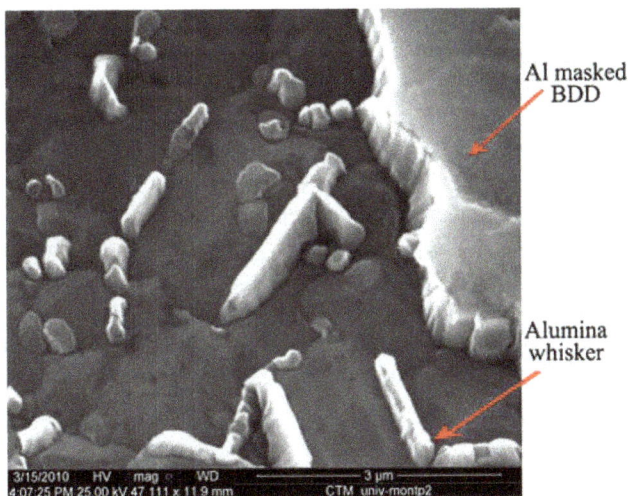

Figure 2.   SEM image of a BDD film etched at 10 mTorr with a ratio Ar/O$_2$ = 5/40 (sccm/sccm) and a RF power 100 W.

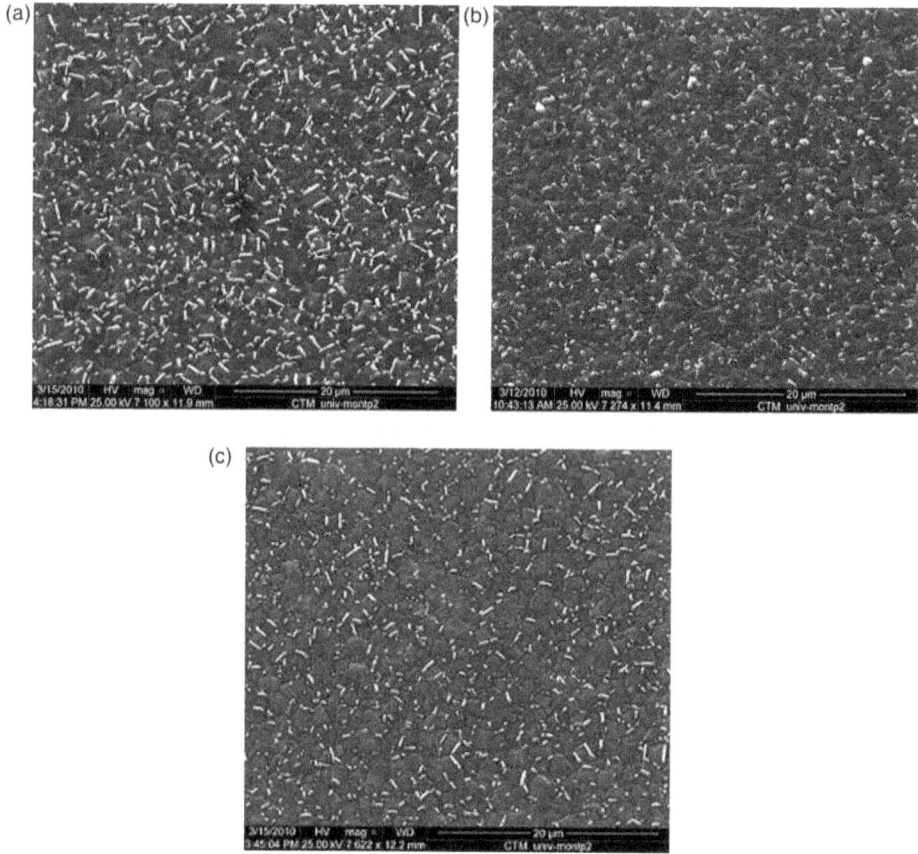

Figure 3. SEM images of a BDD film etched at 10 mTorr and 100 W with (a) $Ar = 11\%$ vol. (ratio $Ar/O_2 = 5/40$), (b) $Ar = 27\%$ vol. (ratio $Ar/O_2 = 15/40$) and (c) $Ar = 43\%$ vol. (ratio $Ar/O_2 = 30/40$ (sccm/sccm)).

to get both a high rate regime and a low whiskers density were then a ratio $Ar/(Ar + O_2)$ of 27% for a RF power of 100 W and a pressure of 10 mTorr.

3.1.2. *RF power effect*

Figure 5 shows the etching rate as the function of the RF power in the range 50–200 W for a chamber pressure of 10 mTorr. The etching rate increased linearly with increase in the RF power which indicated that the production of energetic oxygen species was also increasing as well as the bond breaking efficiency.[12,13]

Micrographies of Figure 6 show that the whiskers density increased with the RF power. As for a constant Ar/O_2 ratio, a power increase led to a higher increase in the chemical etching due to oxygen than of the physical sputtering due to argon, whiskers formation was then mainly related to chemical etching.

3.1.3. *Gas pressure dependence*

The influence of process pressure on the etch rate is depicted in Figure 7 for a ratio Ar/O_2 of 15/40 (sccm/sccm) and a RF power of 100 W. The higher etch rate was obtained at 5 mTorr

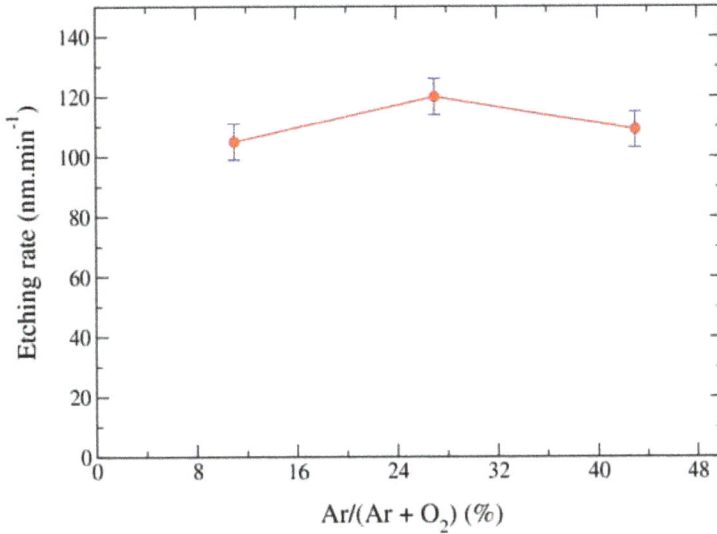

Figure 4.   Effect of Ar content on the etch rate of the BDD film. O_2 flux 40 sccm, RF power 100 W, pressure 10 mTorr, and etching time 10 min.

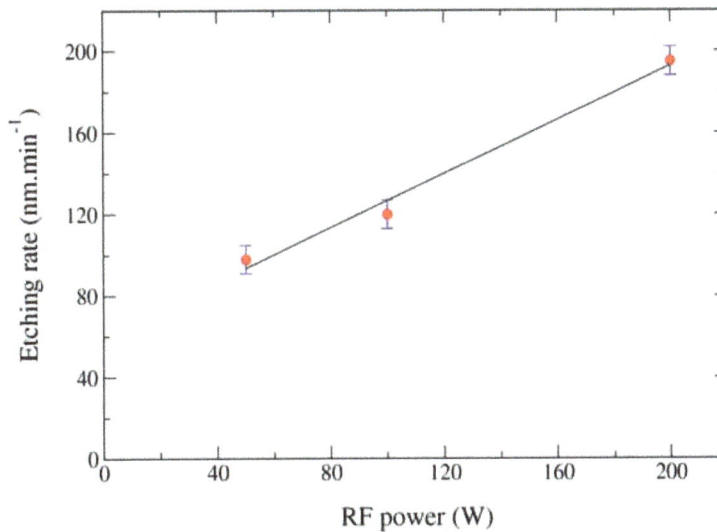

Figure 5.   Effect of the RF power on the etch rate of the BDD film. $Ar/O_2 = 15/40$ (sccm/sccm), pressure 10 mTorr and etching time 10 min.

(130 nm min^{-1}). For higher pressure, a decrease in the etch rate was recorded because of the enhanced ion–ion and ion-electron recombinations which decreased the quantity of accelerated ions with sufficient energy for material removal. The whiskers' densities were also determined at 5, 10 and 20 mTorr and were 8.8%, 9% and 13.8%, respectively. We have previously shown that alumina whiskers formation was related to a high degree of chemical etching. Then as the whiskers density remained high despite a low etching rate, the decrease in the etch rate could be mainly due to a lowering of the physical sputtering due to Ar ions recombination as proposed by Jiang et al.[14]

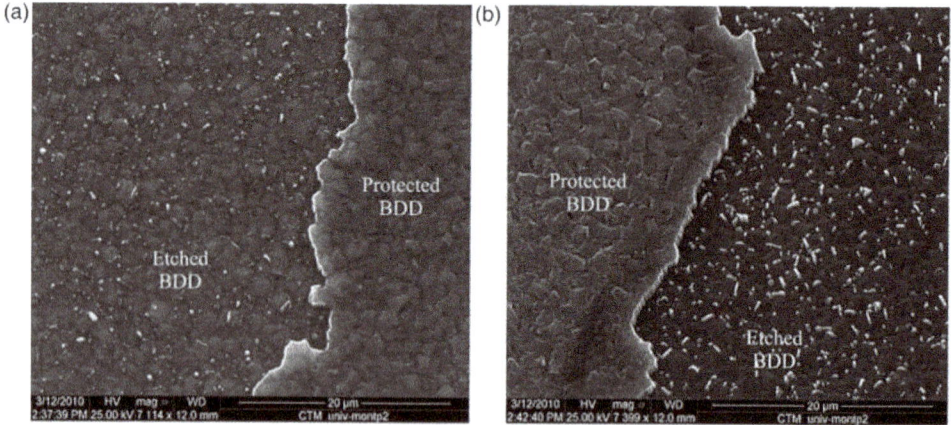

Figure 6.   SEM images of the etched film with Ar/O_2 15/40 (sccm/sccm), pressure 10 mTorr, etching time 10 min and RF power 100 W (a) and 200 W (b).

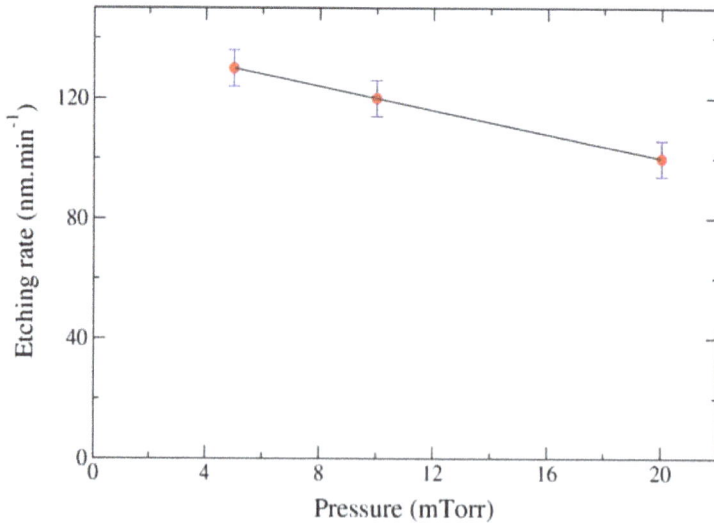

Figure 7.   Effect of the pressure on the etching rate of BDD film (Ar/O_2 15/40 (sccm/sccm), etching time 10 min and an RF power 100 W).

3.1.4.   CHF₃ effect

Fluorocompounds are commonly used in the etching process because of the positive effect on the etch rate of the fluorine radical.[15] The use of a mixture CHF_3/O_2 is particularly of interest to enhance the etch rate. Indeed the concentration of fluorine atoms responsible for the etching is increased by the enhanced dissociation of C−F bonds through reacting with oxygen atoms.[16,17] Unfortunately, the drawback is a decrease in etching selectivity. In this way, Ando et al. [18] have proved that the increase in CF_4/O_2 ratio reduced the selective etching ratio of diamond against Al.

A ratio CHF_3/O_2 = 5/40 (sccm/sccm) led to a etch rate of the BDD film of 130 nm min^{-1} for a pressure of 10 mTorr, an RF power of 200 W and an etching time of 10 min. Such conditions allowed etching without whiskers as proved by the micrography of the surface shown in Figure 8. Note that during the 10-min-etching process, the entire Al mask (i.e. 250 nm) was removed which shows the poor selectivity of the CHF_3/O_2-based etching process (lower that 5) compared with

Figure 8.   SEM image of the etched film with CHF$_3$/O$_2$ 5/40 (sccm/sccm), pressure 10 mTorr, etching time 10 min and an RF power 200 W.

the process carried out in the mixture Ar/O$_2$ for which the selectivity was from 5–10 (i.e. etching rate from 100 to 200 nm min^{-1}). The determined value of 130 nm min^{-1} was then given by default because of the entire removal of Al and the slight etching of the no more protected BDD surface.

The absence of alumina whiskers could be related to both phenomena. First, hydrogen atoms derived from the dissociation of CHF$_3$ reacted with Al to produce alane (AlH$_3$) which are volatile products.[19,20] Second, the presence of CHF$_3$ could also remove more efficiently the deposited alumina than pure O$_2$ (by chemical etching) or than Ar (by physical sputtering).

CHF$_3$ seems then to be a good candidate to etch BDD but with the drawback to decrease the selectivity against BDD face to Al which is unacceptable to pattern thick diamond layers.

Finally very high etching rates (i.e. 200 nm min^{-1}) of BDD were obtained by controlling the input power, the pressure, the nature of the etching gas and their ratio. Nevertheless, whatever were the conditions in the presence of argon, alumina whiskers were identified at the grains boundaries of the films. Contamination of the surface is definitely excluded for MEMS applications and especially in the aim to use the patterned films as the active layer of an electrochemical sensor. The only way to get a sample free of whiskers was to use the mixture CHF$_3$/O$_2$. But this gas mixture induced a high etch rate of Al mask and then a poor process selectivity. Then it was necessary to develop a new kind of mask as presented in the following section.

3.2.   *BDD etching patterned by a hybrid resin*

Most of the studies devoted to the etching of diamond-like materials are based on the use of metals or metallic alloys as masking materials. Very few works in the field used photoresist films to pattern the diamond especially because of the very low resistance of most of the organic-based resins against RIE. To develop an alternative to metallic masks to get both a high selectivity and a high etch rate of BDD without metallic oxides whiskers deposition, a new hybrid resin was investigated here. The resin was synthesized by the sol–gel method from organo-mineral precursors to get a hybrid material in which the organic groups confer the photosensitive properties, while the

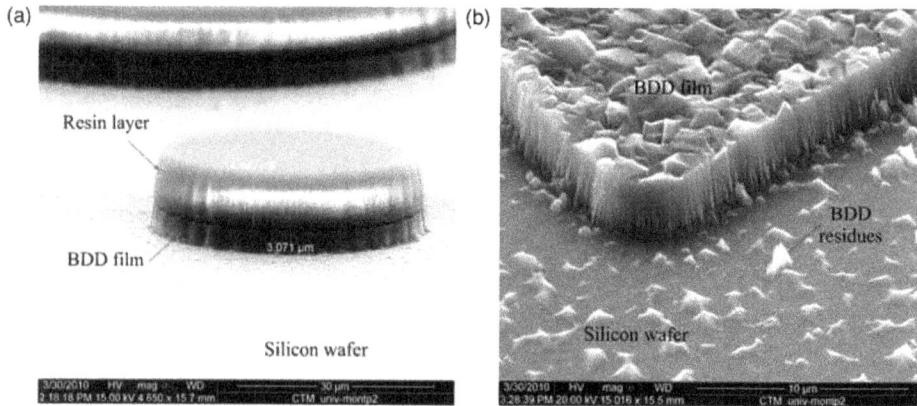

Figure 9. SEM images of different areas of the BDD film etched under a flux of Ar/O_2 15/40 (sccm/sccm), at 10 mTorr with an RF power of 200 W (a) before and (b) after resin dissolution. The thicknesses of the BDD film and of the resin layer are 3.6 and 8.5 μm, respectively.

mineral portion contributes to the mechanical and thermal stability and to the resistance against etching. We did not want to use fluorocompounds as the reactive gas to develop a process as simple as possible. Then the experimental conditions tested were the followings: ratio Ar/O_2 15/40 (sccm/sccm), pressure 10 mTorr and RF power 200 W because the etch rate determined in these conditions was the highest (195 nm min^{-1}). The initial thickness of the protecting layer was 8.5 μm.

Different views obtained by SEM of the etched BDD film before and after dissolution of the protective hybrid layer are shown in Figure 9.

The etching rate of the BDD film, determined after 10 min etching, was 200 nm min^{-1} in accordance with the previous results obtained with the Al mask. As predicted no whiskers were identified on the etched surface. After 18 min of processing, the entire BDD film was removed without the hybrid protective layer etching as depicted in Figure 8. Results show the very high selectivity of the process using the new hybrid mask compared with the Al one for which the selectivity is lower than 10 as previously proved.

4. Conclusion

BDD was successfully etched by ICP-RIE with a high etch rate. The etch rate increased with higher RF power and lower gas pressure and depended also on the gas ratio Ar/O_2. The etch rate decrease with gas pressure increase was related to Ar ions recombination. Best conditions to get the highest etch rate (i.e. 195–200 nm min^{-1}) was a ratio Ar/O_2 of 15/40 (sccm/sccm) (i.e. 27 % of Ar in the mixture), an RF power of 200 W and a pressure of 10 mTorr. Two types of masking were investigated. Basic Al masks induced the formation of alumina whiskers located at the grain boundaries of the polycrystalline BDD film. The whiskers' surface density decreased with increasing Ar content and the formation was mainly related to oxygen chemical etching of the Al mask. The gas mixture CHF_3/O_2 led to an etched surface free of alumina whiskers because of the formation of volatile alane (AlH_3) with the drawback of a very low process selectivity (inferior to 5) which is unacceptable to pattern the thick diamond film. Finally, we have proposed a new commercial hybrid organic/mineral resin to pattern BDD without whiskers formation at a high etching rate (200 nm min^{-1}) and selectivity. In the optimal conditions, a BDD layer of 3.6 μm thick was then etched in 18 min without any removal of the 8.5-μm-thick hybrid protective layer.

References

[1] Reay RJ, Flannery AF, Storment CW, Kounaves SP, Kovacs GTA. Microfabricated electrochemical analysis system for heavy metal detection. Sens Actuat B: Chem. 1996;34:450–455.

[2] Choi JY, Seo K, Cho SR, Oh JR, Kahng SH, Park J. Screen-printed anodic stripping voltammetric sensor containing HgO for heavy metal analysis. Anal Chim Acta. 2001;443:241–247.

[3] Zhu X, Gao C, Choi JW, Bishop PL, Ahn CH. On-chip generated mercury microelectrode for heavy metal ion detection. Lab Chip. 2005;5:212–217.

[4] Zou Z, Jang A, MacKnight E, Wu PM, Do J, Bishop PL, Ahn CH. Environmentally friendly disposable sensors with microfabricated on-chip planar bismuth electrode for *in situ* heavy metal ions measurement. Sens Actuat B Chem. 2008;134:18–24.

[5] McGaw EA, Swain GM. A comparison of boron-doped diamond thin-film and Hg-coated glassy carbon electrodes for anodic stripping voltammetric determination of heavy metal ions in aqueous media. Anal Chim Acta. 2006;575: 180–189.

[6] Madore C, Duret A, Haenni W, Perret A. Microfabricated systems and MEMS V: detection of trace silver and copper at an array of boron-doped diamond microdisk electrodes. New Jersey, NJ: The Elctrochemical Society; 2000. p. 159–168.

[7] Tsunozaki K, Einaga Y, Rao TN, Fujishima A. Fabrication and electrochemical characterization of boron-doped diamond microdisc array electrodes. Chem Lett. 2002;5:502–503.

[8] Pagels M, Hall CE, Lawrence NS, Meredith A, Jones TGJ, Godfried HP, Pickles CSJ, Wilman J, Banks CE, Compton RG, Jiang L. All-diamond microelectrode array device. Anal Chem. 2005;77:3705–3708.

[9] Silva F, Hassouni K, Bonnin X, Gicquel A. Microwave engineering of plasma-assisted CVD reactors for diamond deposition. J Phys: Condens Matter. 2009;21:364202-1–364202-16.

[10] Leech PW, Reeves GK, Holland AS. Reactive ion etching of diamond in CF_4, O_2, O_2 and Ar mixtures. J Mater Sci. 2001;36:3453–3459.

[11] Williams KR, Muller RS. Etch rates for micromachining processing. J Microelectromechanical Syst. 1996;5(4): 256–269.

[12] Lee JM, Chang KM, Kim SW, Huh C, Lee IH, Park SJ. Dry etch damage in *n*-type GaN and its recovery by treatment with an N2 plasma. J Appl Phys. 2000;87:7667–7670.

[13] Feng MS, Guo JD, Lu YM, Chang EY. Reactive ion etching of GaN with BCL_3/SF_6 plasmas. Mater Chem Phys. 1996;45:80–83.

[14] Jiang L, Plank NOV, Blauw MA, Cheung R, Drift EVD. Dry etching of SiC in inductively coupled Cl_2/Ar plasma. J Phys D: Appl Phys. 2004;37:1809–1814.

[15] Kuo Y. Plasma swelling of photoresist. Jpn J Appl Phys. 1993;32:L126–L128.

[16] Lanois F, Planson D, Locatelli M-L, Lassagne P, Jaussaud C, Chante J-P. Chemical contribution of oxygen to silicon carbide plasma etching kinetics in a distributed electron cyclotron resonance (DECR) reactor. J Electron Mater. 1999;28:219–224.

[17] Khan FA, Adesida I. High rate etching of SiC using inductively coupled plasma reactive ion etching in SF_6-based gas mixtures. Appl Phys Lett. 1999;75:2268–2270.

[18] Ando Y, Nishibayashi Y, Kobashi K, Hirao T, Oura K. Smooth and high-rate reactive ion etching of diamond. Diam Relat Mat. 2002;11:824–827.

[19] Pan WS, Steckl AJ. Reactive ion etching of SiC thin films by mixtures of fluorinated gases and oxygen. J Electrochem Soc. 1990;137:212–220.

[20] Yih PH, Saxena V, Steckl AJ. A review of SiC reactive ion etching in fluorinated plasmas. Phys Status Solidi B. 1997;202:605–642.

4

Advantages of using imaged-based fluorescent analysis for nanomaterial studies

Laura K. Braydich-Stolle, Alicia B. Castle, Elizabeth I. Maurer and Saber M. Hussain*

Applied Biotechnology Branch, Human Effectiveness Directorate, Air Force Research Laboratory, Wright-Patterson AFB, OH, USA

Currently, there is no standardised method for making nanomaterials or evaluating their toxicity. Studies show great variability and lack of reproducibility in results, with the discrepancies being caused due to assays used to evaluate exposure toxicity. While the traditional biochemical methods for evaluating toxicity have been employed, there is an issue of how the nanomaterials themselves interact with the dyes, which has created some issues. The current methodologies need to be evaluated to determine the toxicity of the nanomaterials and ensure that there is no interaction between the nanomaterials and the reagents used to complete the assays. This study evaluated the mechanisms of nanomaterial toxicity using mitochondrial function, loss of mitochondrial membrane potential and formation of reactive oxygen species as indicators of toxicity, and compared the data obtained on the BD Pathway 435 confocal to data from a plate reader. Furthermore, nanoparticle uptake and localisation were evaluated using the 3D imaging capabilities on the BD Pathway 435. These studies demonstrated that the fluorescence-based assays analysed on the BD Pathway 435 provided more sensitive detection for the cellular-based assays and was a useful tool for assessing nanoparticle localisation.

Keywords: nanomaterials; assay interference; confocal imaging

1. Introduction

Nanomaterials (NM) are playing an ever-growing role in the consumer product industry. Currently, NM, which are defined as having at least one dimension smaller than 100 nm, are integrated into about 600 consumer products, such as electronic components, cosmetics, cigarette filters, antimicrobial and stain-resistant fabrics and sprays, sunscreens and cleaning products [1,2]. While nanomaterials have greatly improved our everyday life, they are also causing many health problems which can range from granuloma formation in the lung [3] to inflammation at the site of exposure [4]. One of the main mechanisms for toxicity is the formation of reactive oxygen species (ROS), which has been shown in a study examining rat alveolar macrophages. The increase in ROS formation correlated with the decreasing size of the silver nanoparticles [5,6]. Studies such as the one by Carlson et al. show the need for future studies to evaluate the nanomaterials in comparison to their non-toxic bulk counterparts.

Over the past decade, there has been a boom in the nanoparticle research field with around 100 articles written about gold and silver nanoparticles in 1990 to over 5000 articles written in research journals in 2004 and the number is still growing [7]. Several studies and

---

*Corresponding author. Email: saber.hussain@wpafb.af.mil

review papers have shown the nanomaterials to be toxic, but still many questions regarding the health and environmental impact of these new 'wonder' materials remain unanswered. Additionally, review papers have described technical challenges associated with studying the unique properties of nanomaterials and how these properties correlate with the biological effects [8,9].

Currently, there is no standardised method for making nanomaterials or evaluating their toxicity. Studies are showing great variability and lack of reproducibility in results, with the discrepancies as a result of the assays used to evaluate exposure toxicity. At this time, exposure protocols include evaluating mitochondrial membrane function, ROS production, lactate dehydrogenase leakage assays and measuring the inflammatory response. While these are the traditional biochemical methods for evaluating toxicity, the issue of how the nanomaterials themselves interact with the dyes has created some issues. The current methods need to be evaluated to determine the toxicity of the nanomaterials and ensure that there is no interaction between the nanomaterials and the reagents used to complete the assays.

For example, some studies point to the MTT assay as being unreliable due to an interaction between nanomaterials, such as carbon and formazan crystals [10–13]. The formazan crystals bind to the carbon nanotubes, so that the crystals cannot be solubilised, thus leading to inaccurate results. Another study performed to examine the effects of aluminium nanoparticles in mammalian germ-line stem cells found it impossible to assess the nanoparticle effects on mitochondrial function due to particle agglomeration and interference with plate reader measurements [14]. Other articles show that the carbon nanotubes have negligible interaction with the formazan product and they do not influence the results of the study [15–17]. Due to the variability in NM and assays at this time, a method of examining the toxicity of nanomaterials, which can be used across all nanomaterial types and cell models, must be identified.

Additionally, electron microscopy has been the only method used to verify uptake of nano-materials and the sample preparation is labour intensive. Furthermore, the sample preparation for EM requires that the cells be fixed and the preparation can yield artefacts. The confocal 3D imaging of cells labelled with fluorescent organelle dyes, the uptake and localisation of fluo-rescently tagged nanomaterials can be determined in a timely manner. The BD confocal can be utilised for live or fixed cell imaging and the confocal filters out the background fluorescence, so that artefacts are not as readily observed. This study evaluated mechanisms of nanomaterial toxicity using mitochondrial function, loss of mitochondrial membrane potential (MMP), and formation of ROS as indicators of toxicity, and compared the data obtained on the BD Pathway 435 to data from a plate reader, as well as presented nanoparticle uptake and localisation. The HaCaT keratinocytes [18–23] and the A549 lung epithelial cells [24–33] were used for the tox-icity studies since these cell lines are widely used in nanotoxicity studies and carbon and gold nanomaterials were selected for the study since they are also commonly used in nanotoxicity evaluations. In addition to these cell types, germ cells, macrophages and osteoblasts were also used in the uptake and localisation studies to demonstrate the utility of the BD Pathway 435 for multiple cell types. Overall, these studies demonstrated that the BD Pathway 435 provided more sensitive detection for the cellular-based assays and was a useful tool for nanoparticle localisation studies.

2. Materials and methods

2.1. *BD Pathway 435*

The BD Pathway was used for confocal imaging, as well as high content analysis for qualitative data collection. This instrument is a spinning disc confocal and in conjunction with the software

BD Image Explorer, the images can be analysed to determine differences in fluorescent intensity to provide quantitative information.

2.1.1. *Human keratinocytes*

The human keratinocyte cell line (HaCaT) [34] was generously donated by Dr. James Dillman, USAMRICD. The cells were cultured in a flask with RPMI-1640 media (ATCC) supplemented with 10% foetal bovine serum (FBS, ATCC) and 1% penicillin/streptomycin (Sigma) and incubated at 37°C in a humidified incubator with 5% CO_2. For nanoparticle exposure procedures, RPMI-1640 media was supplemented with 1% penicillin/streptomycin without serum.

2.1.2. *Type II alveolar epithelial cells*

The A549 cell line purchased from ATCC (CCL-185) was grown in RPMI-1640 media supplemented with 1% Pen-Strep and 10% heat-inactivated FBS.

2.1.3. *Murine male germline stem cells*

The C18-4 cell line [35] was grown in a 1:1 mixture of Dulbecco's modified Eagle's medium/nutrient F-12 Ham supplemented with 10% FBS (ATCC), 2 mM L-glutamine and 1% penicillin/streptomycin. Cells were incubated at 34°C and 5% CO_2 in a humidified incubator.

2.1.4. *Macrophages*

The U937 cells purchased from ATCC (CRL-1593.2) were grown in RPMI-1640 media supplemented with 1% Pen-Strep and 10% heat-inactivated FBS. Prior to nanoparticle exposures, the U937 cells were stimulated with 100 ng/mL of phorbol-12-myristate-13-acetate (PMA) for 48 h to induce differentiation into macrophages.

2.1.5. *Osteoblasts*

Human foetal osteoblast cells (hFOB 1.19, purchased from ATCC, CRL-11372) were cultured in 25-cm^2 tissue culture flasks at 37°C in a humidified atmosphere of 5% CO_2 in air. A 1:1 mixture of Ham's F12 medium and Dulbecco's modified Eagle's medium was supplemented with 10% FBS and 0.3 mg/mL of G418 disulphate salt.

2.2. *Nanomaterials*

2.2.1. *Multi-walled carbon nanotubes*

Multi-walled carbon nanotubes functionalised with silver alone (designated MWCNT) or silver and a nitrogen functional group (designated MWCNT-CNx) were synthesised and received from Dr Maurico Terrones of The Advanced Materials Department and Laboratory for Nanoscience and Nanotechnology Research, IPICYT, San Louis Potosi, Mexico. These nanomaterials were received dispersed in water at a concentration of 200 μg/mL and were diluted to 25 μg/mL in media for cell exposure studies.

2.2.2. *Gold nanoparticles*

The Au 10 nm and Au 30 nm were synthesised and suspended in sterile water at a concentration of 1 mg/mL from nanoComposix. The nanoparticles were then diluted in media to 25 μg/mL for cellular exposures.

2.2.3. *Aluminium nanoparticles*

The Al_2O_3-40 nm and Al-50 nm were synthesised and generously received in powder form from Dr Karl Martin of NovaCentrix (formerly Nanotechnologies Inc). The nanoparticles were weighed out and dispersed in water to create a stock solution at a concentration of 1 mg/mL.

2.2.4. *Polystyrene nanospheres*

The Nile Red polystyrene nanospheres (cat #: 103125-05) were purchased from Microspheres-Nanospheres, a Corpuscular company, and were 40–60 nm in diameter. The nanoparticles were received at a concentration of 1 mg/mL and diluted to 25 μg/mL in media for exposure experiments. These nanoparticles fluoresce when excited using the TRITC filter settings.

2.2.5. *Carbon foam*

Microcellular carbon foam samples were purchased from Koppers Inc., in large blocks and then cut to fit in a 96-well plate. They were rinsed with phosphate-buffered saline (PBS) to remove any debris and sterilised under ultraviolet light prior to the seeding of cells. Double-sided carbon tape was used to keep the foam adhered to the well when submerged in media during cell culture exposures.

2.3. **Biochemical cellular assays**

2.3.1. *Cell viability assays*

The C18-4 cells were seeded at a density of 50,000 cells/well. Once the cells were 80% confluent in a 96-well plate, different concentrations of aluminium nanoparticles were added to the wells. The cultures were further incubated for 24 h and then washed with PBS to remove any of the nanoparticles that had not been taken up by the cells. MTS was also performed on cells dosed with aluminium nanoparticles that had not been washed. Cell viability was evaluated using the CellTiter 96® AQueous One Solution of Promega, which measures mitochondrial function and directly correlates with cell viability. The relative cell viability (%) related to control wells containing cell culture medium without nanoparticles or PBS as a vehicle was calculated by [A]test/[A] control ×100, where [A]test is the absorbance of the test sample and [A]control the absorbance of control sample. Each experiment was done in triplicate and the data are represented as the mean ± the standard deviation. A Student's t-test was performed and $p < 0.05$ indicated significance. In addition, the Live Cell Dead Cell kit from Biovision was used to evaluate cell viability. The kit contains a green dye that is cell permeable and propidium iodide which is only permeable if the cells are dead. The ratio of red to green cells can be calculated to determine the changes in cell viability.

2.3.2. *Reactive oxygen species*

The HaCaT or A549 cells were seeded on a BD 96-well imaging plate at a density of 100,000 cells/well. Once the cells were 80% confluent, they were dosed with 25 μg/mL of the MWCNT, MWCNT-CNx, Au 10 nm, and Au 30 nm for 24 h. Following exposure, the Image-iT® Live ROS and the MitoSOX™ Red kits (Invitrogen) were used to evaluate the formation of ROS and superoxide, respectively, according to the manufacturer's recommendations.

2.3.3. *Mitochondrial membrane potential*

The HaCaT or A549 cells were seeded on a BD 96-well imaging plate at a density of 100,000 cells/well. Once the cells were 80% confluent, they were dosed with 25 μg/mL of the MWCNT, MWCNT-CNx, Au 10 nm and Au 30 nm for 24 h. Following exposure, the MitE-ψ kit from Biomol International was used to evaluate changes in MMP according to the manufacturer's recommendations.

2.3.4. *Image acquisition and data analysis*

For the endpoint assays, 16 fields in each well were imaged at 20× using the 4 × 4 montage option in the BD Pathway 435 software. For the ROS assays, the FITC and Hoescht filters were used and for the MitoSOX red assay, the Rhodamine and Hoescht filters were used. For the MitE-ψ assay and the Live Cell Dead Cell kit; the FITC, Rhodamine, and Hoescht filters were used. The images were then analysed using the BD AttoVision™ v1.6 and the BD Image Data Explorer, which evaluated differences in fluorescent intensity between the samples. Additionally, for the MMP assay, plates were also read on the SpectraMax plate reader.

2.4. *Cellular uptake and localisation studies*

The macrophages were stimulated with PMA as previously described and seeded in a 96-well imaging plate. The macrophages were exposed to 25 μg/mL of the Nile Red polystyrene nanospheres for 0, 2, 4, 6, 8, 12, 18 and 24 h and then fixed with 4% paraformaldehyde. Initially, time points between 0 and 2 h were also evaluated with no observed uptake occurring, so these time points were not reported. Following fixation, the macrophages were immune labelled with primary antibodies for the lysosome (Santa Cruz: sc-5571 LAMP-2), the mitochondria (Santa Cruz: sc-58348 COX-4), the ER (abcam: ab11420-50 ERp29), the golgi (abcam: ab58826-100 golgi complex) and the peroxisome (abcam 15834-100 catalase antibody). A secondary antibody labelled with an AlexaFluor 488 was used to give the organelle a green fluorescent label and then the nuclei were counterstained using DAPI. Following the staining, the cells were observed in the confocal microscope and the nanospheres fluoresced in the red range and any co-localisation in the organelles appeared yellow, since the organelles were labelled green. To ensure that the NPs were in fact internalised, the z-stack option was used and 10 serial sections of 0.250 μm were taken across 16 fields in each well at 20× using the 4 × 4 montage option in the BD Pathway 435 software. Once the images were obtained, the confocal software allowed for segmenting the cells to measure only the fluorescent intensity within the designated portion. This allowed for eliminating any membrane-bound nanoparticles from analysis or nanoparticles that were stuck to the plate and not taken up. The images were then analysed using the BD AttoVision™ v1.6 and the BD Image Data Explorer, which evaluated the differences in fluorescent intensity between the samples. The red fluorescent intensity was also measured using the SpectraMax Synergy plate reader to compare the uptake trends between the confocal and the plate reader analyses. The collapsed z-stack lysosomal images were selected to qualitatively represent the kinetics of uptake.

2.5. *Cellular morphology*

2.5.1. *Osteoblast actin staining*

For the cell morphology studies with the carbon foam, double-sided carbon tape was placed on one side of the foam prior to inserting the foam into a 24-well plate to help the foam adhere to the plate. Then, 1.5 mL of hFOB suspension at a density of 1×10^6 cells/mL was plated in the wells, and the cells were incubated for 72 h followed by fixation with 4% paraformaldehyde at room temperature for 10 min. The cells were rinsed with PBS and permeabilised with 0.1% Triton X-100 for 5 min, and then the actin filaments were stained with Alexa Fluor 555-phallodin from Invitrogen for 45 min. The samples were then rinsed with PBS and a few drops of Prolong Gold Reagent with DAPI nuclear counterstain were added. Foam samples were then removed from the well, inverted, placed in a 96-well imaging plate and imaged on the BD Pathway 435 confocal microscope.

3. Results and discussion

3.1. *Cellular assays*

3.1.1. *Evaluation of MMP*

The MMP assay used the JC-1 dye to stain non-apoptotic cells red and apoptotic cells green based on how the dye interacts with the membrane potential. MMP changes were evaluated in both HaCaT and A549 cells treated with carbon or gold nanomaterials (Figure 1). The confocal images illustrated that the negative control cells were non-apoptotic and the positive control cells treated with 100 μM tert-butyl hydrogen peroxide showed non-apoptotic and apoptotic cells (Figure 1A–B, 1M–N). In the HaCaT cells, the MWCNT- and CNx-treated cells indicated that a change in MMP was initiated with the CNx cells appearing to have more changes based on the images (Figure 1E and G, respectively). Furthermore, the corresponding light images showed the presence of the nanomaterials in the cell cultures (Figure 1). In comparison, the HaCaT cells that were treated with the Au nanomaterials did not demonstrate any changes in MMP (Figure 1I–L). A similar trend was observed in the A549 cells treated with the same nanomaterials (Figure 1M–X). The MWCNT- and MWCNT-CNx-treated cells displayed a loss of MMP, while the Au-treated cells did not.

Following the image acquisition, the ratio of red/green fluorescent intensity within the images was determined using the BD Data Explorer program (Figure 1Y and Z). In the HaCaT-treated cells, there was a slight reduction in the positive control treatment, as well as both the MWCNT treatments, and there was no reduction for both Au treatments. When the data obtained using the confocal images were compared to the data obtained on a plate reader, there were major discrepancies between the Au treatments. When the fluorescent intensity measurements were obtained on the plate reader, there was significant reduction in the MMP, which did not correlate with what was observed in the images obtained. Furthermore, while there was a reduction in the MWCNT treatments, this reduction appeared greater in the plate reader assays. Previous studies have found that carbon nanomaterials can react and absorb dyes [13] and the MMP assay demonstrated that when a plate reader was used, higher fluorescent intensity measurements were recorded on the FITC setting, which in turn yielded lower red/green ratios, indicating greater loss of MMP (Figure 1Y). These higher intensity measurements could be a result of the carbon nanomaterials absorbing the fluorescent dyes, since this did not correspond to what was observed microscopically, and when the fluorescent intensity was measured from the confocal images that were taken which eliminated the background measurements, a more accurate measurement was obtained. Furthermore, in the A549 cell study, the plate reader assay demonstrated that there was

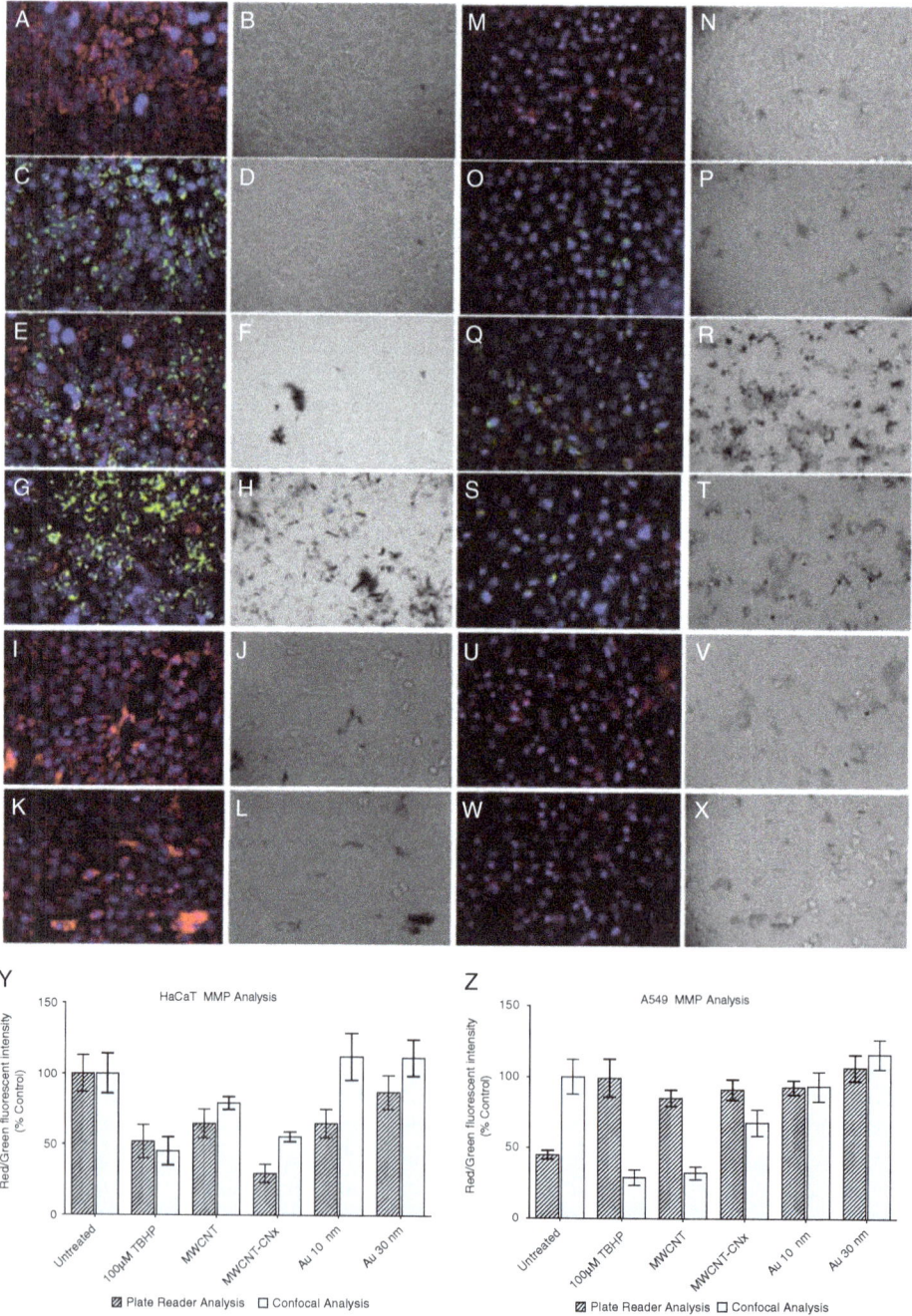

Figure 1.   Measurement of MMP in cells treated with different nanomaterials.
Notes: A–L, confocal and light images of HaCaT cells following exposure to nanomaterials. A–B, negative control; C–D, positive control; E–F, Ag-MWCNT; G–H, Ag-MWCNT-Cx; I–J, 10 nm Au; K–L, 30 nm Au; M–X, confocal and light images of A549 cells following exposure to nanomaterials. M–N, negative control; O–P, positive control; Q–R, Ag-MWCNT; S–T, Ag-MWCNT-Cx; U–V, 10 nm Au; W–X, 30 nm Au; Y–Z, comparison of fluorescent intensity values obtained using the confocal and a plate reader. Y, HaCaT cells; Z, A549 cells. The plate reader graph has higher intensity values due to background measurements. The BD Pathway 435 evaluated the intensity from the confocal images which eliminated the background fluorescence and gave a more accurate measurement.

no significant change in MMP for any of the treatments (Figure 1Z); yet the images and confocal-based analysis demonstrated significant changes in MMP for both the MWCNT treatments. When evaluating the light microscopy images, there were clumps of dye that did not wash away despite the repeated washes and these clumps would have given off enough of a red fluorescent signal to mask the green fluorescent readings. Overall, the confocal analysis provided a more sensitive and accurate assessment of MMP measurements.

3.1.2. *Detection of ROS*

The production of ROS was (Figure 2A–L) evaluated in the HaCaT cells treated with the different nanomaterials. Previous work had demonstrated that the MWCNT disrupted the cell mono-layer while the other treatments did not (data not shown). The negative control (Figure 2A) showed little to no production of ROS while the positive control induced formation of ROS (Figure 2B). The MWCNTs and the MWCNT-CNx generated ROS formation as well (Figure 2C and D, respectively). In comparison, the Au nanoparticle treatments did not induce ROS pro-duction (Figure 2E and F). Moreover, when the formation of superoxide was evaluated using the MitoSox™ Red kit (Figure 2G–L), the negative control showed little to no production (Figure 2G), the positive control demonstrated production of superoxide (Figure 2H) and the Ag-MWCNTs induced formation of superoxide (Figure 2I) while the Ag-MWCNT-CNx and both Au treatments formed minimal amounts of superoxide (Figure 2J–L). In comparison, when the A549 cells were exposed to the nanomaterials, both MWCNT treatments produced sig-nificant production of ROS and superoxide (Figure 2O, P, U, V), while the Au treatments did not have a significant effect (Figure 2Q, R, W, X). Since the lung epithelial cells by design are more sensitive than keratinocytes, this difference was not surprising. However, when evaluating the formation of ROS, the levels of ROS observed by the Ag-MWNT-CNx were surprising since this treatment had not disrupted the morphology. The MitoSOX™ Red assay demonstrated that there was formation of superoxide in the presence of the Ag-MWCNTs but not the Ag-MWCNT-CNx, indicating that the nitrogen functional groups were capable of interacting with the ROS probe from the Image-iT® Live ROS kit (Figure 2). Therefore, the MitoSOX™ Red kit is a more sensitive method of detection when functional groups are present on nanomaterials.

3.1.3. *Cellular viability assays*

Carbon nanotubes are not the only nanomaterials that have interfered with biochemical assays. A previous study demonstrated that Al-NPs interfered with the MTS assay and that accurate readings could not be obtained [14]. In this study, the C18-4 cells were treated with the aluminium nanoparticles and then morphology was assessed. In the control cells, there was expression of actin (Figure 3A) and at this nanoparticle dose, the expression was drastically reduced in the Al_2O_3 treatment (Figure 3B) and non-existent in the Al treatment (Figures 3C). These morphological changes indicated that the nanoparticles were disrupting the cellular environment. When the C18-4 cells were treated with the Al-NPs and cellular viability was assessed after 24 h using the MTS assay, there did not appear to be a disruption in cellular proliferation (Figure 3D), which was contradictory to the morphology assay. However, when the excess nanomaterials were washed prior to performing the MTS assay, a reduction in cell viability was observed (Figure 3E), and this reduction occurred in a dose-dependent manner with the Al_2O_3 nanoparticles, demonstrating the least toxicity. The validity of this washing technique was verified using the live cell/dead cell assay from Biovision, which yielded results comparable to the MTS assay, performed post-washing (Figure 3F). The cellular viability assays confirmed that the excess Al-NPs interfered with

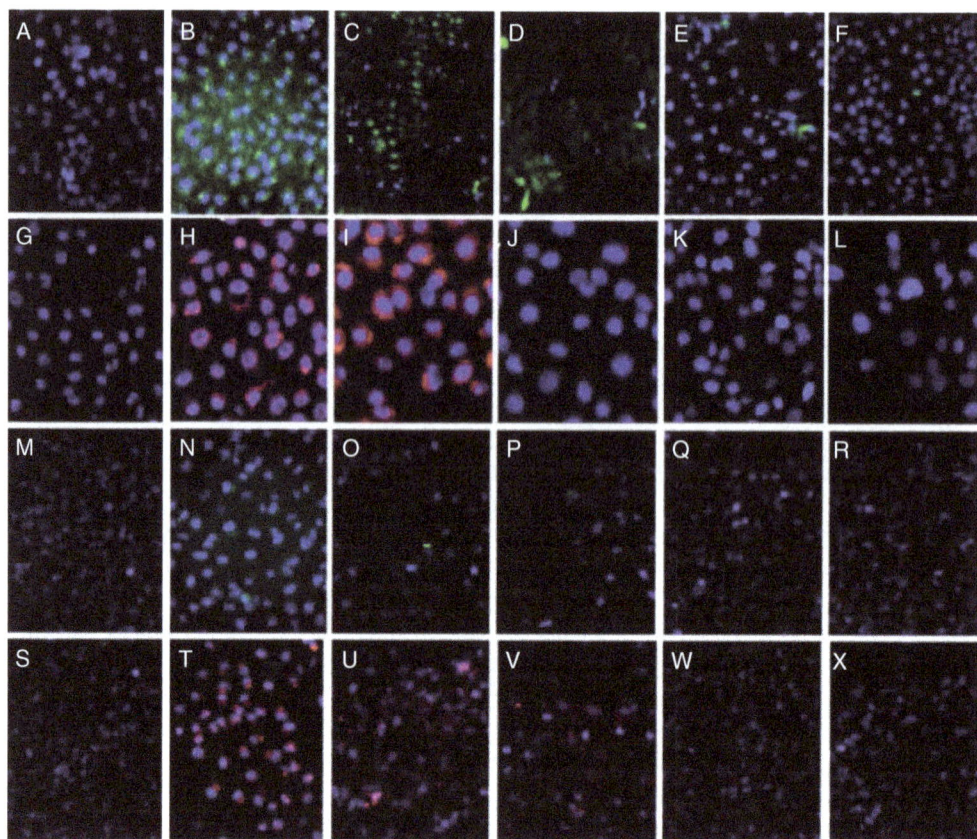

Figure 2.   Measurement of ROS formation in cells following exposure to nanomaterials.
Notes: A–F, evaluation of ROS in HaCaT cells. A, negative control; B, positive control; C, Ag-MWCNT;
D, Ag-MWCNT-CNx; E, 10 nm Au, F 30 nm Au. The MWCNTs and CNx nanotubes had similar levels of
ROS production when the Image-iT Live ROS detection kit was used, which was not expected based on
morphology results (data not shown), which showed that there was no disruption with the Ag-MWCNT-CNx.
The Au nanoparticles did not induce the production of ROS. G–L, evaluation of superoxide formation in
HaCaT cells. G, negative control; H, positive control; I, MWCNT; J, CNx; K, 10 nm Au; L, 30 nm Au. When
formation of superoxide was detected using the MitoSox red kit, the MWCNTs produced superoxide while
there was minimal expression of superoxide in the CNx-treated cells which correlated with the morphology
data. These data suggested that the nitrogen components of the functional group added to the carbon nanotubes
were reacting with the ROS Image-iT Live assay to produce nitrogen reactive species. M–R, evaluation of
ROS in A549 cells. M, negative control; N, positive control; O, Ag-MWCNT; P, Ag-MWCNT-CNx; Q,
10 nm Au; R, 30 nm Au. Both the MWNCT- and the CNx-treated cells generated ROS. S–X, evaluation of
superoxide formation in A549 cells. S, negative control; T, positive control; U, MWCNT; V, CNx; W, 10 nm
Au; X, 30 nm Au. When superoxide formation was evaluated, this confirmed that both the MWCNT- and
the CNx-treated cells were under oxidative stress.

plate readings when the MTS assay was performed (Figure 3D–F). However, a simple washing
step prior to running the MTS assay eliminated this problem (Figure 3E). Since nanomaterials can
interfere with assays, verification of the data using another assay is critical, and since fluorescent
assays tend to yield high levels of background on plate readers, the BD Pathway 435 provides a
method for eliminating fluorescent background and verifying other assays using fluorescent-based
technology.

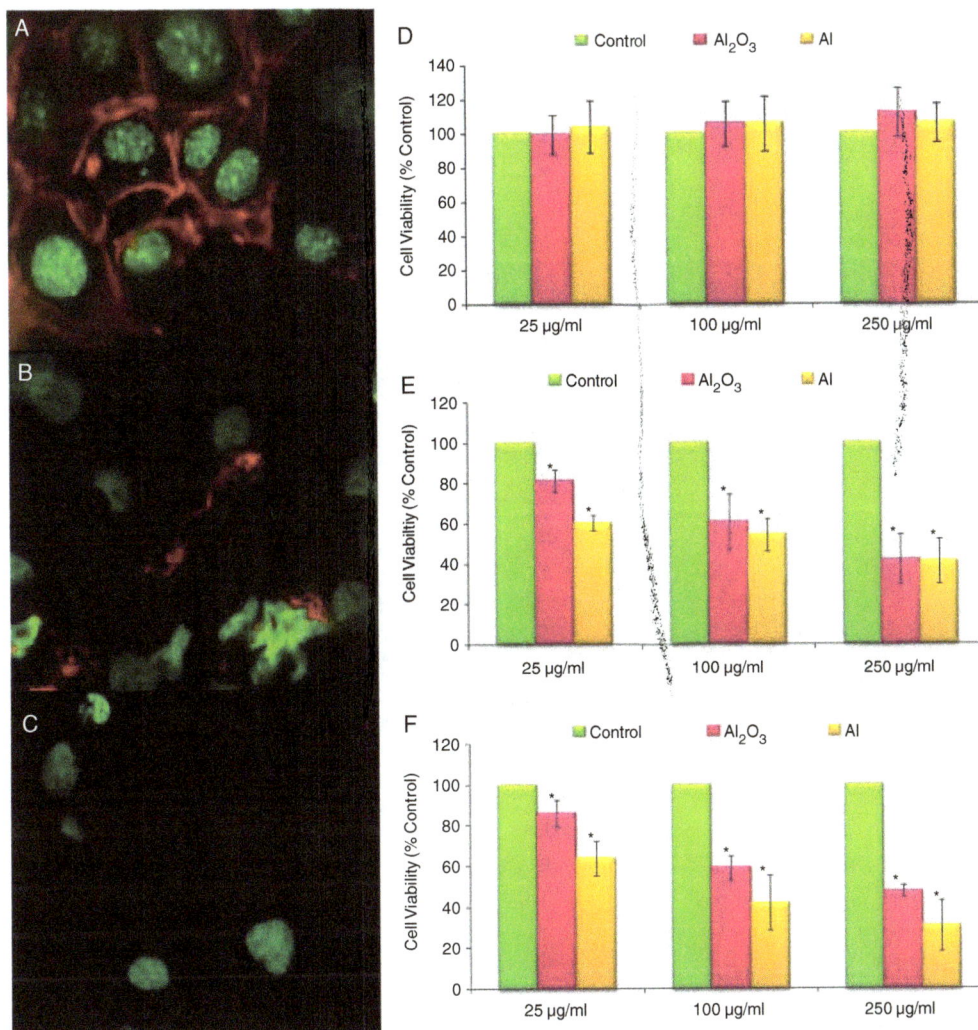

Figure 3.   Cell viability measurements in germ-line stem cells treated with Al nanoparticles.
Notes: A–C, confocal images of C18-4 cells with actin and nuclear staining. A, control; B, 100 μg/mL Al$_2$O$_3$; C, 100 μg/mL Al; D, MTS measurement without washing following exposure. E, MTS measurement with washing following exposure. F, live cell dead cell staining analysed on the BD Pathway 435. The Al MTS that was run without washing was highly inaccurate and did not correspond to observed morphology changes (A–C). When the excess Al nanoparticles were washed away prior to running the MTS assay, this greatly improved the measurement. Furthermore, the results from the washing technique were confirmed using the LCDC assay.

3.1.4.   *Cellular uptake and localisation studies*

The kinetics of nanoparticle uptake was evaluated using macrophages since they are known to actively engulf foreign material. The level of uptake was assessed at multiple time points within a 24-h period and is represented qualitatively in Figure 4A–L. Beginning at 2 h, the nanoparticles begin to adhere to the cell membrane (Figure 4B) and then they are gradually internalised and accumulate within the cell over time (Figure 4C–E). Once the cells reached the 12- and 18-h time points, the appearance of the cytoplasm was almost completely red, suggesting continued

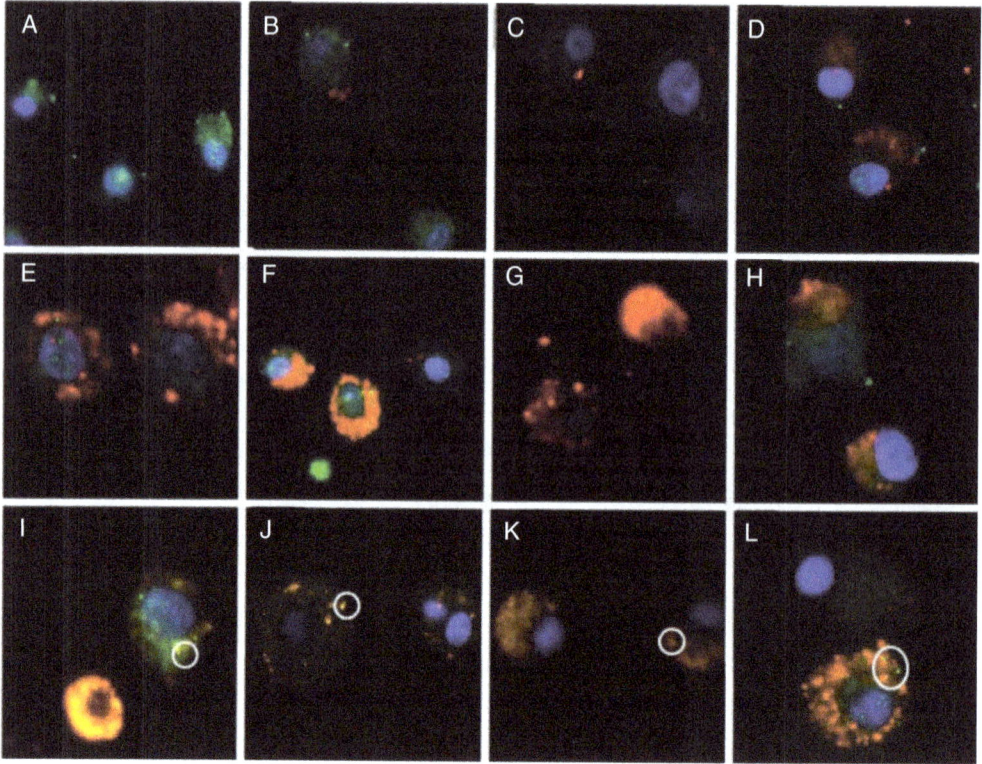

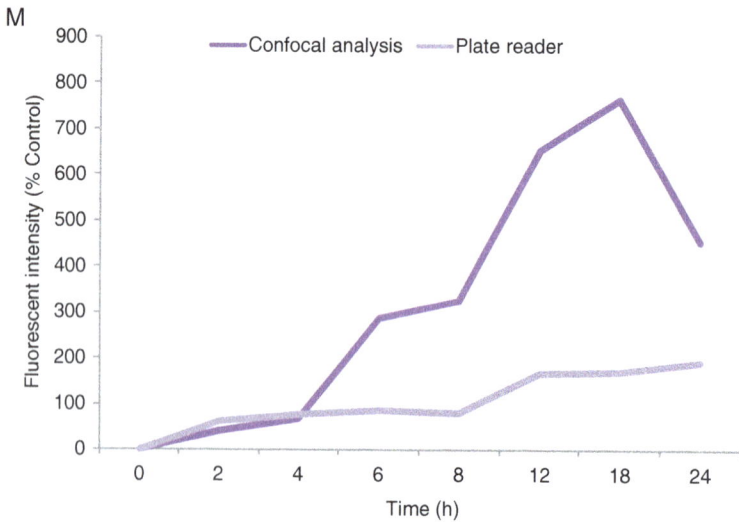

Figure 4.   Nanoparticle uptake and localisation in human alveolar macrophages.
Notes: A–H, kinetics of nanoparticle uptake. A, 0 h; B, 2 h; C, 4 h; D, 6 h; E, 8 h; F, 12 h; G, 18 h; H, 24 h.
I–L, organelle co-localisation at 12 h. I, mitochondria; J, golgi; K, ER; L, peroxisome; and M, comparison
of uptake using the BD 435 Pathway and a SpectraMax plate reader.

accumulation (Figure 4F and G, respectively). Interestingly, at the 24-h time point, there appeared
to be less nanoparticle internalisation indicating that the nanoparticles may be shuttled out of the
cell potentially via exocytosis. When the red fluorescent intensity was measured on the plate

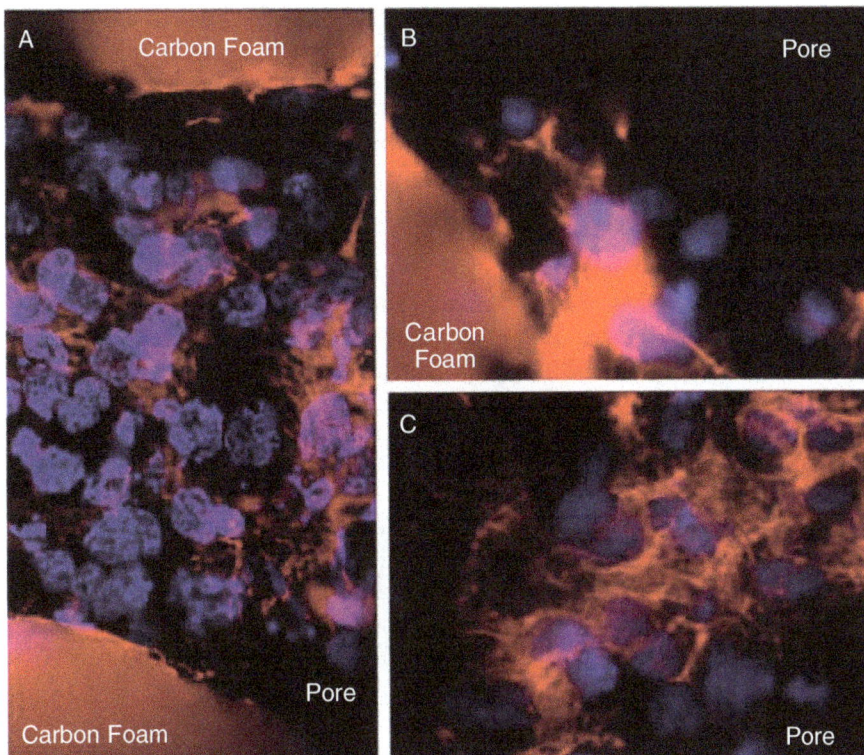

Figure 5.   Osteoblast growth on carbon foam.
Notes: The osteoblasts were stained with Alexa 555-phallodin to visualise the actin in the cytoplasm and counterstained with the nuclear dye Hoescht 3334. The carbon foam absorbed the phalloidin dye but using the confocal imaging to filter out some of the background, cells were seen colonising the pores of the carbon foam. These osteoblasts expressed actin and can be seen attaching to the edges of the carbon foam pore and spreading out into the pore to form a cellular network.

reader, there only appeared to be a steady increase over time and this peak at 18 h was not observed (Figure 4M). However, when the analysis was performed using the images obtained on the BD Pathway, this trend was observed. One reason for these discrepancies is that for the plate reader assays, any nanoparticles that were adhering to the membrane or to the bottom of the plate would contribute to the intensity measurement and since the BD Pathway allows for sectioning the cytoplasm of the cell and measuring only the intensity within that sectioned sample, any membrane-bound or plate-bound nanoparticles were not factored into the analysis. These settings allowed for more control over the uptake measurements and yielded a more sensitive and accurate assessment of nanoparticle uptake. In addition to evaluating the kinetics of uptake, the localisation of the nanoparticles was also evaluated. For the localisation studies, antibodies targeting the specific organelles were used to stain the lysosome, mitochondria, ER, Golgi and peroxisomes. Since the 12- and 18-h time points demonstrated the maximum levels of uptake, the 12-h time point was reported for the localisation studies. Figure 4F demonstrated that after 12 h, there was some lysosomal localisation; however, there was significant localisation within the mitochondria (Figure 4I), which confirms many studies that have suggested endocytosis as the uptake mechanism and reported changes in mitochondrial function. In addition, there was also some localisation within the ER and Golgi, indicating that nanoparticles have the capability to impact protein expression through the synthesis and post-translation processes. Furthermore, the localisation in the peroxisomes suggests that the nanoparticles can exit via exocytosis and this supports the trend observed in the kinetic data where there is a peak at 18 h and then a drop at 24 h. Taken together, these localisation experiments demonstrated that the nanoparticles have the potential to enter a cell and impact a wide range of cellular processes and these interactions require further investigation.

3.1.5. *Osteoblast colonisation of carbon foam*

Electron microscopy has been the only method available to demonstrate osteoblast growth on the carbon foam, and while this technique provides information on the osteoblast attachment and growth, there are limitations. Using fluorescent microscopy, the osteoblasts can be labelled using immunofluorescence to detect markers of osteoblasts and osteoblast differentiation; however, the propensity for the carbon foam to absorb fluorescent dyes makes this type of analysis problematic. The actin staining demonstrated that the confocal can filter out the background levels of fluorescence to distinguish the cells in the pores and on the edges of the foam, making more in-depth cellular analysis possible. The carbon foam did absorb the Alexa-Fluor 555-phallodin dye (Figure 5A and B); however, the confocal was able to filter out some of the background from absorption so that the cells could be visualised and distinguished from the carbon foam (Figure 5). The carbon foam supported osteoblast growth in the pore (Figure 5A) and along the edges of the carbon foam (Figure 5B). Furthermore, it was demonstrated that the osteoblast cells were able to expand throughout the pore and form a network of cells with strong actin expression (Figure 5A and C).

4. Conclusion

This study has demonstrated that the properties of nanomaterials can make assessment of the biological responses to these materials difficult. Some of the carbon nanomaterials absorbed the dyes or reagents used, resulting in inaccurate readings. In addition, functional groups added to the nanomaterials interacted with reagents to yield false positive readings. The BD Pathway 435 provided a platform on which cellular-based fluorescent assays were used and the problems associated with the false reading were corrected for since the evaluations were based on the images and what is being analysed was controlled for. Furthermore, the confocal option on the BD Pathway 435 corrected the high levels of background associated with some of the fluorescence assays and nanomaterials. One suggestion has been to run just the nanomaterials with the dyes and subtract out the background in order to run plate reader assays. While this technique definitely would improve the background measurements, it would not be as accurate since the porosity between materials can differ and once the material has been taken up by a cell, this could also change the interaction/absorption of the dye. Therefore, the BD pathway provides a more sensitive method of measurement when compared to a plate reader, and with the evaluation of MMP and ROS production in both the HaCaT and A549 cell lines using the same nanomaterials, the applicability of this device was demonstrated. Additionally, background from metallic-based nanoparticles was lessened through washing techniques which was confirmed using the BD Pathway 435. The BD Pathway 435 confocal microscope also provided information on nanoparticle uptake using the 3D imaging and Z stack option, as well as, more information on nanoparticle localisation in conjunction with organelle labelling. This microscopy technique is less labour intensive when compared to electron microscopy and the use of cellular stains and antibodies can provide more detailed information on cellular profiles and cellular interactions of the nanomaterials. Overall, the use of fluorescence-based technology on a platform such as the BD Pathway 435 provides researchers with tools for assessing the biological interactions of nanomaterials that take into account some of the problems encountered with studying these materials with unique properties.

Acknowledgements
We thank Col. Patricia Reily and Dr John Schlager for their support of this study. Dr Braydich-Stolle is funded through the Henry Jackson Foundation, Ms Castle and Ms Maurer through the Consortium Research Fellows Program. This study was cleared for public release 88ABW-2011-1013.

References

[1] Woodrow Wilson International Center for Scholars, *The project on emerging nanotechnologies*. Available at: http://www.nanotechproject.org

[2] S.F. Hansen, E.S. Michelson, A. Kamper, P. Borling, F. Stuer-Lauridsen, and A. Baun, *Categorization framework to aid exposure assessment of nanomaterials in consumer products*, Ecotoxicology 17 (2008), pp. 438–447.

[3] D.B. Warheit, B.R. Laurence, K.L. Reed, D.H. Roach, G.A. Reynolds, and T.R. Webb, *Comparative pulmonary toxicity assessment of single-wall carbon nanotubes in rats*, Toxicol. Sci. 77 (2004), pp. 117–125.

[4] P. Hoet, I. Briusle-Hohlfeld, and O. Salata, *Nanoparticles-known and unknown health risks*, J. Nanobiotechnol. 2 (2004), 15pp. DOI: 10.1186/1477-3155-2-12.

[5] Nel, T. Xia, L. Madler, and N. Li, *Toxic potential of materials at the nanolevel*, Science 311 (2006), pp. 622–627.

[6] C. Carlson, S.M. Hussain, A.M. Schrand, L.K. Braydich-Stolle, K.L. Hess, R.L. Jones, and J.J. Schlager, *Unique cellular interaction of silver nanoparticles: Size-dependent generation of reactive oxygen species*, J. Phys. Chem. B 112 (2008), pp. 13608–13619, (35 (2008), pp. 209–217).

[7] S. Eustis and M.A. El-Sayed, *Why gold nanoparticles are more precious than pretty gold: Noble metal surface plasmon resonance and its enhancement of the radiative and nonradiative properties of nanocrystals of different shapes*, Chem. Soc. Rev. 35 (2006), pp. 209–217.

[8] S.T. Stern and S.E. McNeil, *Nanotechnology safety concerns revisited*, Toxicol. Sci. 101 (2008), pp. 4–21.

[9] S.M. Hussain, L.K. Braydich-Stolle, A.M. Schrand, R.C. Murdock, K.O. Yu, D.M. Mattie, J.J. Schlager, and M. Terrones, *Toxicity evaluation for safe use of nanomaterials: Recent achievements and technical challenges*, Adv. Mater. 2 (2009), pp. 1–11.

[10] R.H. Hurt, M. Monthioux, and A. Kane, *Toxicology of carbon nanomaterials: Status, trends, and perspectives on the special issue*, Carbon 44 (2006), pp. 1028–1033.

[11] N.A. Monteiro-Riviere and A.O. Inman, *Challenges for assessing carbon nanomaterial toxicity to the skin*, Carbon 44 (2006), pp. 1070–1078.

[12] M. Davoren, E. Herzog, A. Casey, B. Cottineau, G. Chambers, H.J. Byrne, and F.M. Lyng, *In vitro toxicity evaluation of single walled carbon nanotubes on human A549 lung cells*, Toxicol. Vitro. 21 (2007), pp. 438–448.

[13] J.M. Worle-Knirsch, K. Pulskamp, and H.F. Krug, *Oops they did it again! Carbon nanotubes hoax scientists in viability assays*, Nano. Lett. 6 (2006), pp. 1261–1268.

[14] L. Braydich-Stolle, S. Hussain, J.J. Schlager, and M.C. Hofmann, *In vitro cytotoxicity of nanoparticles in mammalian germline stem cells*, Toxicol. Sci. 88 (2005), pp. 412–419.

[15] C.M. Grabinski, S.M. Hussain, K. Lafdi, L. Braydich-Stolle, and J.J. Schlager, *Effect of particle dimension on biocompatibility of carbon nanomaterials*, Carbon 45 (2007), pp. 2828–2835.

[16] A.M. Schrand, H. Huang, C. Carlson, J.J. Schlager, E. Omacr Sawa, S.M. Hussain, and L. Dai, *Are diamond nanoparticles cytotoxic?*, J. Phys. Chem. B 111 (2007), pp. 2–7.

[17] P. Wick, P. Manser, L.K. Limbach, U. Dettlaff-Weglikowska, F. Krumeich, S. Roth, W.J. Stark, and A. Bruinink, *The degree and kind of agglomeration affect carbon nanotube cytotoxicity*, Toxicol. Lett. 168 (2007), pp. 121–131.

[18] E. Herzog, A. Casey, F.M. Lyng, G. Chambers, H.J. Byrne, and M. Davoren, *A new approach to the toxicity testing of carbon-based nanomaterials: The clonogenic assay*, Toxicol. Lett. 174 (2007), pp. 49–60.

[19] B. Zhao, Y.-Y. He, P.J. Bilski, and C.F. Chignell, *Pristine (C60) and hydroxylated [C60(OH)24] fullerene phototoxicity towards HaCaT keratinocytes: Type I vs Type II mechanisms*, Chem. Res. Toxicol. 21 (2008), pp. 1056–1063.

[20] B.C. Schanen, A.S. Karakoti, S. Seal, D.R. Drake III, W.L. Warren, and W.T. Self, *Exposure to titanium dioxide nanomaterials provokes inflammation of an in vitro human immune construct*, ACS Nano 3 (2009), pp. 2523–2532.

[21] W. Lu, D. Senapati, S. Wang, O. Tovmachenko, A.K. Singh, H. Yu, and P.C. Ray, *Effect of surface coating on the toxicity of silver nanomaterials on human skin keratinocytes*, Chem. Phys. Lett. 487 (2010), pp. 92–96.

[22] C. Gong, G. Tao, L.Q. Yang, J.J. Liu, Q.C. Liu, and Z.X. Zhuang, *SiO_2 nanoparticles induce global genomic hypomethylation in HaCaT cells*, Biochem. Biophys. Res. Commun. 397 (2010), pp. 397–400.

[23] M.C. Moulton, L.K. Braydich-Stolle, M.N. Nadagouda, S. Kunzelman, S.M. Hussaina, and R.S. Varma, *Synthesis, characterization and biocompatibility of "green" synthesized silver nanoparticles using tea polyphenols*, Nanoscale 2 (2010), pp. 763–770.

[24] M. Huang, Z. Ma, E. Khor, and L.-Y. Lim, *Uptake of FITC-chitosan nanoparticles by A549 cells*, Pharm. Res. 19 (2002), pp. 1488–1494.

[25] J.S. Kim, T.J. Yoon, K.N. Yu, M.S. Noh, M. Woo, B.G. Kim, K.H. Lee, B.H. Sohn, S.B. Park, J.K. Lee, and M.H. Cho, *Cellular uptake of magnetic nanoparticle is mediated through energy dependent endocytosis in A549 cells*, J. Vet. Sci. 7 (2006), pp. 321–326.
[26] W. Lin, Y.W. Huang, X.D. Zhou, and Y. Ma, *Toxicity of cerium oxide nanoparticles in human lung cancer cells*, Int. J. Toxicol. 25 (2006), pp. 451–457.
[27] W. Lin, Y.-W. Huang, X.-D. Zhou, and Y. Ma, *In vitro toxicity of silica nanoparticles in human lung cancer cells*, Toxicol. Appl. Pharmacol. 217 (2006), pp. 252–259.
[28] J.M. Hillegass, A. Shukla, S.A. Lathrop, M.B. MacPherson, N.K. Fukagawa, and B.T. Mossman, *Assessing nanotoxicity in cells in vitro*, Rev. Nanomed. Nanobiotechnol. 2 (2010), pp. 219–231.
[29] H.K Patra, S. Banerjee, U. Chaudhuri, P. Lahiri, and A.K. Dasgupta, *Cell selective response to gold nanoparticles*, Nanomedicine 3 (2007), pp. 111–119.
[30] Y. Jin, S. Kannan, M. Wu, and J.X. Zhao, *Toxicity of luminescent silica nanoparticles to living cells*, Chem. Res. Toxicol. 20 (2007), pp. 1126–1133.
[31] S. Park, Y.K. Lee, M. Jung, K.H. Kim, N. Chung, E.-K. Ahn, Y. Lim, and K.-H. Lee, *Cellular toxicity of various inhalable metal nanoparticles on human alveolar epithelial cells*, Inhal. Toxicol. 19 (2007), pp. 59–65.
[32] A. Casey, E. Herzog, F.M. Lyng, H.J. Byrne, G. Chambers, and M. Davoren, *Single walled carbon nanotubes induce indirect cytotoxicity by medium depletion in A549 lung cells*, Toxicol. Lett. 179 (2008), pp. 78–84.
[33] Simon-Deckers, B. Gouget, M. Mayne-L'Hermite, N. Herlin-Boime, C. Reynaud, and M. Carrière, *Cellular uptake of magnetic nanoparticle is mediated through energy-dependent endocytosis in A549 cells*, Toxicology 253 (2008), pp. 137–146.
[34] P. Boukamp, R.T. Petrussevska, D. Breitkreutz, J. Hornung, A. Markham, and N.E. Fusenig, *Normal keratinisation in a Spontaneously immortalized aneuploid human keratinocyte cell line*, J. Cell Biol. 106 (1988), pp. 761–771.
[35] M.C. Hofmann, L.K. Braydich-Stolle, L. Dettin, E. Johnson, and M. Dym, *Immortalization of mouse germ line stem cells*, Stem Cells 23 (2005), pp. 200–210.

5

Hydrothermal synthesis and luminescent properties of $Ce_xGd_{1-x}F_3$:Ln^{3+} nanocrystals

Xiaoqing Zhang, Xianping Fan*, Xvsheng Qiao and Qun Luo

State Key Laboratory of Silicon Materials, Department of Materials Science and Engineering, Zhejiang University, Hangzhou 310027, P.R. China

The $Ce_xGd_{1-x}F_3$:Ln^{3+} ($x = 0$, 0.5 and 1, $Ln^{3+} = Tb^{3+}$, Dy^{3+} and Eu^{3+}) colloidal nanocrystals have been synthesised by the hydrothermal procedure. The $Ce_{0.5}Gd_{0.5}F_3$:Tb^{3+} nanocrystals consisted of a well-crystallised hexagonal CeF_3 phase and formed a (Ce, Gd)F_3:Tb^{3+} solid solution. The nanocrystals have a near-spherical shape with diameters between 30 and 60 nm. The excitation and emission spectra of the $Ce_xGd_{1-x}F_3$:Ln^{3+} nanocrystals confirmed the procedure of energy transfer from Ce^{3+} ions via Gd^{3+} lattices to Ln^{3+} ions was more efficient than that of direct energy transfer from Ce^{3+} ions to Ln^{3+} ions. The colloidal solutions of $Ce_{0.5}Gd_{0.5}F_3$:Tb^{3+} and CeF_3:Tb^{3+} nanocrystals showed intense green luminescence upon short wavelength ultraviolet irradiation.

Keywords: lanthanide; rare earth fluorides; nanocrystals; luminescence

1. Introduction

Recently, lanthanide (Ln^{3+})-doped nanocrystals have attracted great interest due to a wide variety of applications in high-resolution displays, electroluminescent devices, lasers and biological labels [1–4]. Among the most host materials, rare earth fluorides were good candidates for biological labelling, because of their attractive chemical and optical features, such as low toxicity, large efficient Stokes shifts, as well as high resistance to photobleaching, blinking and photochemical degradation [5]. CeF_3 was an ideal host for the luminescent material, because this material has a very low phonon energy, which made the quenching of the excited state of the Ln^{3+} ions minimal. Ce^{3+} ions have strong absorption in ultraviolet (UV) region and could transfer energy to activator ions (Tb^{3+}, Dy^{3+}), which emitted bright visible light. Great efforts have been devoted to synthesise the Ln^{3+}-doped CeF_3 nanocrystals. There were also many reports about the synthesis of CeF_3 nanostructures, such as nanocrystals, nanodisks and nanowires [6–8]. Guo [9] reported the luminescent properties of CeF_3:Tb^{3+} nanodiskettes prepared by hydrothermal microemulsion. Kong et al. [4] used diethyleneglycol (DEG) as the solvent to prepare functionalised CeF_3:Tb^{3+} nanocrystals with a SiO_2–NH_2 layer. These nanocrystals could be conjugated with biotin molecules, but the biofunctionalisation of the nanocrystals had

---

*Corresponding author. Email: fanxp@zju.edu.cn

less effect on their luminescence properties. Therefore, the $CeF_3:Ln^{3+}$ nanocrystals had great potential as biological fluorescence probes.

In order to increase the emission intensity of $CeF_3:Tb^{3+}$ nanocrystals, Li et al. [10] synthesised $CeF_3:Tb^{3+}@LaF_3$ (core–shell) nanoplates by an effective hydrothermal process. The core–shell structure made the luminescent intensity and the lifetime of the Tb^{3+} greatly enhanced in comparison to the bare $CeF_3:Tb^{3+}$. Although the structure of $CeF_3:Tb^{3+}@LaF_3$ core–shell could increase the luminescent intensity of Tb^{3+} ions, the synthesis process was usually complicated. Ce^{3+} ions and other Ln^{3+} co-doped gadolinium compounds might effectively increase the luminescent intensity of Ln^{3+} ions because the energy transfer from Ce^{3+} ions to Ln^{3+} ions via the Gd^{3+} sublattice was more efficient than the direct energy transfer from Ce^{3+} ions to Ln^{3+} ions [5,11–14]. The synthesis process and luminescence properties of Ce^{3+} and Ln^{3+} co-doped $NaGdF_4$ nanocrystals have been reported [5,11,15]. However, little attention has been paid to the synthesis process and luminescence properties of the Ln^{3+}-doped (Ce, Gd)F_3 solid solution nanocrystals. In this article, the hydrothermal synthesis and luminescent properties of $Ce_xGd_{1-x}F_3:Ln^{3+}$ ($x = 0$, 0.5 and 1, $Ln^{3+} = Tb^{3+}$, Dy^{3+} and Eu^{3+}) colloidal nanocrystals are described.

2. Experimental

To prepare the 5 mol% Ln^{3+}-doped $Ce_xGd_{1-x}F_3$ colloidal nanocrystals, 1 mmol $CeCl_3$ and/or $GdCl_3$, 0.05 mmol $LnCl_3$ and 1 mmol CTAB (cetyltrimethylammonium bromide), which was used as surfactant, were added to a beaker containing 100 mL of water. After being stirred magnetically for about 10 min at 50°C, the mixture became a colourless transparent solution. Then, slight excess sodium fluoride, 100 mL of 0.05 M NaF solution, was added to the beaker. After being stirred for 15 min at 50°C, the mixture was poured into a Teflon-lined autoclave and was heated subsequently to 180°C (the lower hydrothermal synthesis temperature would result in the formation of $NaGdF_4$ nanocrys-tals) for 2 h under a nitrogen atmosphere. Vigorous stirring was continuously applied throughout the hydrothermal process. The obtained nanocrystals were collected by centrifugation, washed with ethanol and DI water several times, and dried in a vacuum at ambient temperature.

X-ray diffraction (XRD) measurements were performed in a DIMAX-RA X-ray diffractometer (Rigaku Corporation, Tokyo, Japan) with Cu-Kα radiation at 4°/min scanning rate. Transmission electron microscopy (TEM) measurements were carried out on a JEOL2010 TEM operating at an acceleration voltage of 200 kV. Excitation and emission spectra were taken using an F-4500 fluorescence spectrophotometer (Hitachi Ltd., Tokyo, Japan) equipped with a 150 W xenon lamp as the excitation source. Luminescence decay curves were measured with a PLS920P spectrometer (Edinburgh Instruments Ltd., Livingston, UK), using excitation of microsecond flashlamps. All the measurements were performed at room temperature.

3. Results and discussion

Figure 1 shows the XRD patterns of the 1 mmol $CeCl_3$ and/or $GdCl_3$, 0.05 mmol $TbCl_3$ and 5 mmol NaF mixtures which were hydrothermal synthesised for 2 h at 180°C.

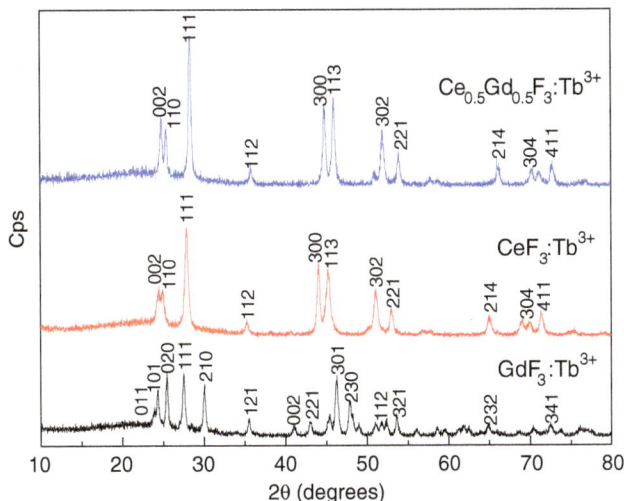

Figure 1. XRD patterns of the $Ce_xGd_{1-x}F_3$:Tb^{3+} ($x = 0$, 0.5 and 1) nanocrystals.

The peak positions of the nanocrystals prepared by hydrothermal synthesis of (1 mmol $GdCl_3 + 0.05$ mmol $TbCl_3 + 5$ mmol NaF) mixtures agreed well with the data reported in the JCPDS standard card (12-0788) for orthorhombic GdF_3 crystals. The nanocrystals prepared by hydrothermal synthesis of (1 mmol $CeCl_3 + 0.05$ mmol $TbCl_3 + 5$ mmol NaF) mixtures exhibited prominent peaks well-accordant with JCPDS standard card (08-0045) of the hexagonal CeF_3 crystal with no second phase. The XRD patterns of the nanocrystals prepared by hydrothermal synthesis of (0.5 mmol $CeCl_3 + 0.5$ mmol $GdCl_3 + 0.05$ mmol $TbCl_3 + 5$ mmol NaF) mixtures also showed prominent peaks corresponding to hexagonal CeF_3 crystal, which indicated formation of $Ce_{0.5}Gd_{0.5}F_3$:Tb^{3+} solid solution. From Figure 1, however, the position of all diffraction peaks of the $Ce_{0.5}Gd_{0.5}F_3$:Tb^{3+} nanocrystals could be observed to shift towards higher diffraction angles, which indicated decrease in the lattice parameters. The radius of six-coordinate Gd^{3+} ions (0.938 Å) is slightly smaller than that of six-coordinate Ce^{3+} ions (1.01 Å) [16]. This decrease in the lattice parameters indicated that the Gd^{3+} ions have substituted the Ce^{3+} sites and a (Ce, Gd)F_3 solid solution with hexagonal phase of CeF_3 was formed. The solid solution of (Ce,Gd)F_3 adopting the hexagonal phase of CeF_3 against the orthorhombic phase of GdF_3 may be explained by dopant-controlled crystallisation processes. It seems that lighter Ln^{3+} component in the solid solution usually dominates the crystal phase of the products [17]. Figure 2 shows the TEM images of $Ce_xGd_{1-x}F_3$:Tb^{3+} ($x = 0$, 0.5 and 1) nanocrystals. The CeF_3:Tb^{3+} and GdF_3:Tb^{3+} nanocrystals have an elongated spherical shape, with diameters between 30 and 60 nm. The $Ce_{0.5}Gd_{0.5}F_3$:Tb^{3+} solid solution nanocrystals were spherical with diameters about 40 nm.

Figure 3 gives the excitation and emission spectra of the $Ce_xGd_{1-x}F_3$:Tb^{3+} ($x = 0$, 0.5 and 1) nanocrystals in aqueous solutions. The emission spectra consisted of some sharp emission peaks corresponding to the transitions of the excited $^5D_4$ energy level to $^7F_6$ (491 nm), $^7F_5$ (543 nm), $^7F_4$ (584 nm) and $^7F_3$ (621 nm) levels of the Tb^{3+} ions. The very weak broad emission band centred at 360 nm could be ascribed to 4f–5d transition of

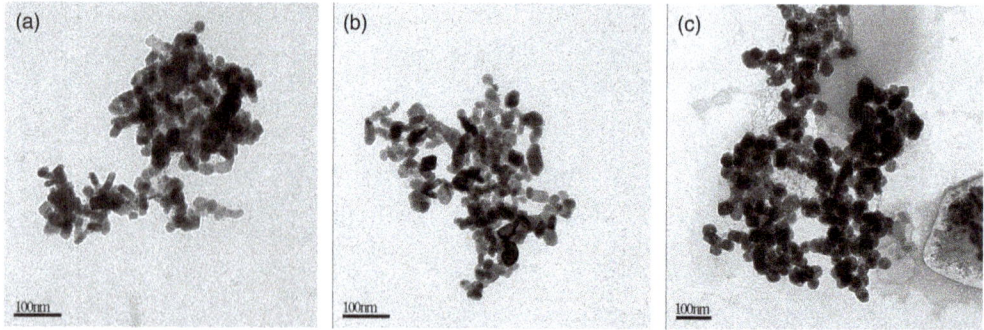

Figure 2. TEM images of the (a) $GdF_3:Tb^{3+}$, (b) $CeF_3:Tb^{3+}$ and (c) $Ce_{0.5}Gd_{0.5}F_3:Tb^{3+}$ nanocrystals.

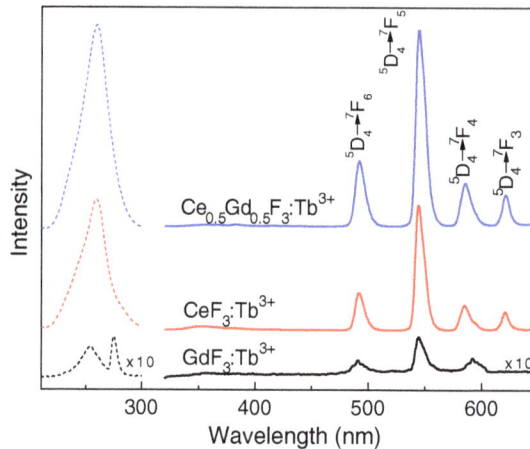

Figure 3. Excitation (monitored at 543 nm) and emission (excited at 260 nm for $x = 0$, 0.5 and at 275 nm for $x = 1$) spectra of the $Ce_xGd_{1-x}F_3:Tb^{3+}$ ($x = 0$, 0.5 and 1) nanocrystals in 5 mmol/L aqueous solutions.

Ce^{3+} ions [7]. The excitation spectra of both $CeF_3:Tb^{3+}$ and $Ce_{0.5}Gd_{0.5}F_3:Tb^{3+}$ nanocrystals exhibited only a broad band at around 260 nm corresponding to the 4f–5d transition of Ce^{3+} ions. The appearance of the Ce^{3+} transitions in the excitation spectra and Tb^{3+} transitions in the emission spectra implied that there was an efficient energy transfer from Ce^{3+} ions to Tb^{3+} ions. According to Förster's theory, the energy transfer efficiency through multipolar interaction mainly depended on the overlapping extent between the emission of sensitiser and the absorption of activator [18]. Because the wide band emission of Ce^{3+} ions and many absorption lines of Tb^{3+} ions were all at the UV region, which led to the overlapping between Ce^{3+} ions emission and Tb^{3+} absorption, Ce^{3+} ions could efficiently transfer energy to Tb^{3+} ions. On the other hand, the emission intensity of the $Ce_{0.5}Gd_{0.5}F_3:Tb^{3+}$ nanocrystals could be found to be stronger than that of the $CeF_3:Tb^{3+}$ nanocrystals, which indicated different energy transfer and migration

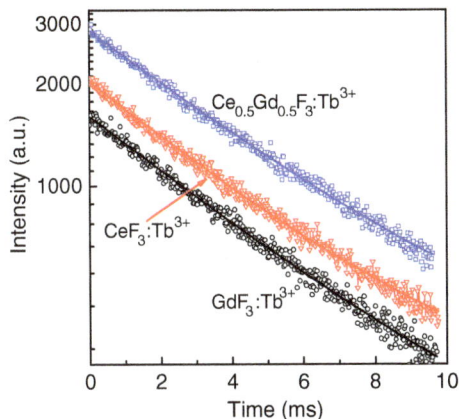

Figure 4. Luminescence decay curves and fitted curves for $^5D_4 \rightarrow {}^7F_5$ transition of the $Ce_xGd_{1-x}F_3$:Tb^{3+} ($x = 0$, 0.5 and 1) nanocrystals in 5 mmol/L aqueous solutions.

processes in both CeF_3:Tb^{3+} and $Ce_{0.5}Gd_{0.5}F_3$:Tb^{3+} nanocrystals. Wang et al. [5] have described the energy transfer mechanism in Ce^{3+} and Ln^{3+} activated gadolinium compounds. The luminance process started with the excitation of Ce^{3+} via the allowed Ce^{3+} 4f–5d transition. The excitation energy was then transferred to the Gd^{3+} and migrates over the Gd^{3+} sublattice. Finally, trapping of the migrating energy occurred at Ln^{3+} ions, followed by emission from Ln^{3+} ions. In contrast, the excitation energy of Ce^{3+} ions could be directly transferred to Ln^{3+} ions in the CeF_3:Ln^{3+} nanocrystals. The stronger emission intensity of the $Ce_xGd_{1-x}F_3$:Tb^{3+} nanocrystals in comparison with the CeF_3:Tb^{3+} nanocrystals implied that the procedure of energy transfer from Ce^{3+} ions via Gd^{3+} lattices to Tb^{3+} ions was more efficient than that of direct energy transfer from Ce^{3+} ions to Tb^{3+} ions. In addition, the excitation spectrum of GdF_3:Tb^{3+} nanocrystals presented a broad band at 255 nm corresponding to the 4f $\rightarrow$ 5d absorption of Tb^{3+} [19,20] and a sharp peak at 275 nm corresponding to the transition from $^8S_{7/2} \rightarrow {}^6I_J$ transition of Gd^{3+} [21], respectively. Because of the very small absorption coefficients of the Tb^{3+} ions [22], the emission intensity of the GdF_3:Tb^{3+} nanocrystals could be found to be much weaker than that of the CeF_3:Tb^{3+} and $Ce_{0.5}Gd_{0.5}F_3$:Tb^{3+} nanocrystals.

Figure 4 shows luminescence decay curves and fitted curves of $^5D_4$ state of Tb^{3+} in the $Ce_xGd_{1-x}F_3$:Tb^{3+} ($x = 0$, 0.5 and 1) nanocrystals (monitoring 543 nm emission line). Luminescence decay curve was best fitted into a single exponential function as $I(t) = I_0 \exp(-t/\tau)$ [23]. The average lifetime of the $^5D_4$ state of Tb^{3+} ions in $Ce_{0.5}Gd_{0.5}F_3$:Tb^{3+}, CeF_3:Tb^{3+} and GdF_3:Tb^{3+} was calculated to be about 5.17, 4.78 and 5.12 ms, respectively. The structure of nanocrystals seems to have no influence on the lifetimes of $^5D_4$ state of Tb^{3+} ions.

Due to the strong absorption in the wavelength range between 200 and 300 nm resulting from the allowed 4f $\rightarrow$ 5d transition of Ce^{3+} ions, the $Ce_{0.5}Gd_{0.5}F_3$:Tb^{3+} and CeF_3:Tb^{3+} nanocrystals showed intense emission upon short wavelength UV irradiation. Figure 5 shows photographs of visible luminescence in 5 mmol/L aqueous solutions of $Ce_xGd_{1-x}F_3$:Tb^{3+} ($x = 0$, 0.5 and 1) nanocrystals, excited with a hand-held UV

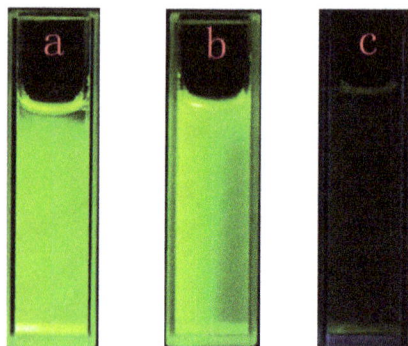

Figure 5. Photographs showing luminescence from (a) $Ce_{0.5}Gd_{0.5}F_3$:Tb^{3+}, (b) CeF_3:Tb^{3+} and (c) GdF_3:Tb^{3+} nanocrystals in 5 mmol/L aqueous solutions. All the samples were excited at 254 nm with a hand-held UV lamp.

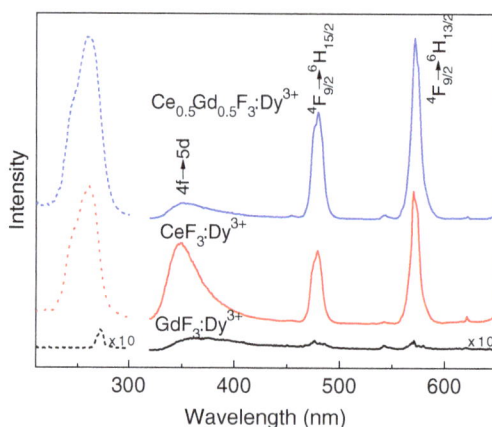

Figure 6. Excitation (monitored at 571 nm) and emission (excited at 260 nm for $x = 0$, 0.5 and at 275 nm for $x = 1$) spectra of the $Ce_xGd_{1-x}F_3$:Dy^{3+} ($x = 0$, 0.5 and 1) nanocrystals in 5 mmol/L aqueous solutions.

lamp at 254 nm. As can be seen, the aqueous solution of the $Ce_{0.5}Gd_{0.5}F_3$:Tb^{3+} and CeF_3:Tb^{3+} nanocrystals exhibited bright green luminescence.

The same procedure has been used to synthesise $Ce_xGd_{1-x}F_3$:Dy^{3+} and $Ce_xGd_{1-x}F_3$:Eu^{3+} colloidal nanocrystals. Figure 6 shows the excitation and emission spectra of the $Ce_xGd_{1-x}F_3$:Dy^{3+} ($x = 0$, 0.5 and 1) nanocrystals in aqueous solutions. The emission intensity of the CeF_3:Dy^{3+} and $Ce_{0.5}Gd_{0.5}F_3$:Dy^{3+} nanocrystals was much stronger than that of the GdF_3:Dy^{3+} nanocrystals. The sharp emission peaks at 481 and 571 nm could be assigned to the $^4F_{9/2} \rightarrow {}^6H_{15/2}$ and $^4F_{9/2} \rightarrow {}^6H_{13/2}$ transitions of Dy^{3+} ions, respectively [24,25]. In addition, a strong broad band emission between 300 and 400 nm corresponding to 4f–5d transition of Ce^{3+} ions could be observed in the CeF_3:Dy^{3+} nanocrystals. It was obvious that the energy transfer from Ce^{3+} ions to Dy^{3+} ions was incomplete. The broad band emission corresponding to 4f–5d

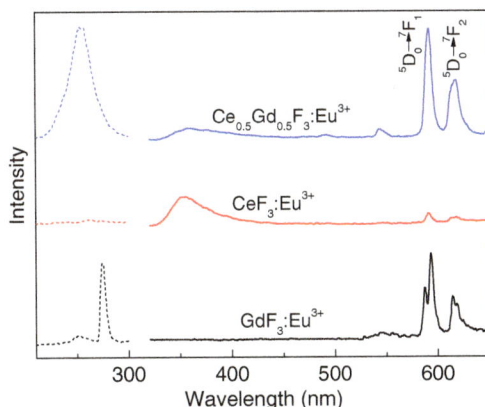

Figure 7. Excitation (monitored at 591 nm) and emission (excited at 260 nm for $x = 0$, 0.5 and at 275 nm for $x = 1$) spectra of the $Ce_xGd_{1-x}F_3:Eu^{3+}$ ($x = 0$, 0.5 and 1) nanocrystals in 5 mmol/L aqueous solutions.

transition of Ce^{3+} ions obviously decreased in $Ce_{0.5}Gd_{0.5}F_3:Dy^{3+}$ nanocrystals, which indicated the more effective $Ce^{3+} \to Dy^{3+}$ energy transfer via the Gd^{3+} sublattice.

Figure 7 shows the excitation and emission spectra of the $Ce_xGd_{1-x}F_3:Eu^{3+}$ ($x = 0$, 0.5 and 1) nanocrystals in aqueous solutions. Ce^{3+} and Eu^{3+} ions were known to quench each other's emission due to electron transfer quenching [26,27] and direct sensitisation of Eu^{3+} ions by Ce^{3+} ions was not applicable. Thus $Ce^{3+} \to Eu^{3+}$ energy transfer has to occur via the Gd^{3+} sublattice [26,28]. Therefore, the luminescence of Eu^{3+} ions could hardly be observed in the $CeF_3:Eu^{3+}$ nanocrystals under 260 nm excitation. In contrast, the emission intensity of the $Ce_{0.5}Gd_{0.5}F_3:Eu^{3+}$ nanocrystals was much stronger than that of the $CeF_3:Eu^{3+}$ nanocrystals, which indicated that the energy transfer procedure from Ce^{3+} to Eu^{3+} ions via Gd^{3+} lattices was very efficient.

4. Conclusions

The $Ce_xGd_{1-x}F_3:Ln^{3+}$ ($x = 0$, 0.5 and 1, $Ln^{3+} = Tb^{3+}$, Dy^{3+} and Eu^{3+}) colloidal nano-crystals have been synthesised by the hydrothermal procedure. Similar to $CeF_3:Tb^{3+}$ nanocrystals, the $Ce_{0.5}Gd_{0.5}F_3:Tb^{3+}$ nanocrystals showed well-crystallised hexagonal CeF_3 phase. The nanocrystals have a near-spherical shape with diameters between 30 and 60 nm. The emission intensity of the $Ce_{0.5}Gd_{0.5}F_3:Ln^{3+}$ colloidal nanocrystals was obviously stronger than that of the $CeF_3:Ln^{3+}$ colloidal nanocrystals, which indicated that the procedure of energy transfer from Ce^{3+} ions via Gd^{3+} lattices to Ln^{3+} ions was more efficient than that of direct energy transfer from Ce^{3+} ions to Ln^{3+} ions.

Acknowledgements

The authors gratefully acknowledge the support provided for this research by the Science and Technology Department of Zhejiang Province (2008C21051), the Research Fund of the Doctoral

Program of Higher Education of China (20070335012) and Program for Changjiang Scholars and Innovative Research Team in University.

References

[1] T. Jüstel, H. Nikol, and C. Rondad, *New developments in the field of luminescent materials for lighting and displays*, Angew. Chem. Int. Ed. 37 (1998), pp. 3085–3103.

[2] K. Kawano, K. Arai, H. Yamada, N. Hashimoto, and R. Nakata, *Application of rare-earth complexes for photovoltaic precursors*, Sol. Energy Mater. Sol. Cells 48 (1997), pp. 35–41.

[3] D.B. Barber, C.R. Pollock, L.L. Beecroft, and C.K. Ober, *Amplification by optical composites*, Opt. Lett. 22 (1997), pp. 1247–1249.

[4] D.Y. Kong, Z.L. Wang, C.K. Lin, Z.W. Quan, Y.Y. Li, C.X. Li, and J. Lin, *Biofunctionalization of $CeF_3:Tb^{3+}$ nanoparticles*, Nanotechnology 18 (2007), p. 075601.

[5] F. Wang, X.P. Fan, and M.Q. Wang, *Multicolour $PEI/NaGdF_4:Ce^{3+}$, Ln^{3+} nanocrystals by single-wavelength excitatio*, Nanotechnology 18 (2007), p. 025701.

[6] K. Kompe, H. Borchert, J. Storz, A. Lobo, S. Adam, T. Moller, and M. Haase, *Green-emitting $CePO_4:Th/LaPO_4$ core-shell nanoparticles with 70% photoluminescence quantum yield*, Angew. Chem. Int. Ed. 42 (2003), pp. 5513–5516.

[7] L. Zhu, Q. Li, X.D. Lin, J.Y. Li, Y.F. Zhang, J. Meng, and X.Q. Cao, *Morphological control and luminescent properties of CeF_3 nanocrystals*, J. Phys. Chem. C 111 (2007), pp. 5898–5903.

[8] Z.Y. Wang, Z.B. Zhao, and J.S. Qiu, *Carbon nanotube templated synthesis of CeF_3 nanowires*, Chem. Mater. 19 (2007), pp. 3364–3366.

[9] H. Guo, *Photoluminescent properties of $CeF_3:Tb^{3+}$ nanodiskettes prepared by hydrothermal microemulsion*, Appl. Phys. B 84 (2006), pp. 365–369.

[10] C.X. Li, X.M. Liu, P.P. Yang, C.M. Zhang, H.Z. Lian, and J. Lin, *LaF_3, CeF_3, $CeF_3:Tb^{3+}$, and $CeF_3:Tb^{3+}@LaF_3$ (core-shell) nanoplates: Hydrothermal synthesis and luminescence properties*, J. Phys. Chem. C 112 (2008), pp. 2904–2910.

[11] J.C. Boyer, J. Gagnon, L.A. Cuccia, and J.A. Capobianco, *Synthesis, characterization, and spectroscopy of $NaGdF_4:Ce^{3+}$, $Tb^{3+}/NaYF_4$ core/shell nanoparticles*, Chem. Mater. 19 (2007), pp. 3358–3360.

[12] C.X. Guo and S.Y. Bai, *Luminescence and energy transfer of $Gd_xY_{1-x}P_5O_{14}:Ce,Tb.1.$ Preparation, structure and spectral properties*, J. Rare Earths 13 (1995), pp. 257–261.

[13] H.S. Kiliaan, J.F. Kotte, and G. Blasse, *Energy-transfer in the luminescent system $Na(Y,Gd)F_4\_Ce,Tb$*, J. Electrochem. Soc. 134 (1987), pp. 2359–2364.

[14] G. Blasse and B.C. Grabmaver, *Luminescent Materials*, Springer, Berlin, 1994.

[15] C.H. Liu, H. Wang, X.R. Zhang, and D.P. Chen, *Morphology- and phase-controlled synthesis of monodisperse lanthanide-doped $NaGdF_4$ nanocrystals with multicolor photoluminescence*, J. Mater. Chem. 19 (2009), pp. 489–493.

[16] H.C. Yang, C.Y. Li, and Y. Tao, *The luminescence of $CaYBO_4: RE^{3+}$ (RE = Eu, Gd, Tb, Ce) in VUV-visible region*, J. Lumin. 126 (2007), pp. 196–202.

[17] F. Wang, Y. Han, C.S. Lim, Y.H. Lu, J. Wang, J. Xu, H.Y. Chen, C. Zhang, M.H. Hong, and X.G. Liu, *Simultaneous phase and size control of upconversion nanocrystals through lanthanide doping*, Nature 263 (2010), pp. 1061–1065.

[18] F. Wang, H.W. Song, G.H. Pan, and L.B. Fan, *Luminescence properties of Ce^{3+} and Tb^{3+} ions codoped strontium borate phosphate phosphors*, J. Lumin. 128 (2008), pp. 2013–2018.

[19] J.C. Wang, Q. Lin, and Q.F. Liu, *Synthesis and luminescence properties of Eu or Tb doped Lu_2O_3 square nanosheets*, Opt. Mater. 29 (2007), pp. 593–597.

[20] E. Zych, *On the reasons for low luminescence efficiency in combustion-made $Lu_2O_3:Tb$*, Opt. Mater. 16 (2001), pp. 445–452.

[21] F.T. You, Y.X. Wang, J.H. Liu, and Y. Tao, *Hydrothermal synthesis and luminescence properties of NaGdF₄:Eu*, J. Alloys Compounds 343 (2002), pp. 151–155.

[22] X.P. Fan, X.H. Lv, S.B. Li, F. Wang, and M.Q. Wang, *The in-situ synthesis process and luminescence behavior of a p-hydroxybenzoic acid-terbium complex in sol-gel derived host materials*, J. Mater. Chem. 12 (2002), pp. 3560–3564.

[23] C.X. Li, J. Yang, Z.W. Quan, P.P. Yang, D.Y. Kong, and J. Lin, *Different microstructures of ss-NaYF₄ fabricated by hydrothermal process: Effects of pH values and fluoride sources*, Chem. Mater. 19 (2007), pp. 4933–4942.

[24] B.V. Rao and S. Buddhudu, *Emission analysis of RE³⁺ (Dy³⁺ or Tb³⁺): Ca$_s$Ln(=Y,Gd)(VO₄)₃ powder phosphors*, Mater. Chem. Phys. 111 (2008), pp. 65–68.

[25] G.F. Wang, W.P. Qin, and D.S. Zhang, *Enhanced photoluminescence of water soluble YVO₄:Ln³⁺ (Ln = Eu, Dy, Sm, and Ce) nanocrystals by Ba²⁺ doping*, J. Phys. Chem. C112 (2008), pp. 17042–17045.

[26] M. Yu, J. Lin, Z. Wang, J. Fu, S. Wang, H.J. Zhang, and Y.C. Han, *Fabrication, patterning, and optical properties of nanocrystalline YVO₄:A (A = Eu³⁺, Dy³⁺, Sm³⁺, Er³⁺) phosphor films via sol-gel soft lithography*, Chem. Mater. 14 (2002), pp. 2224–2231.

[27] R.T. Wegh, H. Donder, K.D. Oskam, and A. Meijerink, *Visible quantum cutting in LiGdF₄:Eu³⁺ through downconversion*, Science 283 (1999), pp. 663–665.

[28] M. Karbowiak, A. Mech, A. Bednarkiewicz, W. Strek, and L. Kepinski, *Comparison of different NaGdF₄:Eu³⁺ synthesis routes and their influence on its structural and luminescent properties*, J. Phys. Chem. Solids 66 (2005), pp. 1008–1019.

6

Polyelectrolyte-mediated self-assembly of polystyrene nano-spheres into honeycomb-patterned microbeads

Haifeng Guo[a,b]* and Feng Ye[a]

[a]*Institute for Advanced Ceramics, School of Materials Science and Engineering, Harbin Institute of Technology, Harbin 150080, China;* [b]*PetroChina Pipeline R&D Center, Langfang, Hebei 065000, China*

Honeycomb-patterned polystyrene (PS) microbeads were prepared through the self-assembly of PS nano-spheres induced by solvent evaporation in the presence of a cationic polyelectrolyte – poly(diallyl dimethyl ammonium chloride) (PDAD). Irregular particles were synthesised when the suspension was added with a single anionic electrolyte – sodium dodecyl sulphate (SDS), and large beads without regular patterns were obtained with the addition of both of SDS and PDAD. The building blocks – PS nano-spheres were synthesised by a nano-emulsion method using SDS as the emulsifier. The surfaces of the synthetic PS nano-spheres were negatively charged by sulphate radicals of SDS. The interaction between positively charged PDAD and negatively charged sulphate radicals of PS nano-spheres plays the crucial role in the formation of honeycomb patterns. Also, the honeycomb-patterned PS microbeads have a potential application in acid-sensitive drug delivery system.

Keywords: polystyrene; nano-sphere; self-assembly; honeycomb-pattern; drug delivery system

1. Introduction

Recently, many methods have been explored to direct the bottom-up assembly of nanoparticles and one-dimensional nanoscaled building blocks, such as nanotubes, nanowires and nanorods, into various novel superstructures (1–5). Evaporation-induced self-assembly is a kind of system far from equilibrium that can exhibit complex transitory structures, even when equilibrium fluctuations are limited (6). The relatively weak attractions between nanocrystals which are efficiently screened in solution become manifest as the solvent evaporates, initiating intricate assembly (7).

Carefully designed interfaces will lead to desirable surface properties for nanoscaled building blocks (5,8). In recent reports (9–11), an extensive group of nanocrystals with tunable sizes and modified surfaces was prepared by properly tuning the chemical reactions at the interfaces, and then these nanocrystals were assembled into desirable hierarchical structures induced by solvent evaporation. However, most of these evaporation-induced assemblies are mainly two-dimensional arranged structures, but those of three-dimensional regular patterns are rarely reported.

Herein, honeycomb-patterned PS microbeads were synthesised through the self-assembly of PS nano-spheres induced by evaporation in the presence of polyelectrolyte-poly(diallyl dimethyl ammonium chloride) (PDAD) for the first time. The samples were characterised by scanning electron microscopy (SEM), energy-dispersive spectroscopy (EDS), Fourier transform infrared spectroscopy (FTIR) and transmission electron microscopy (TEM).

---

*Corresponding author. Email: guo_iac@yahoo.com.cn

2. Materials and methods

2.1. *Synthesis of PS nano-spheres*

All the chemicals were of analytical grade and used as purchased without further purification. First, PS nanoparticles were synthesised using a modified emulsion polymerisation method (12). The emulsifier, sodium dodecyl sulphate (SDS) (2 g); monomer, styrene (30 mL); dispersing stabiliser, trisodium phosphate (0.15 g) and the initiator, ammonium persulphate (1.56 g) were used. The polymerisation reaction was carried out at 75°C. After polymerisation for several hours, the resulting sol suspension was centrifuged (20,000 r/min) and thoroughly washed with de-ionised water, and then PS gel was prepared. One half of the gel was dried at room temperature and the other half was re-dispersed in water (0.05 g/mL, the mass of the gel was directly measured), and colloid suspension formed. The colloid suspension was further divided into several parts, one was added with aqueous $CaCl_2$ solution (10 wt%) (suspension-I), and one (12 mL) diluted 10 times the volume with de-ionised water and designated as suspension-II. The precipitate of suspension-I was centrifuged and thoroughly washed.

2.2. *Self-assembly of PS nano-spheres and characterisations*

The above suspension-II (120 mL), which was polymerised for 2 h, was divided into four parts after sonicating for 10 min in an ultrasonic bath. The first one (30 mL) was added with 3 mL aqueous PDAD solution (1 wt%), and then dispersed by ultrasonication for 10 min (suspension-II-1); the second one was added with SDS solution (1 wt%) (suspension-II-2). The third one was first added with PDAD and then with SDS solution (suspension-II-3), and the rest was suspension-II (was polymerised for 2 h). The washed precipitate of suspension-I was re-dispersed in de-ionised water with a sonication of 10 min and PS suspension solution was obtained (suspension-I-1). Then, these PS suspensions (had the same pH ranged in 7–8) were individually evaporated within an aluminium container in a natural environment (18°C, for about 30 h). The as-dried products upon the container inner surfaces were slightly scraped, collected together and gently ground into powders. The obtained PS powders were observed by SEM (FEISirion200, 20 kV) and TEM (EM20, Philips, 200 kV). The zeta potential of suspensions was determined by a zetasizer 3000HSa. The elemental information was investigated by EDS. Also, the chemical composition of the synthetic PS particles was characterised by FTIR.

2.3. *PS assemblies collapsed in weak acid environment*

The as-dried PS powders were re-dispersed in phosphate buffer solution (10 mM, pH range 4–5). After being incubated for 0.5 h, the suspensions were characterised by SEM.

3. Results and discussion

Figure 1 shows the SEM and TEM micrographs of the synthetic PS nano-spheres. The PS nanospheres of the suspensions without adding anything (suspension-II) were well-dispersed on the aluminium chips. These PS particles had a uniform diameter, about 15 nm for 2-h polymerisation and 40 nm for 5-h polymerisation. But the particles were seriously aggregated when the PS suspension was added with aqueous $CaCl_2$ solution.

Figure 2 illustrates the SEM micrographs of PS products obtained from different suspensions. With the addition of PDAD, micro-scaled honeycomb patterned beads were obtained. The diameter of the microbeads was in the range 1–5 μm. At larger magnifications, regular honeycomb patterns were observed upon the surface of the microbeads. These honeycomb patterns were in

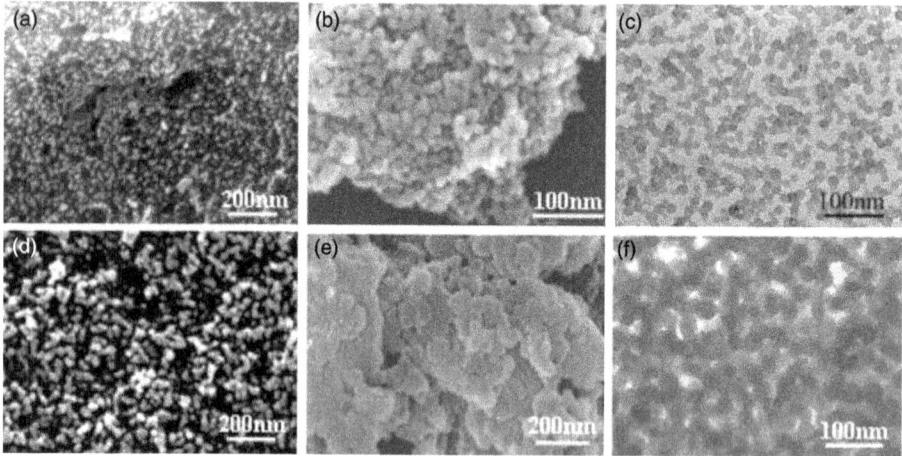

Figure 1.   SEM micrographs of PS nanoparticles (suspension-II): (a) dispersed particles polymerised for 2 h, (b) particles polymerised for 2 h, (d) dispersed particles polymerised for 5 h and (e) aggregated particles polymerised for 5 h; TEM micrographs of PS nanoparticles (suspension-II) (c) dispersed particles polymerised for 2 h and (f) aggregated particles polymerised for 5 h.

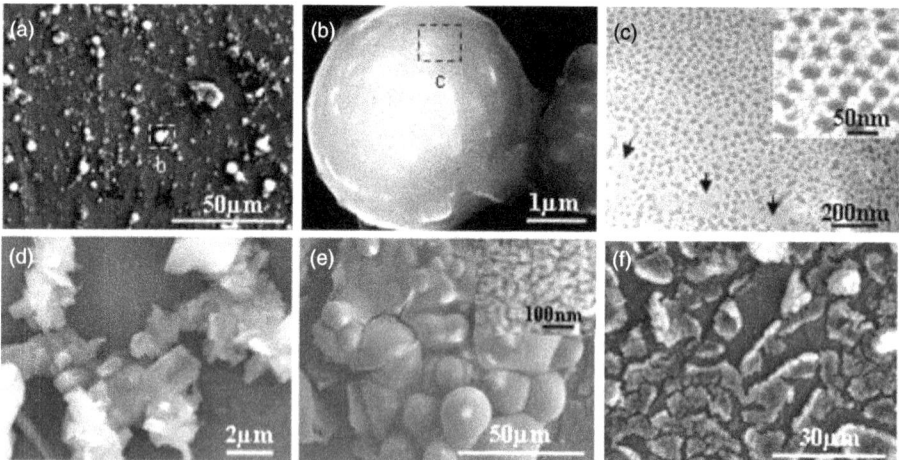

Figure 2.   SEM micrographs of honeycomb-patterned PS microbeads with different magnifications (from suspension-II-1) (a) 1250×, (b) 2500× and (c) 100,000× (inset: 500,000×); (d) irregular cylinder-like particles from suspension-II-2; (e) large irregular sphere-like particles from suspension-II-3 (inset shows the densely aggregated PS nano-spheres); (f) scroll-like particles from suspension (was polymerised for 2 h) at a high concentration.

hexagon-like morphology and were composed of uniform nanospheres which had the same size and morphology as the PS nanoparticles shown in Figure 1(a) and (c). The voids of the honeycomb had a scale of around 30 nm, and two neighbouring voids shared a linear-like boundary consisting of 2–4 nanoparticles. But there existed some dense nanoparticles patches among the honeycomb patterns (arrow denoted). The large micrographs of these PS beads demonstrate the details, as shown in Figure S1 of support information. But when the suspension was added with SDS or mixtures of SDS and PDAD, irregular cylinder-like or large sphere-like particles without regular patterns were formed, respectively (Figure 2d and e). Also, when the initial suspension-II

(a)   (b)

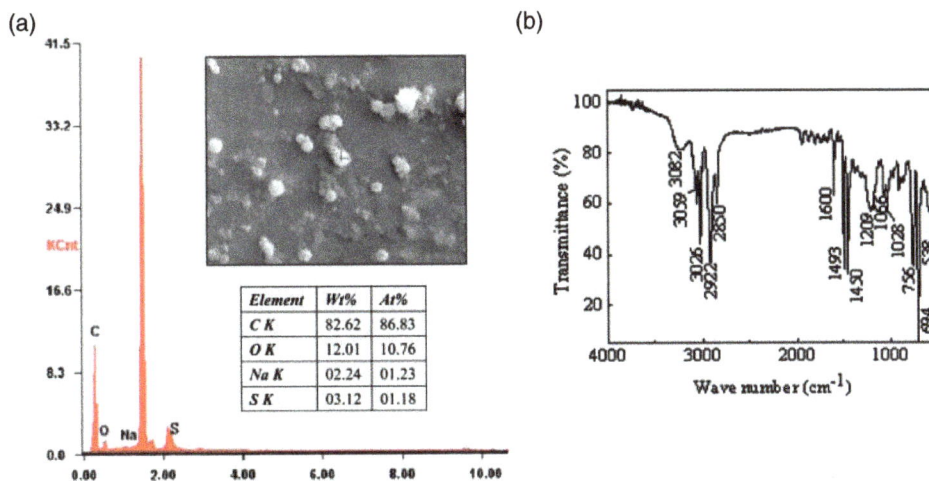

Figure 3.   (Colour online) (a) EDS analysis of PS microbead: the bead contained S and Na which came from SDS (inset micrograph shows the position for collection of EDS data), and (b) FTIR spectrum of the as-dried PS gel product which was polymerised for 2 h.

(not being diluted) was slowly dried, scroll-like products were observed with no microbeads, as shown in Figure 2(f).

The EDS spectrum of the PS microbead (Figure 3a) indicates that the microbeads mainly contained C and O together with a small amount of S and Na. This result shows that SDS molecules have incorporated into the PS microbeads, which was well-consistent with the result of FTIR. Figure 3(b) gives the FTIR spectrum of the dried gel-like product which was polymerised for 2 h. The bands at 3082, 3059, 3026, 2922, 2850, 1600, 1493, 1450, 756, 694, 538 and 1670–1944 cm^{-1} were of polystyrene, and the band at 1200 cm^{-1} was attributed to the ν(S=O) stretch vibration of sulphate radicals ($-SO_3^-$). The results indicated that the synthetic PS nanoparticles were negatively charged by SDS hydrophilic groups on the surfaces. The zeta potential of suspension-II (was polymerised for 2 h, pH ~9) was measured to be −42 eV. This further supported the result of FTIR.

In the usual evaporation-induced self-assembly (9–11), the surface of the building blocks was modified or coated with hydrophobic or hydrophilic groups at first, and then directly assembled these modified building blocks into various structures. Polystyrene nano-spheres were polymerised inside SDS nano-emulsion which had a hydrophobic core and hydrophilic outer surface. During the process of polymerisation, hydrophobic styrene monomer continuously entered the hydrophobic parts of SDS emulsion. Consequently, PS nano-spheres gradually grew up with the polymerisation time with negatively charged SDS hydrophilic groups on the surfaces. Compared to the traditional method, herein, PDAD – a kind of cationic polyelectrolyte which is positively charged in aqueous solution, or SDS, or both, was added into the suspensions before dryness, separately. With the addition of PDAD, PS nanoparticles (~15 nm) self-assembled into three-dimensional honeycomb-patterned microbeads. This was possibly due to negatively charged PS nanoparticles which were absorbed onto the twisted PDAD chains due to the electrostatic attractions, and the PDAD–PS complex acted as the nuclei for the assembly of PS nano-spheres into microbeads. As the solvent evaporated, the relatively weak interactions among the negatively charged PS nano-spheres, positively charged PDAD chains and water molecules became manifest, initiating continuous self-assembly of PS nano-spheres upon the nuclei, resulting in honeycomb-patterned microbeads. For the SDS-added sample, the repulsive interaction produced no nuclei or cores, which could absorb the PS nano-spheres around and promote the regular self-assembly when the solvent evaporated, resulting in the formation of irregular cylinder-like particles, and for

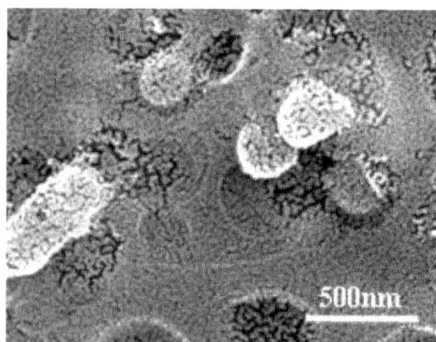

Figure 4.   SEM micrograph of honeycomb-patterned PS microbeads was incubated in a weak environment for 0.5 h.

the sample with the addition of both SDS and PDAD, the firstly added PDAD interacted with PS nano-spheres and formed PDAD–PS complexes, and the SDS added later interacted with PDAD–PS complexes forming large complexes. When the solvent evaporated, these large complexes had weak interactions with each other, and therefore were only aggregated, resulting in the formation of large irregular sphere-like particles. The PS nano-spheres with negatively charged hydrophilic groups could easily disperse in the diluted aqueous solution due to the repulsive interaction between two nanoparticles. But when the suspension was added with aqueous $CaCl_2$ solution, the negative charge was neutralised by Ca^{2+}, and therefore, the obtained PS nanoparticles were seriously aggregated. Bormashenko et al. (13) reported that the solution concentration plays an important role in these kinds of self-assembly. In our experiments, a high-solution concentration (0.05 g/mL) led to irregular aggregations. It is worth mentioning that the obtained honeycomb-patterned PS microbeads collapsed into single PS nanoparticles when being incubated in a weak acid environment for 0.5 h, as shown in Figure 4. But the mechanism is very complicated and might be achieved by a further computation simulation. This study can be extended to self-assembly for other kinds of materials and allow us to have a better understanding of evaporation-induced self-assembly. The obtained honeycomb-patterned PS microbeads have the potential application in acid-sensitive drug delivery system because the microbeads possess a large amount of 30 nm voids on the surfaces which are favourable in adsorption of drug molecules, and have a pH-responsive property.

4.   Conclusion

Self-assembly of polystyrene nano-spheres with negative hydrophilic surfaces into honeycomb-patterned microbeads was successfully achieved, induced by evaporation in the presence of PDAD. The building blocks – PS nano-spheres had a uniform diameter distribution of around 15 nm for 2-h polymerisation and 40 nm for 5-h polymerisation. The particles could be well-dispersed on the aluminium chips when nothing was added, but they were seriously aggregated when the solution was added with aqueous $CaCl_2$ solution. The resultant honeycomb-patterned microbeads were in the range 1–5 μm. When anionic electrolyte SDS or both SDS and PDAD were added, the resulting product was irregular cylinders or large beads.

References

C.B. Murray, S. Sun, W. Gaschler, H. Doyle, T.A. Betley, and C.R. Kagan, *Colloidal synthesis of nanocrystals and nanocrystal superlattices*, IBM J. Res. Dev. 45 (2001), pp. 47–55.

Y. Xia, P. Yang, Y. Sun, Y. Wu, B. Mayers, B. Gates, Y. Yin, F. Kim, and H. Yan, *One-dimensional nanostructures: Synthesis, characterization, and applications*, Adv. Mater. 15 (2003), pp. 353–389.

A.L. Rogach, D.V. Talapin, E.V. Shevchenko, A. Kornowski, M. Haas, and H. Weller, *Organization of matter on different size scales: Monodisperse nanocrystals and their superstructures*, Adv. Funct. Mater. 12 (2002), pp. 653–664.

J. Polleux, N. Pinna, M. Antonietti, and M. Niederberger, *Ligand-directed assembly of preformed titania nanocrystals into highly anisotropic nanostructures*, Adv. Mater. 16 (2004), pp. 436–439.

F. Kim, S. Kwan, J. Akana, and P. Yang, *Langmuir-Blodgett nanorod assembly*, J. Am. Chem. Soc. 123 (2001), pp. 4360–4361.

E. Rabani, D.R. Reichman, P.L. Geissler, and L.E. Brus, *Drying-mediated self-assembly of nanoparticles*, Nature 426 (2003), pp. 271–274.

J. Tang, G. Ge, and L.E. Brus, *Gas-liquid-solid phase transition model for two-dimensional nanocrystal self-assembly on graphite*, J. Phys. Chem. B 106 (2002), pp. 5653–5658.

P.D. Yang, *Wires on water*, Nature 425 (2003), pp. 243–244.

H.F. Chen, B.H. Clarkson, K. Sun, and J.F. Mansfield, *Self-assembly of synthetic hydroxyapatite nanorods into an enamel prism-like structure*, J. Colloid Interface Sci. 288 (2005), pp. 97–103.

X. Wang, J. Zhuang, Q. Peng, and Y.D. Li, *Liquid–solid–solution synthesis of biomedical hydroxyapatite nanorods*, Adv. Mater. 18 (2006), pp. 2031–2034.

X. Chen, Z.M. Chen, B. Yang, G. Zhang, and J.C. Shen, *Regular patterns generated by self-organization of ammonium-modified polymer nanospheres*, J. Colloid Interface Sci. 269 (2004), pp. 79–83.

A.Y. Men'shikova, T.G. Evseeva, B.M. Shabsel's, I.V. Balanina, A.K. Sirotkin, and S.S. Ivanchev, *Synthesis of monodisperse polystyrene particles in the presence of sodium dodecyl sulfate and carboxyl-containing initiator*, Russ. J. Appl. Chem. 78 (2005), pp. 1008–1012.

E. Bormashenko, R. Pogreb, A. Musin, O. Stanevsky, Y. Bormashenko, G. Whyman, O. Gendelman, and Z. Barkay, *Self-assembly in evaporated polymer solutions: Influence of the solution concentration*, J. Colloid Interface Sci. 297 (2006), pp. 534–540.

7

Exploring the insertion of ethylenediamine and bis(3-aminopropyl)amine into graphite oxide

A.B. Dongil[a], B. Bachiller-Baeza[a,b*], I. Rodríguez-Ramos[a,b] and A. Guerrero-Ruiz[b,c]

[a]*Instituto de Catálisis y Petroleoquímica, CSIC, c/Marie Curie No. 2, Cantoblanco, 28049 Madrid, Spain;* [b]*Grupo de Diseño y Aplicación de Catalizadores Heterogéneos, Unidad Asociada UNED-CSIC (ICP), Spain;* [c]*Dpto. Química Inorgánica y Técnica, Fac. de Ciencias, UNED, C/ Senda del Rey nº 9, 28040, Madrid, Spain*

The influence of ethylenediamine and bis(3-aminopropyl)amine insertion into graphite oxide using toluene and ethanol–water media was studied by elemental analysis, X-ray diffraction, X-ray photoelectron spectroscopy, thermogravimetry, temperature-programmed desorption and diffuse reflectance infrared Fourier transform spectroscopy. In every case, amine was incorporated between the basal planes and on the edges of the material through different interactions with epoxy, hydroxyl, carboxyl and carbonyl groups resulting in new-layered materials. Results show that incorporation was homogeneous for ethylenediamine irrespective of the solvent and that the higher level of intercalation in non-polar media was probably a consequence of the inhibition of competitive reactions with the solvent. For bis(3-aminopropyl)amine, the higher reaction times required to obtain satisfactory degrees of intercalation in the non-polar solvent suggests that with this solvent the process is kinetically controlled for longer chain amines.

Keywords: graphite oxide; amine-functionalized; intercalation; surface reaction; expanded graphite oxide

1. Introduction

Graphite oxide (GO) is a pseudo-two-dimensional layered carbonaceous material rich in oxygen-containing functional groups whose composition depends on the level of oxidation. It is generally accepted that it consists of randomly distributed intact graphitic regions and regions of aliphatic six-membered rings.[1] Epoxy and hydroxyl functionalities lie above and below these carbon layers while carbonyl groups and carboxylic acid groups having strong acidic character decorate the sheet edges.[2] This variety of functional groups bound to the carbon sheets makes the material strongly hydrophilic and gives rise to interesting properties like cation exchange capacity and rich intercalation ability. In this sense, great attention in recent years has been paid to the preparation of intercalated GO composites by intercalation of polar organic molecules, polymers or inorganic materials inside the nanospace between the layers by different methods.[3–6] Furthermore, the exfoliation and subsequent reduction of GO has been proposed as a promising route for large-scale production of graphene sheets and graphene-based materials.[7,8] The electrical, optical and mechanical properties have urged the application of these materials in the fabrication of microelectronic devices.

---

*Corresponding author. Email: b.bachiller@icp.csic.es

Most of the reported works are based on the ability of GO to form stable colloidal suspensions in aqueous media.[9,10] However, in some applications, it would be more favourable to improve the dispersion of the material in organic media, as polar aprotic solvents, and to increase its compatibility with polymeric matrices. Therefore, several methods have been explored to attach some molecules to GO as a previous step of chemical functionalization.[6,11,12] Similar approaches have also been applied to reduce the aggregation and restacking of graphene sheets due to the van der Waals interactions. These strategies provide a path to improve the synthesis of the quasi-two-dimensional carbon nanosheets and to tune the properties or even the number of graphene sheets. On the other hand, the use of bifunctional molecules can provide materials with good dispersibility in both aqueous and organic media[13] and could lead to materials with free functional groups or pillared structures which open its applicability as catalytic supports or adsorbents.

With this in mind, we have explored the incorporation of two different amines, a diamine and a triamine, inside the nanospace of a lab-prepared GO. In order to get a better knowledge of the solvent effect on the intercalation process, the preparation of the materials was carried out in two solvents with different polarities, ethanol and toluene, and the materials were characterized by a variety of complementary techniques.

2. Experimental

GO was synthesized from natural graphite powder (99.999% stated purity, −200 mesh, Alfar Aesar) following a modification of Brodie's method.[14] The graphite was added to a reaction flask containing fuming nitric acid (HNO_3, 20 ml/g of support) which was previously cooled to 273 K in an iced bath, after that potassium chlorate (8 g/g of support) was slowly added. The reaction was left to proceed for 21 h under stirring, and the final solid was filtered, extensively washed with deionized water until neutral pH and dried under vacuum at 323 K overnight.

Incorporation of ethylenediamine (E) and bis(3-aminoethyl)amine (T) was performed according to a reported method.[15] For each experiment, 0.02 mol amine per gram of GO was dissolved in ethanol and added dropwise to a suspension of GO in water at room temperature. The mixture was stirred for 24 h and then the solution was filtered off, the solid washed with ethanol/water and finally dried under vacuum at 333 K overnight. Samples obtained by this procedure were named GOE and GOT. A second procedure was also studied where the amine and the GO were dispersed in toluene (t). The mixture was stirred for 24 h for GOE-t and GOT-t and for 96 h for GOT-t-96, the solution was filtered off, and the solid washed with toluene and finally dried under vacuum at 333 K.

C, H and N contents of the samples were determined by elemental analysis with a LECO CHNS-932 system. The crystalline phases present in these samples were determined from the X-ray diffraction patterns (XRD). The diffractograms were recorded on a Polycristal X'Pert Pro PANalytical apparatus using Ni-filtered Cu Kα radiation ($\lambda = 0.15406$ nm) and a graphite monochromator. For each sample, Bragg's angles between 4° and 90° were scanned at a rate of 0.04 deg/s.

The amount of oxygen groups present on the surface was determined by thermogravimetric analysis (TG) in a CI Electronics microbalance (MK2-MC5). The sample was purged under flowing He for 2 h and then heated at a 5 K·min^{-1} rate up to 1023 K. The chemical nature of these functional groups was evaluated by temperature-programmed desorption (TPD) experiments under vacuum in a conventional volumetric system connected to a SRS RGA-200 mass spectrometer. The sample (0.010 g) was evacuated for 30 min at room temperature and ramped to 1023 K at a 5 K·min^{-1} rate.

Infrared spectra were collected by using a VARIAN 670 spectrometer equipped with a diffuse reflectance accessory. The diffuse reflectance infrared Fourier transform (DRIFT) spectra were recorded by a mercury–cadmium–telluride detector from 256 scans and with a resolution of

$4\,cm^{-1}$. GO and its derivatives were thoroughly ground and mixed with pre-dried potassium bromide to a final concentration of approximately 1% (w/w).

The surface modification was also analysed by X-ray photoelectron spectroscopy (XPS). The spectra were obtained on a ESCAPROBE P spectrometer from OMICROM equipped with a EA-125 hemispherical multichannel Electronics analyzer. The pressure in the analysis chamber was kept below 10^{-9} Pa. The excitation source was the Mg Kα line ($hv = 1253.6\,eV$, 150 W). The binding energy was referenced to the C 1s line at 284.6 eV. The error in determination of electron binding energies and the line widths did not exceed 0.2 eV.

3. Results

3.1. *Structural characterization*

Elemental analysis of the samples (Table 1) corroborates the incorporation of amine. In ethanol, the degree of molecular incorporation was similar for both amines, and comparing with parent GO, a reduction in the amount of oxygen groups was evident. In toluene, the amount of triamine incorporated was much lower than that of diamine, and the oxygen groups were reduced to a lower extent than for the ethanol series.

Alterations in the structure of GO after the modification treatment are observed in the XRD patterns of the amine-modified derivatives shown in Figure 1. The layer-to-layer distance or d-spacing for the prepared GO, obtained from the position of the 001 reflection, was 0.55 nm. Since interlayer spacing depends on the degree of oxidation and hydration level, this low value compared with those reported in the literature might suggest a relatively low degree of oxidation and hydration.[15,16] The interlayer distance increased from 0.55 nm for the parent GO to 0.62 and 0.70 nm for GOE and GOT samples, respectively. This is consistent with previous results reporting that the longer the chain of alkylamines, the longer the interlayer spacing.[15] For samples prepared in toluene, the patterns showed the presence of several peaks consequence of layers with different d-spacings. The 001 reflection peak for GOE-t and GOT-t-96 was shifted to lower angles compared to those for samples prepared in ethanol (GOE and GOT), which implied higher interlayer spacing (0.75 and 0.85 nm). The shoulder at 13.3° could be due to residual solvent molecules intercalated between the layers. Moreover, the patterns showed the peak at 15.8°, the same value that for the parent GO, indicating that some galleries remained unoccupied. The relative intensity of this peak can be related to the incorporation degree of the amine molecule and therefore, for longer chain molecules and in apolar media, longer intercalation times are needed. On the other hand, the broad hump that appeared in the range $2\theta = 20$–26° for some of the patterns, and that corresponds to a d-spacing of 0.36 nm, was close to the value for natural graphite, $2\theta = 26.6°$ (Figure 1(a)), and suggested the presence of random or disordered graphitic platelets (turbostratic platelets). For OGT-t-96, the presence of this hump could also be related

Table 1.   Elemental analysis.

Sample	%C	%O	%N
GO	66.81	32.03	0.02
GOE	78.08	15.83	5.09
GOT	77.09	13.86	7.52
GOE-t	69.62	17.51	10.12
GOT-t	68.51	25.91	3.74
GOT-t-96	64.58	25.86	6.68

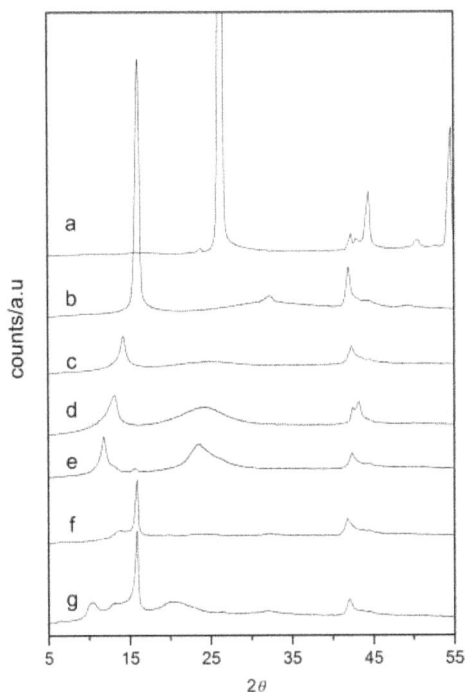

Figure 1.   XRD patterns for (a) natural graphite, (b) GO (c) GOE, (d) GOT, (e) GOE-t, (f) GOT-t and (g) GOT-t-96.

to the contraction of unoccupied layers caused by the expansion of adjacent layers where amine molecules are intercalated [17] and not only to GO reduction.

3.2. *Thermal analyses*

The original GO and the modified samples were characterized by thermal analyses, and Figure 2 displays the TG curves. The profile for GO sample showed a strong weight loss of 30% at 535 K that corresponded to the release of different gases, H_2O, CO and CO_2, during the reductive exfoliation of GO and that was a consequence of the decomposition of the oxygen-containing functionalities, mainly epoxy and OH groups. For higher temperatures, the weight loss followed a slower rate, and a total mass loss of about 40% was eventually reached. The thermograms for the amine-modified samples also showed a pronounced drop but at lower temperature, 430 K, and a continuous or slower decay for higher temperatures. This shift of the main weight loss to lower temperatures is consistent not only with the increase in the interlayer distance observed by XRD, but also with the reducing character of the intercalated molecules. It has been reported that the higher the reducing character of the intercalated molecule, the lower the exfoliation temperature.[17] The total weight loss was in all cases lower than that for GO sample, although the one for the GOT-t-96 comparing with that for GOT-t was consequent with a higher amount of amine incorporated.

The analysis of the gases evolved during the TPD experiments was also carried out. The TPD profiles for GO and the amine-modified samples are displayed in Figures 3 and 4, respectively. The profile of the GO sample showed intense signals at 530 K for $m/z = 18$, 44 and 28, which are assigned to H_2O, CO_2 and CO, coming from the decomposition of the oxygen functional groups. For the amine-modified samples, the fragmentation patterns of the studied amine were considered, and therefore, some other mass/charge ions were monitored during the TPD to study

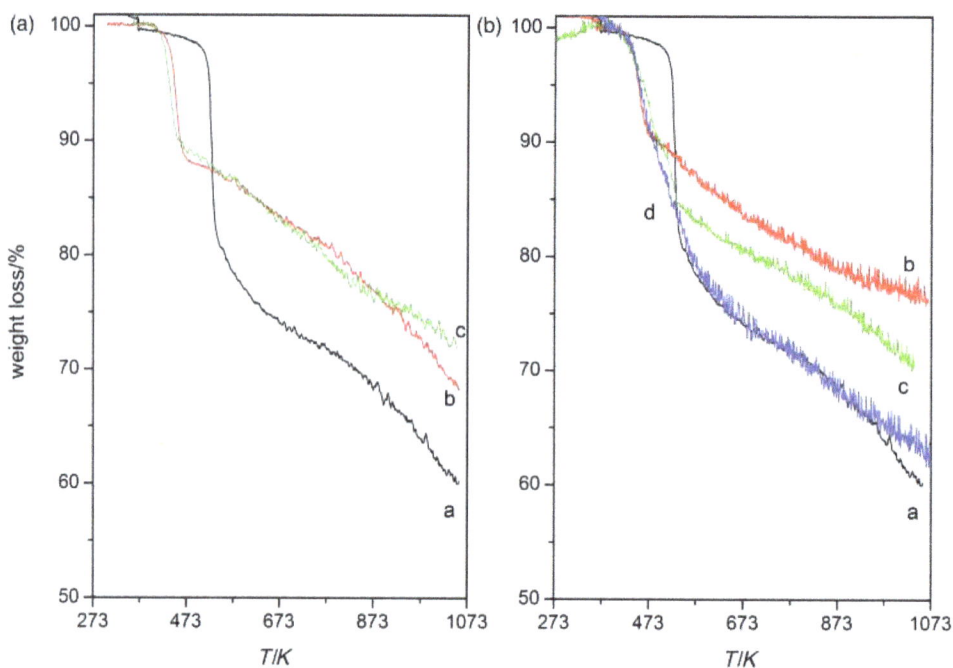

Figure 2.   Thermograms in He for the amine-modified samples (A) prepared in ethanol (a) GO, (b) GOE and (c) GOT, and (B) prepared in toluene, (a) GO, (b) GOE-t, (c) GOT-t and (d) GOT-t-96.

their contribution to the general spectra. To asses for evolution of molecularly adsorbed amine, we have followed signals $m/z = 30$ and 73 for diamine and triamine, respectively. Mass 73 was selected because major fragment for the triamine molecule is $m/z = 44$ and can be masked by the CO_2 contribution. The lack of m/z 30 and 73 could be taken as a proof that other kinds of interactions were responsible of the amine attachment. The first point to remark when analysing the TPD profiles was that an intense peak at 470 K and a small shoulder at higher temperatures, 573 K, were observed for samples prepared in ethanol (Figure 3). Contributions to the main peak of $m/z = 18$, 17, 16 (not shown), 44 and 28 were identified. The intense and wide peak obtained for $m/z = 18$ was assigned to water desorption. The presence of masses $m/z = 44$ and 28 seemed to indicate that the amine thermal decomposition contributed to the profile. In addition, the appearance of a shoulder at 723 K with contributions of m/z 17 and 44 seemed to reflect the additional linkage of the amine through stronger bonds. This peak was probably due to decomposition of the amine in NH_3 and fragments leading to $m/z = 44$. The evolution of $m/z = 2$ assigned to H_2 was observed at high temperatures and could be in line with the dehydrogenation of the amine or of its decomposition products.

The general spectra were clearly different for samples prepared in toluene (Figure 4). While for the diamine-modified sample, the TPD showed one broad peak at 450 K and a small one at 560 K, for the triamine samples they showed three main peaks at 440, 490 and 530 K. Again, the masses that mainly contributed to the profile were 18, 17, 28 and 44 whose relative contribution increased considerably in the two peaks at higher temperatures. The temperature range of the latter peak, i.e. 530–540 K, corresponded to the temperature of GO exfoliation. Thus, this peak must be attributed to the decomposition of functional groups in the residual not intercalated GO structures, which were also detected by XRD. In addition, as occurred for samples prepared in ethanol, a small shoulder at 680 K with contributions of masses 17 and 44 was detected. This

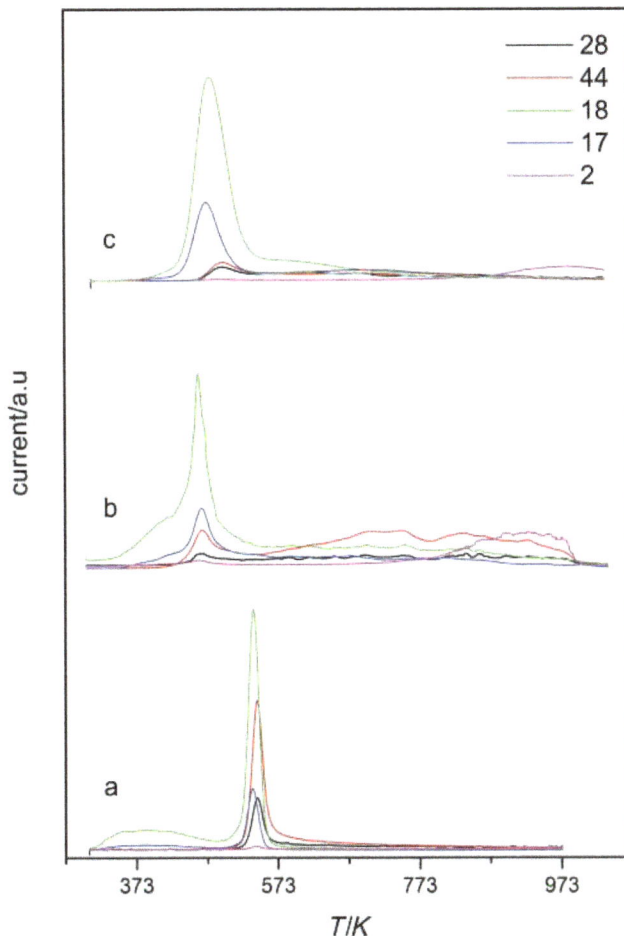

Figure 3.   TPD profiles for amine-modified samples prepared in ethanol (a) GO, (b) GOE and (c) GOT.

indicated that amine was strongly adsorbed on other type of oxygen-surface groups different from hydroxyl and epoxy groups. An increase in $m/z = 2$ for temperatures starting at 873 K can reveal, as suggested above, the dehydrogenation of the adsorbed amine.

3.3. *DRIFT*

The infrared spectra of GO also changed after modification (Figure 5). Peaks appearing in the range 3000–3600 and at 1380 cm^{-1} which were assigned to the stretching and bending vibrations of the acidic $-$OH groups [12,15,18] seemed to be significantly attenuated for both modified series. For samples prepared in ethanol, the band at 1713 cm^{-1} due to C=O stretching vibration of carbonyl and/or carboxyl groups situated at the edges of the lamellae disappeared, while the peak at 1578 cm^{-1} could indicate the presence of un-oxidized graphitic domains.[18] New bands at 1410 and 1216 cm^{-1} could be assigned to δCH$_2$ and ρNH$_2$, respectively,[19] and would confirm the presence of amine. On the other hand, new features related to the presence of amine could be detected in the samples prepared in toluene, i.e. peaks at 1667 and 1565 cm^{-1}.[12] The appearance of these two peaks added to the fact that the C=O stretching vibration was absent or shifted to lower wavenumber could indicate the formation of an amide bond. Therefore, these peaks

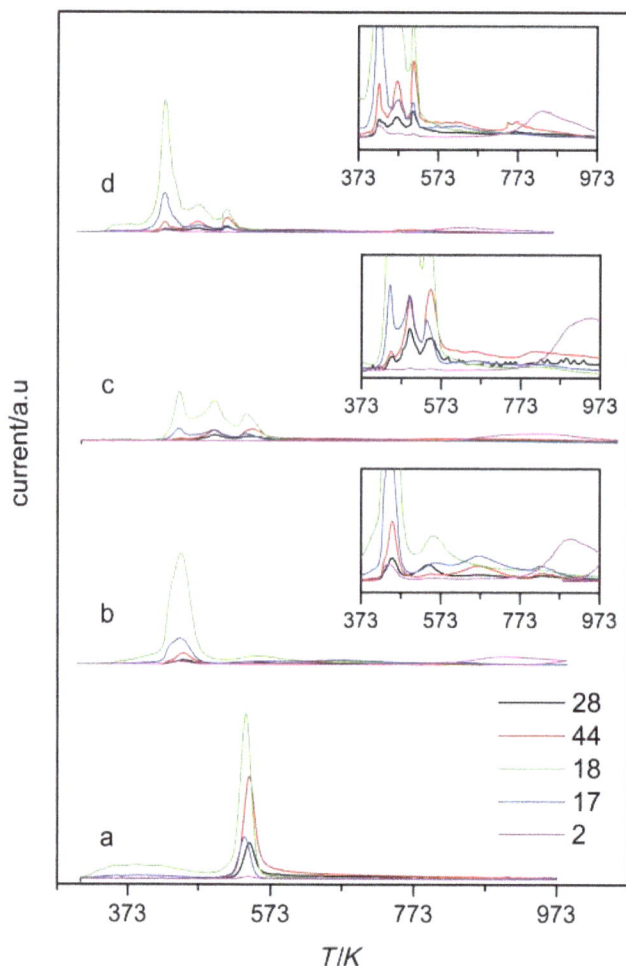

Figure 4.   TPD profiles for amine-modified samples prepared in toluene (a) GO, (b) GOE-t, (c) GOT-t and (d) GOT-t-96.

would correspond to the amide carbonyl stretching mode and the coupling of the C−N stretching vibration with the CHN deformation vibration, respectively.[15] However, these two bands could also be assigned to asymmetric and symmetric deformation bands of hydrogen-bonded NH_2.[20] This would indicate that amine molecules reacted with acidic hydroxyl groups forming hydrogen-bonded NH_2 groups. Therefore, DRIFT spectra of the functionalized-samples are not conclusive concerning amidation, but they have allowed the identification of amine functionalities and of differences in the interaction depending on the solvent used in the synthesis.

3.4. *XPS analysis*

XPS analyses were carried out to asses the type of functional groups on the prepared samples. The relative amounts of the different species, C, O and N, were calculated from the corresponding peak areas divided by the sensitivity factors (1.00 for C, 2.85 for O and 1.77 for N) and are shown in Table 2. The O/C ratios obtained for the GO series showed a small decrease in the oxygen content after the functionalization with amines at expenses of an increase in the N/C ratio, which

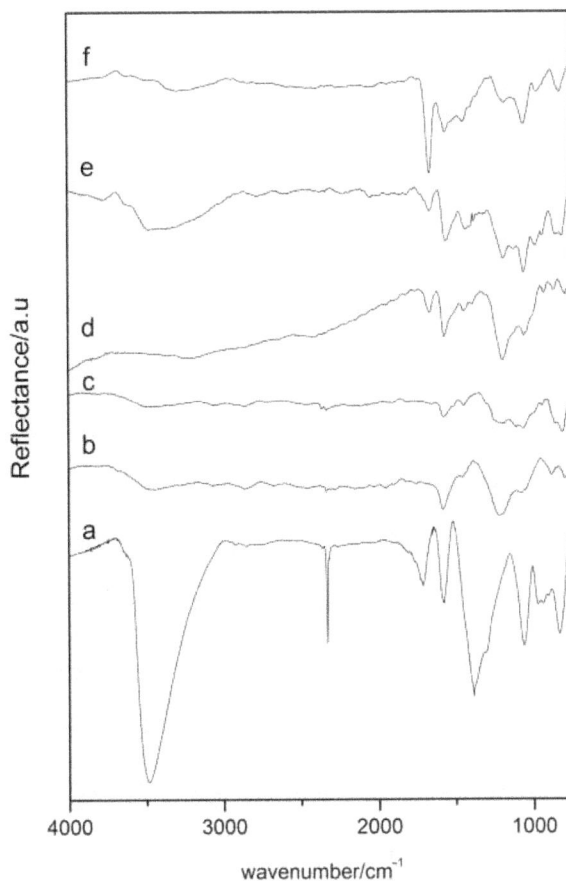

Figure 5. DRIFT spectra for amine-modified samples (a) GO, (b) GOE, (c) GOT, (d) GOE-t, (e) GOT-t and (f) GOT-t-96.

Table 2. XPS results.

Sample	C_{arom}/C_{defect}	Atomic ratio			Relative peak area O 1s (%)			
		O/C	N/C	N1/N2	Peak A	Peak B	PeakC	PeakD
GO	0.16	0.22	—	—	17	47	25	11
GOE	0.90	0.20	0.04	3.8	24	62	12	2
GOT	0.75	0.18	0.06	3.0	21	63	15	1
GOE-t	0.38	0.14	0.11	1.2	28	64	8	—
GOT-t	0.41	0.13	0.09	1.1	33	56	10	—
GOT-t-96	0.49	0.15	0.27	1.1	66	28	6	—

suggested the incorporation of the amine. Moreover, the N/C ratios for the toluene series were higher than for the ethanol series implying a greater proportion of amines over the surface of the samples of that series. Comparing the data of the two amines in each series, the higher N/C ratios for the triamine samples were consequent taking into account the number of amine groups in the molecule. The N/C ratio obtained for GOT-t-96 was also remarkable.

According to the possible components contributing to the C 1s region in carbon materials, the position of the maximum for the high-resolution C 1s spectra of the GO sample (supplementary material, Figure SI 1), 285.1 eV and the absence of the $\pi \rightarrow \pi^*$ peak characteristic of large polyaromatic structures were coherent with the structure of the GO. Thus, the creation of epoxy and hydroxyl groups induced this loss in the aromatic character of the graphene sheets. Besides, the asymmetry of the peak was also consequence of the high proportion of oxygen groups. After incorporation of the amines, the maxima of the C 1s spectra shifted to lower BE, which indicated that the contribution due to graphitic structures had increased. Furthermore, a shoulder at intermediate BE has developed and, in fact, was maximum for GOT-t-96 sample, and could be ascribed to the presence of C—N structures and to a change in the concentration of the different type of oxygen groups.

The modification in the type of groups that occurred after intercalation of both amines was reflected by the shift of the maximum of the O 1s spectra to lower BE. Therefore, information about the nature of the surface oxygen groups was obtained by deconvolution of the peak in four components: carbonyl oxygen (peak A, 531.4); oxygen atoms in C—O bonds like hydroxyl and ether groups (peak B, 532.6); oxygen atoms in acidic carboxyl groups (peak C, 534.6) and adsorbed water (peak D, 535.6). Table 2 shows the relative area of the different peaks. The data suggested a decrease in the proportion of carboxyl groups that was slightly higher for samples prepared in toluene. This went along with an increase in the C=O type groups.

The N1s peak could also be deconvoluted in two components at 399.5 (N1) and 400.5 eV (N2) (supplementary material, Figure SI 2). The assignation of species was complicated but a tentative assignment from the literature could be hydrogen-bonded and hydrogen-bonded protonated amines, respectively, species that seemed to coexist and that were in equilibrium.[19] For the samples prepared in toluene, the N1/N2 ratio decreased, which would point to a higher proportion of protonated species.

4. Discussion

The incorporation of amine molecules to the GO interlayer has been assessed by all the characterization techniques applied in this study and seems to depend on the solvent used in the preparation and on the size or length chain of the molecule. The relative orientation of molecules intercalated on GO depends on their structure and functionalities, and on the type of interaction with the GO.[19,21,22] According to these variables, some of the possible conformations are shown in Scheme 1. Briefly, with the alkyl chain parallel to the layer forming a one- or double-layer structure; with the chain titled leading to a tail conformation; or bridge or loop conformations in the case of bifunctional molecules.[18] The d-spacing values of our samples vary in the range 0.62–0.85 nm and seem to indicate that amine molecules are probably arranged in a bilayer or tail orientation to yield such values. The amount of amine molecules incorporated when the samples were prepared in the polar solvent was similar for both ethylenediamine (E) and bis(3-aminoethyl)amine (T) and their distribution very homogeneous as the single peak detected in the XRD spectra suggested. In apolar media where interferences due to the solvent–solid interactions should be absent, the level of incorporation of E was higher, while it decreased for T. In addition, the remaining peak at $2\theta = 15.8°$ in the XRD spectra of OGT-t and OGT-t-96 reflects the existence of unoccupied galleries. Furthermore, the discrepancy between the N content from elemental analysis and the XPS results, where the N/C ratios showed that T amine was incorporated in a higher extent on the samples prepared in toluene, also confirms the heterogeneity in OGT-t and OGT-t-96 samples. All these facts can only be rationalized if it is considered that T molecules are located close to or at the edges of GO layers in the toluene samples. Consequently, these amine molecules could

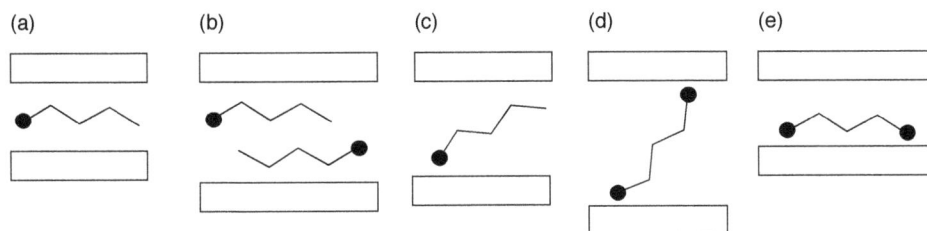

Scheme 1. Representation of possible conformations for intercalated mono- and bifunctional molecules on graphite oxide: a) monolayer, b) bilayer, c) tail, d) bridge e) loop. •: functional group

be hindering the access of other molecules to the interlayer. Besides, the higher levels of incorporation obtained for sample treated during 96 h would be indicating that the intercalation is a kinetically controlled process that depends on the chain length and number of functionalities of the amine and on the solvent.

On the other hand, the elemental analysis and the O/C ratios obtained by XPS prove that the incorporation of amine reduces the amount of oxygen groups. The decrease in the DRIFT bands associated to the oxygen groups, added to the decrease in the weight loss observed for all the TG profiles of the modified samples, seems to confirm the elimination of some of the oxygen-containing groups due to reaction with the amine or to the partial reduction forming disordered carbon. The C_{arom}/C_{defect} ratios obtained by XPS (Table 1) have also pointed to partial reduction of the GO structure that is higher in the samples prepared in ethanol. The small development in the XP spectra of the 291 eV peak that is associated with the polyaromatic character and the D/G intensity ratios obtained by Raman (not shown) seem to indicate that the reduction implies not the generation of a continuous sp^2 phase, but most likely the creation of sp^2 domains isolated by disordered domains.[23]

The question of the type of interaction between the amine and the oxygen groups is not trivial to explain. The amine must be mainly adsorbed either by hydrogen bonding interactions between amine moieties and the OH groups or by nucleophilic attack on the epoxy groups, although the exchange of protons with acidic groups at the periphery of the GO layers cannot be disregarded. Furthermore, contrary to what it has been reported for graphite and carbon nanofibers series also modified with bis(3-aminoethyl)amine,[24] $m/z = 27$ was absent in the TPD profile for the amine-modified GO samples. The presence of mass $m/z = 27$, which is assigned to desorption of HCN during the TPD, was associated in that case to the interaction of the amine with the carboxylic acid groups. Then, its absence in the profiles of amine-modified GO samples suggests a small number of amide species, consistent with the low amount of carboxylic groups, or that the amine undergoes different chemical transformations during the TPD run and leads to desorption of other products. The formation of amide bonds has been reported after $SOCl_2$ activation of the carboxylic groups.[25] So, other groups present on the GO different from carboxylic groups seem to be involved in the attachment of the amine molecule. Several groups based in XRD and infrared analysis have proposed the nucleophilic attack to the epoxy groups as the main route of insertion.[15,26] Our results seem to be more in agreement with the interaction via hydrogen bonding forming hydrogen-bonded and hydrogen-bonded protonated species. Furthermore, the deconvolution of the N 1s region of the XPS has shown that the solvent used in the synthesis influences the ratio between both types of species. The lower N1/N2 ratios for samples prepared in toluene would also reflect a higher proportion of hydrogen-bonded protonated species that would be in line with the favoured acid–base interaction between the amine and the surface acid groups, i.e. the hydroxyl groups and/or the carboxylic groups located on the borders of the graphene sheets. This interaction is favoured in this case because the solvent used for the preparation does not interfere or react with the solid surface. However, it is also important to highlight that those assignments were suggested for alkylamines,[19] and in our case, we are studying amines with two and three NH_2 groups. Therefore, the presence of some additional amine groups should also be considered.

5. Conclusion

In conclusion, the method of preparation of amine-modified GO derivatives is crucial in determining the properties of the material, i.e. the amount of amine incorporated and the type of interaction. Specifically, the temperature and time of reaction, the solvent, the alkyl chain of the amine molecule to be incorporated and the previous reduction or not of the GO nanosheets are the main parameters to be controlled. The incorporation of amines into GO with a polar solvent, ethanol–water, implies a homogeneous distribution, but the partial reduction of the GO structure. The resulting material does not contain large size sp^2 domains, but abundant and isolated by disordered domains. The degree of incorporation when the apolar solvent, toluene, is used depends on the amine chain length: the larger the molecule the lower the amount of amine incorporated and the homogeneity. Comparing both solvents, toluene gives better levels of intercalation than ethanol for ethylendiamine due to the lack of interferences caused by the solvent and the GO. Finally, the amine molecules inserted in the GO interlayers seem to be mainly retained by hydrogen bonding interactions with the OH species.

Acknowledgement

The authors acknowledge the MICINN of Spain (Projects CTQ-2011-29272-C04-01 and -03) for financial support.

References

[1]   He H, Klinowski J, Forster M, Lerf A. A new structural model for graphite oxide. Chem Phys Lett. 1998;287:53–56.
[2]   Jeong H-K, Lee YP, Lahaye RJWE, Park M-H, An KH, Kim IJ, Yang C-W, Park CY, Ruoff RS, Lee YH. Evidence of graphitic AB stacking order of graphite oxides. J Am Chem Soc. 2008;130:1362–1366.
[3]   Hu Z-L, Aizawa M, Wang Z-M, Yoshizawa N, Hatori H. Synthesis and characteristics of graphene oxide-derived carbon nanosheet-Pd nanosized particle composites. Langmuir. 2010;26(9):6681–6688.
[4]   Matsuo Y, Higashika S, Kimura K, Miyamoto Y, Fukutsuka T, Sugie Y. Synthesis of polyaniline-intercalated layered materials via exchange reaction. J Mater Chem. 2002;12:1592–1596.
[5]   Wang Z-M, Hoshinoo K, Shishibori K, Kanoh H, Ooi K. Surfactant-mediated synthesis of a novel nanoporous carbon-silica composite. Chem Mater. 2003;15:2926–2935.
[6]   Matsuo Y, Fukunaga T, Fukutsuka T, Sugie Y. Silylation of graphite oxide. Carbon. 2004;42:2113–2130.
[7]   Pei S, Cheng H-M. The reduction of graphene oxide. Carbon. 2012;50:3210–3228.
[8]   Guardia L, Villar-Rodil S, Paredes JI, Rozada R, Martínez-Alonso A, Tascón JMD. UV light exposure of aqueous graphene oxide suspensions to promote their direct reduction, formation of graphene–metal nanoparticle hybrids and dye degradation. Carbon. 2012;50:1014–1024.
[9]   Park S, Ruoff RS. Chemical methods for the production of graphenes. Nat Nanotechnol. 2009;4:217–224.
[10] Stankovich S, Dikin D, Piner RD, Kohlhaas KA, Kleinhammes A, Jia Y, Wu Y, Nguyen SBT, Ruoff RS. Synthesis of graphene-based nanosheets via chemical reduction of exfoliated graphite oxide. Carbon. 2007;45:1558–1565.
[11] Georgakilas V, Bourlinos AB, Zboril R, Steriotis TA, Dallas P, Stubos AK, Trapalis C. Organic functionalisation of graphenes. Chem Commun. 2010;46:1766–1768.
[12] Stankovich S, Piner RD, Nguyen ST, Ruoff RS. Synthesis and exfoliation of isocyanate-treated graphene oxide nanoplatelets. Carbon. 2006;44:3342–3347.

[13] Xu C, Wu X, Zhu J, Wang X. Synthesis of amphiphilic graphite oxide. Carbon 2008;46(2):386–389.

[14] Brodie BC. On the atomic weight of graphite. Philos Trans R Soc Lond. 1859;149:249–259.

[15] Bourlinos AB, Gournis D, Petridis D, Szabó T, Szeri A, Dékány I. Graphite oxide: chemical reduction to graphite and surface modification with primary aliphatic amines and amino acids. Langmuir. 2003;19:6050–6055.

[16] You S, Luzan SM, Szabo T, Talyzin AV. Effect of synthesis method on solvation and exfoliation of graphite oxide. Carbon. 2013;52:171–180.

[17] Barroso-Bujans F, Cerveny S, Verdejo R, del Val JJ, Alberdy JM, Alegría A, Colmenero J. Permanent adsorption of organic solvents in graphite oxide and its effect on the thermal exfoliation. Carbon. 2010;48:1079–1087.

[18] Szabó T, Berkesi O, Forgó P, Josepovits K, Sanakis Y, Petridis D, Dékány I. Evolution of surface functional groups in a series of progressively oxidized graphite oxides. Chem Mater. 2006;18:2740–2749.

[19] Herrera-Alonso M, Abdala AA, McAllister MJ, Aksay IA, Prud'homme RK. Intercalation and stitching of graphite oxide with diaminoalkanes. Langmuir. 2007;23:10644–10649.

[20] Matsuo Y, Miyabe T, Fukutsuka T, Sugie Y. Preparation and characterization of alkylamine-intercalated graphite oxides. Carbon. 2007;45:1005–1012.

[21] Matsuo Y, Niwa T, Sugie Y. Preparation and characterization of cationic surfactant intercalated graphite oxide. Carbon. 1999;37:897–901.

[22] Dékány I, Krüger-Grasser R, Weiss A. Selective liquid sorption properties of hydrophobized graphite oxide nanostructures. Colloid Polym Sci. 1998;276:570–576.

[23] Singh V, Joung D, Zhai L, Das S, Khondaker SI, Seal S. Graphene based materials: past, present and future. Progress Mater Sci. 2011;56:1178–1271.

[24] Dongil AB, Bachiller-Baeza B, Guerrero-Ruiz A, Rodríguez-Ramos I, Martínez-Alonso A, Tascón JMD. Surface chemical modifications induced on high surface area graphite and carbon nanofibers using different oxidation and functionalization treatments. J Col Inter Sci. 2011;355:179–189.

[25] Niyogi S, Bekyarova E, Itkis ME, McWilliams JL, Hamon MA, Haddon RC. Solution properties of graphite and graphene J Am Chem Soc. 2006;128:7720–7721.

[26] Wang S, Chia P-J, Chua L-L, Zhao L-H, Png R-Q, Sivaramakrishnan S, Zhou M, Goh RG-S, Friend RH, Wee AT-S, Ho PK-H. Band-like transport in surface-functionalized highly solution-processable graphene nanosheets. Adv Mater. 2008;20:3440–3446.

8

Experimental evaluation of the penetration of TiO_2 nanoparticles through protective clothing and gloves under conditions simulating occupational use

Ludwig Vinches[a]*, Nicolas Testori[a], Patricia Dolez[a,b], Gérald Perron[a], Kevin J. Wilkinson[c] and Stéphane Hallé[a]

[a]École de technologie supérieure, 1100 Notre-Dame Ouest, Montréal, QC, Canada H3C 1K3; [b]CTT Group, 3000 rue Boullé, Saint-Hyacinthe, QC, Canada J2S 1H9; [c]Department de chimie, Université de Montréal, C.P. 6128, succ. Centre-ville, Montréal, QC, Canada H3C 3J7

Titanium dioxide nanoparticles ($nTiO_2$) are found in numerous manufactured products. While a few studies have been carried out to measure the efficiency of chemical protective clothing and gloves against nanoparticles (NPs), they have generally not considered the conditions prevailing in occupational settings. This study was designed to evaluate the resistance of protective clothing against NPs under conditions simulating occupational use. Nitrile and butyl rubber gloves, as well as cotton/polyester woven and polyolefin non-woven clothing samples were placed into contact with $nTiO_2$ in the form of powders or colloidal solutions. Simultaneously, mechanical deformations were applied to the samples. Preliminary results showed that $nTiO_2$ may penetrate some of the materials after prolonged dynamic deformations and/or when the NPs are in colloidal solutions. The effect was partly attributed to modifications in the physical and mechanical properties of protective materials that were induced by repetitive mechanical deformations.

Keywords: titanium dioxide nanoparticles; protective clothing; protective gloves; occupational use

1. Introduction

The National Science Foundation has estimated that the number of nano-products manufactured around the world will double every three years until 2020 [1]. At the same time, the number of workers and researchers in the industry will attain six million people. Nanoparticles (NPs), particularly titanium dioxide nanoparticles ($nTiO_2$), are increasingly present in commercial products such as paints, varnishes and sunscreens [2,3] and as such are a new potential source of hazard. Indeed, increasing numbers of studies are cautioning their likely harmful effects on health. For example, a small increase in the number of cancers among workers in contact with $nTiO_2$ has been reported [4]. Moreover, studies conducted on rats and mice, exposed to 250-nm TiO_2 pigment particles and 20-nm $nTiO_2$, observed a greater occurrence of pulmonary effects [5]. In addition, the observed inflammatory response was higher with $nTiO_2$ [6,7]. Based on these results, the International Agency for Research on Cancer has classified nanosized titanium dioxide in the 2B-group, as being possibly carcinogenic to humans [8]. In response to this classification, several agencies have recommended the application of the precautionary principle [9]. While this implies the use of protective gloves and clothing, for the most part, they have not yet been validated for use with NPs.

*Corresponding author. Email: ludwig.vinches.1@ens.etsmtl.ca

With respect to the limited number of studies that have examined the efficiency of protective clothing and gloves exposed to NPs, most involved aerosols. Furthermore, conclusions reached on the efficiency of protective gloves appear to be conflicting. For example, while the penetration of 30 and 80 nm graphite NPs has been reported for nitrile, vinyl, latex and neoprene commercial glove samples [10], no penetration was observed for the same glove models for 40 nm graphite, 10 nm TiO_2 or Pt NPs [11,12]. In the case of air-permeable fabrics, tests have been performed with oleic acid, KCl, NaCl, graphite, TiO_2 and Pt nano-aerosols with diameters as small as 10 nm [11,13]. Several authors have determined the variation of NP penetration through fabrics to be function of the particle diameter and air flow rate, in agreement with the filtration theory [14,15]. Nonetheless, a much higher efficiency against NP penetration was measured with a thin high-density polyethylene (PE) non-woven membrane than with other, thicker fabrics, both with and without air flow [11,12].

In occupational settings, exposure to NPs may also involve both colloidal solutions and powders. This situation is especially relevant to protective clothing and gloves. For example, Ahn et al. [16] exposed nitrile, latex and cotton glove samples to nanoclay and alumina nanopowder. Scanning electron microscope (SEM) observations revealed the accumulation of the NPs in micrometer-size pores at the surface of the nitrile and latex gloves. Furthermore, large quantities of NPs were found on and within the fibres of the cotton glove samples. Penetration through glove materials is likely to be facilitated by the presence of a liquid carrier. In addition, the polymers forming the gloves are often sensitive to solvents such that some of them may sustain significant swelling [17,18]. Finally, when they are in service, protective clothing and gloves are subjected to mechanical constraints, for example, biaxial strains at interphalangeal joints, elbows and knees, which may also affect the penetration of NPs. Indeed, elastomer resistance to organic solvents has been shown to be sensitive to small elongations (less than 20%) [19]. In addition, prehension forces of up to 500 N have been induced by adult hands [20]. Such a compression of the membrane may induce an effect similar to the application of a high pressure with the NPs in an aerosol, i.e. reduction of the barrier properties.

This paper reports some preliminary results on the penetration of NPs through protective clothing and gloves obtained with commercial 15 nm $nTiO_2$ in powder and colloidal solutions. The tested protective clothing and glove materials consisted of nitrile and butyl rubber as well as of cotton/polyester weaves and polyolefin non-woven fabrics. Repeated biaxial deformations were applied to samples exposed to NPs in order to simulate typical use conditions of protective equipment under an occupational setting. Complementary analyses on the dynamic biaxial deformations and the contact of the protective materials with colloidal solutions of the NPs were also carried out.

2. Materials

2.1. *Protective gloves and clothing*

Two models of protective gloves, corresponding to two families of elastomers, were selected for this study: disposable nitrile rubber gloves (100 μm thick) and non-supported butyl rubber gloves (700 μm thick). Samples were taken from the palm section of the gloves.

2.2. *Nanoparticles*

$nTiO_2$ (Nanostructured & Amorphous Materials, Inc., Houston, TX) was been selected for this study due to its use in numerous applications (paints, cosmetics, electronic compounds, etc.) [3]. $nTiO_2$ is labelled as 99.7% pure anatase with an average particle size of 15 nm. Colloidal solutions of $nTiO_2$ were also selected. Both were prepared from a 15 nm diameter anatase: a stock solution of 15 wt% in water (Nanostructured & Amorphous Materials, Inc., Houston, TX) and a stock solution of 20 wt% in 1,2-propanediol (MK Impex, Mississauga, ON).

2.3. *Ultra-pure solvents*

Solvents corresponding to the liquid carriers of the two colloidal solutions of $nTiO_2$ were also tested for comparison purposes. Ultra-pure 1,2-propanediol (99.5 wt%) was obtained from Sigma Aldrich and high-performance liquid chromatography water was obtained from Acros Organics.

3. Methods

3.1. *NP penetration experimental set-up*

A test set-up has been developed for the purpose of this study [21] (Figure 1). It includes an exposure chamber and a sampling chamber, which are separated from each other by the sample. Both chambers as well as all components in contact with NPs are made of ultrahigh molecular weight PE in order to limit the adsorption of $nTiO_2$. The set-up has been designed to the exposure of glove samples to powders or colloidal solutions while simultaneously subjecting them to static or dynamic mechanical constraints. NPs are introduced in the exposure chamber and placed into contact with the external surface of the glove sample. In the case of powder NPs, a thin circular nitrile rubber membrane is placed on top of the sample to prevent the NP from being dispersed in the exposure chamber. As shown in Figure 1, the test system is also equipped with a probe mounted on a pneumatic system for deforming the sample. The system is computer controlled and includes a 500 N load cell and a position detector. The whole set-up is enclosed in a glove box in order to ensure operator safety during assembly, dismounting and clean-up operations, as well as during tests.

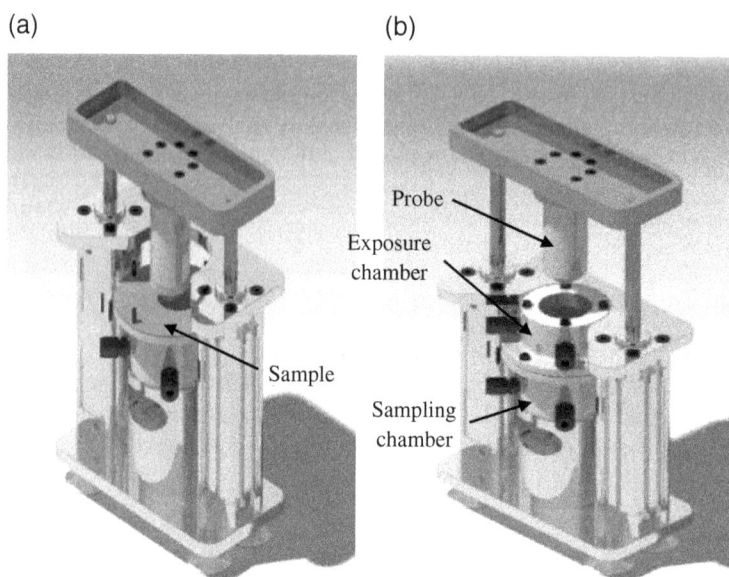

Figure 1.   Set-up isometric view (a) without the exposure chamber and (b) with the exposure chamber.

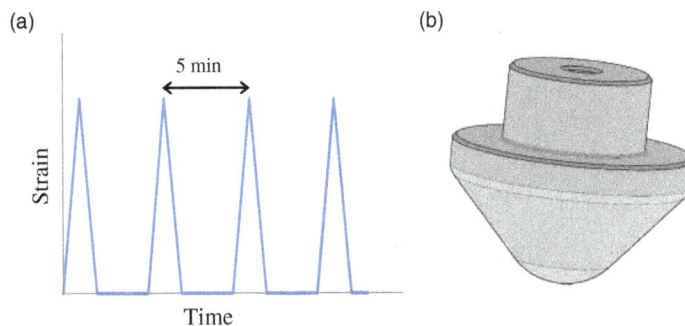

Figure 2.   (a) Schematic representation of the variation of the sample deformation as a function of time during a dynamic deformation event and (b) schema of a conical-spherical probe head.

The time profile of the applied sample deformations is illustrated in Figure 2(a). They consist in the application of a 50% deformation every 5 min for the glove samples and a 6% deformation every 5 min for the textile materials. The probe head used in this study is shown in Figure 2(b). It corresponds to a conical-spherical geometry that has been shown to successfully simulate biaxial deformations that are produced, for example, in gloves during hand flexion [22].

3.2. NP sampling protocol and detection methods

A sampling protocol was developed to collect NPs that crossed from the exposure chamber into the sampling chamber. Sample collection was facilitated by the use of a sampling solution, which was placed in the sampling chamber during the set-up. The sampling solution consisted of ultra-pure water acidified with 1% nitric acid [23]. Before the beginning of each test, contamination of the sampling chamber by $nTiO_2$ was verified by analysing a control sample that was produced by rinsing the sampling chamber. If a trace of contamination in the test control was measured, the result of the following test was discarded.

During the experiment, the maximum sample deformation height was such that no contact occurred between the sampling solution and the sample surface. Once the test was over and before dismounting the test setup, the chamber assembly was gently tilted and rotated so that the sampling solution could rinse the walls of the sampling chamber. The sampling solution was then transferred into a vial. Titanium concentrations in the sampling solution were analysed by inductively coupled plasma mass spectrometry (ICP–MS, Varian 820) for glove materials or by atomic absorption spectroscopy (Perkin Elmer AAnalyst 800) for protective clothing materials. Moreover, a small quantity of sampling solution was centrifuged on mica for qualitative analysis by Atomic Force Microscopy (Veeco-DI).

3.3. Quantification of characteristic features on the gloves surface

Some characteristic features can be observed on the gloves surface: micrometer-size pores for nitrile rubber and platelets for butyl rubber gloves [16]. The surface morphology of five samples, for each elastomer, was analysed by SEM (Hitachi S3600N $- V_{acc} = 15\,kV - magnification \times 300$) and the quantification of the surface area of these features was performed using the software ImageJ (image processing).

3.4. Swelling measurements

Gravimetric measurements were performed on glove samples in order to evaluate the effect of the $nTiO_2$ solutions and especially their liquid carriers on the glove materials. Indeed, elastomers

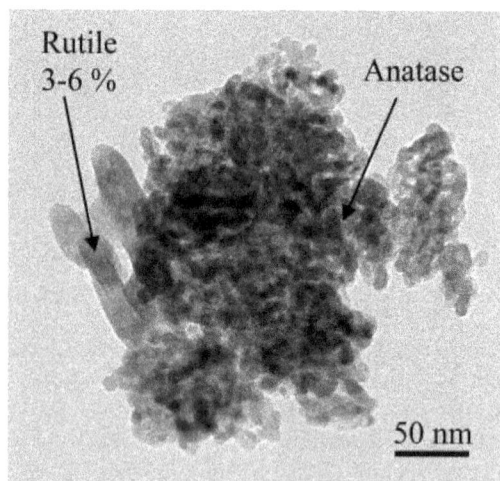

Figure 3.   TEM image of the nTiO$_2$ powder.

may be sensitive to swelling in some solvents. Rectangular samples (4×50 mm) were taken from the palm section of the gloves. To obtain statistically significant data, triplicate measurements were performed for each solution. The measurements were performed by immersing the samples in the commercial nTiO$_2$ colloidal solutions and in the ultra-pure solvents corresponding to the colloidal carriers. At regular intervals, samples were removed from the liquid, their surface gently blotted with a paper towel and their mass determined using a precision balance (± 0.1 mg). The gain in mass was computed using Equation (1):

$$\Delta M(t) = \frac{M_t - M_0}{M_0} \tag{1}$$

with M_t being the mass at time t and M_0 the corresponding value before immersion.

4.   Results and discussion

4.1.   *Characterization of the nTiO$_2$*

The size distribution of the nTiO$_2$ was verified by statistical analysis of 174 particles imaged by transmission electron microscopy (TEM, JEM-2100F). Two allotropic forms of TiO$_2$ were observed: a spherical anatase and a rod-like rutile (Figure 3). Analysis by X-ray diffraction (Philips X'PERT) confirmed the presence of 3–6% rutile in the nTiO$_2$. Moreover, these NPs seldom behaved as individual particles when in air [2]. In fact, only two particles with a diameter lower than 20 nm were counted among the 174 particles. Figure 4 displays the size distribution of the analysed sample in terms of circular diameter, which is the diameter of a circle having the same surface as the particle [24]. It can be seen that the average dimension of the particles/aggregates was situated around 100 nm, although even some micrometric-size agglomerates were recorded.

Analysis of the aqueous nTiO$_2$ stock solution (following dilution to 10 mg/L) by fluorescence correlation spectroscopy gave a hydrodynamic diameter for the particles of 21 ± 2 nm. The same analysis was not possible for the nTiO$_2$ in 1,2-propanediol due to an incompatibility between the cell material and the liquid carrier. In addition, no characterization of the colloidal solutions was achievable by microscopy due to the formation of a μm-thick viscous film on the NPs following

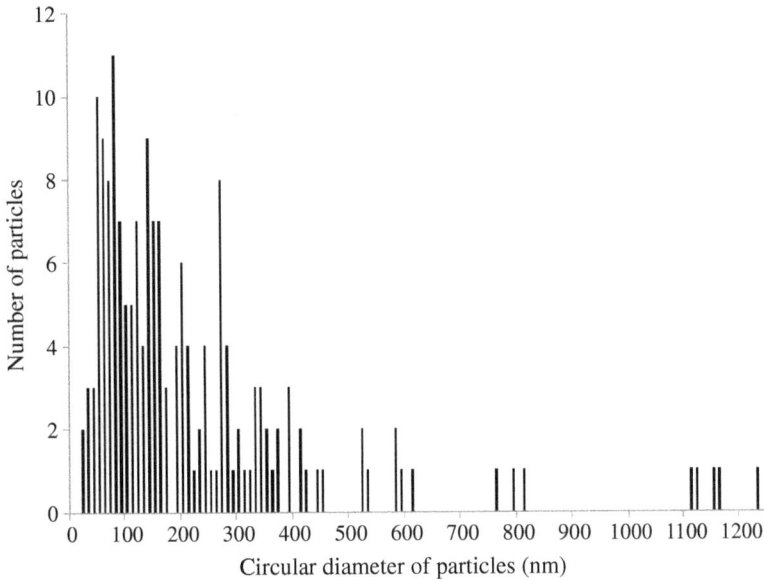

Figure 4.   Size distribution of the nTiO$_2$ powder.

solvent evaporation. This film was attributed to the presence of additives in the colloidal solutions, which are used as stabilizing agents [25].

4.2.   *Characterization of the gloves and protective clothing*

Figure 5(a) shows the micrometer-size pores which can be observed on the outer surface of the nitrile rubber glove. For its part, the outer surface of butyl rubber gloves shows platelets (Figure 5(b)). The same features were observed on the inner surface of both types of gloves. Two types of protective clothing were also studied: a 35/65 cotton/polyester plain woven fabric used to produce laboratory coats and a polyethylene–polypropylene (PE–PP) non-woven used for coveralls. Figure 5(c) and 5(d) show, respectively, the outer surface of the PE–PP non-woven and the cotton/polyester woven materials. In both cases, the surfaces displayed porosities ranging from 1 to 10 μm for the PE–PP non-woven and from 10 to 100 μm for the cotton/polyester woven.

4.3.   *Protective glove materials and* nTiO$_2$ *powder*

Glove samples were exposed to the nTiO$_2$ powder using a dynamic exposure (one biaxial deformation every 5 min for up to 7 h). Control experiments were carried out under identical conditions except that no NP was introduced into the exposure chamber. Controls ensured that the presence of titanium in the sampling solutions was related to the passing of NPs through the materials rather than simply a contamination of the sampling solution by the materials. No significant differences were observed for measurements performed with and without NPs when the butyl rubber was tested (Figure 6). This result indicated that the butyl rubber was not permeable to the nTiO$_2$ powder, even after 7 h of 50% deformations to the glove material.

On the other hand, a much higher concentration in titanium was recorded in the sampling solution after 7 h of dynamic deformations for nitrile rubber gloves. Based upon its thermodynamic constants, TiO$_2$ is unlikely to dissolve significantly and thus the ICP–MS results suggest that nTiO$_2$ was penetrating the nitrile rubber gloves. The presence of NPs nTiO$_2$was confirmed by atomic

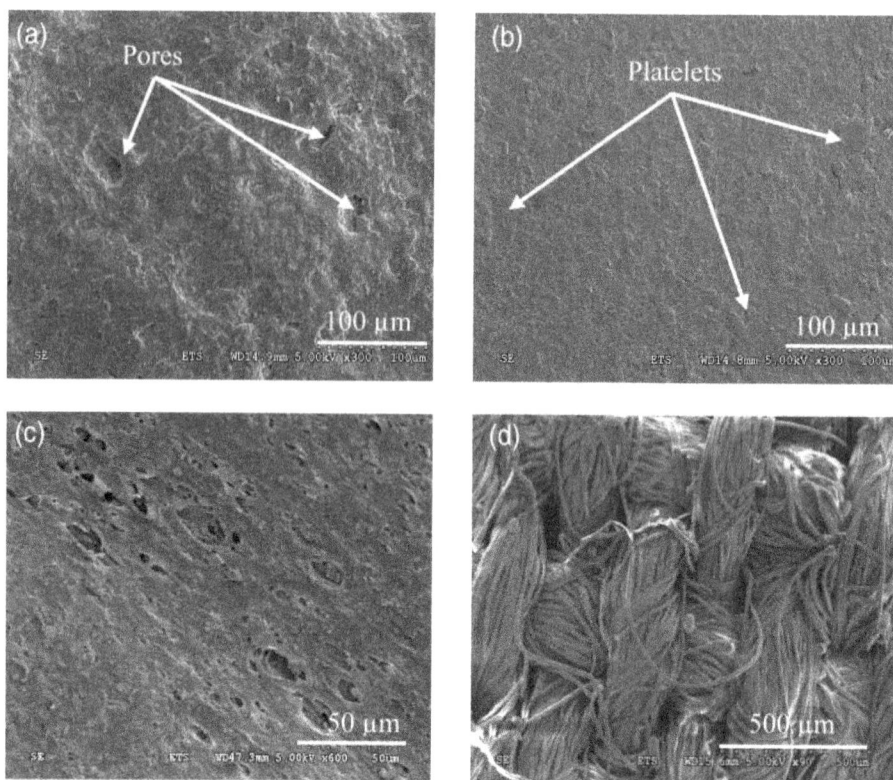

Figure 5. (a) SEM images of the surfaces of nitrile rubber, (b) butyl rubber, (c) PE–PP non-woven coverall and (d) laboratory coat.

force microscopy (Figure 7). These results indicated the passage of $nTiO_2$ through the nitrile gloves after dynamic deformations of more than 5 h.

In order to find explanations for the possible passage of the NPs through the nitrile rubber gloves, an analysis of the effect of repeated mechanical deformations on the glove surface morphology was carried out. For that purpose, the variation in the surface features identified for each elastomer (pores for nitrile rubber and platelets for the butyl rubber, see Figure 5) was determined as a function of time. Dynamic biaxial deformations were applied to the glove samples for up to 7 h using the test set-up described in Section 3.1. The surfaces of the nitrile and butyl rubber samples after 7 h of dynamic biaxial deformations are given in Figure 8. In comparison with corresponding pictures for the non-deformed samples (Figure 5), a significant change in the surface was apparent in the nitrile rubber samples following 7 h of dynamic deformations. In particular, the number, diameter and depth of the pores appear to be largely increased, possibly indicating a weakening of the membrane. It is possible to speculate that NPs may have accumulated in the deeper imperfections.

Figure 9 displays the variation in these surface features as a function of the duration of the applied deformations. A very slow increase can be seen for both elastomers during the first 5 h of deformations. The deterioration of the materials could be attributed to abrasion of the sample surface by the probe. This abrasion phenomenon might possibly be amplified in the presence of NPs. Indeed, TiO_2 particles, which are much harder than the elastomer, could act as an abrasive powder. It is especially interesting to note that the nitrile rubber samples experienced a sharp increase in the surface area of pores after 7 h of repeated deformations. Such an observation could

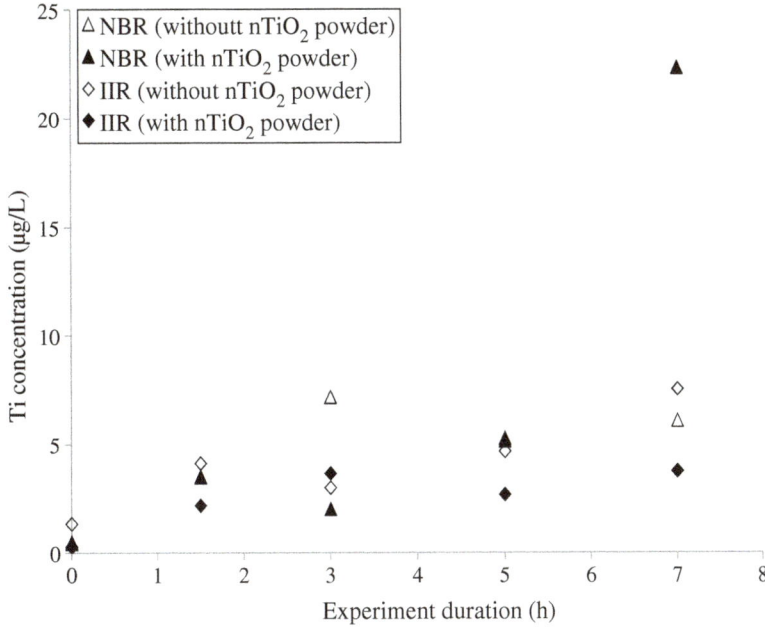

Figure 6.   Variation in concentration of titanium in the sampling solutions as a function of the duration of the 50% biaxial deformations for nitrile (nitrile butadiene rubber (NBR)) and butyl (isobutylene isoprene rubber (IIR)) rubber samples exposed to nTiO$_2$ powder.

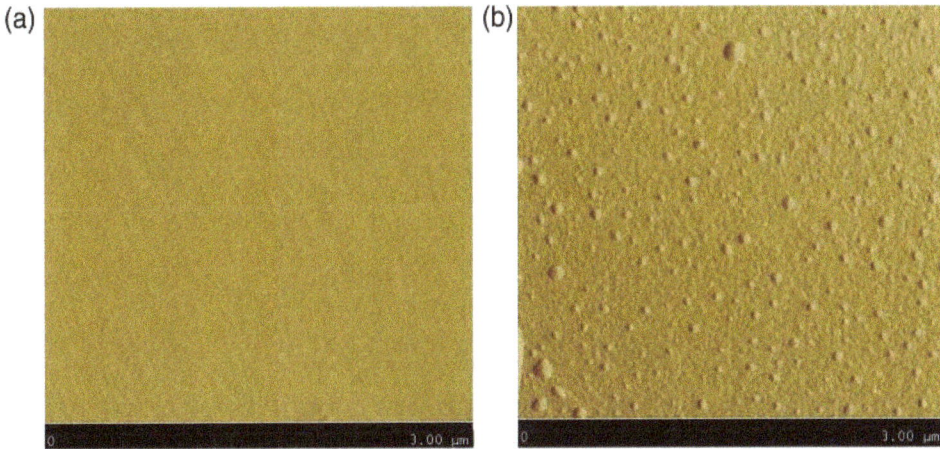

Figure 7.   (a) Atomic force microscopy pictures of native mica substrate and (b) after centrifugation of the sampling solution for the 7 h dynamic biaxial deformations/nTiO$_2$ powder/nitrile rubber condition.

explain the jump in the Ti concentration in the sampling solution that was measured for the nitrile rubber after these longer times.

The effect of the repeated deformations on the agglomeration of the nTiO$_2$ powder was investigated. In this case, the sample surface was analysed by a field emission gun SEM (FEG–SEM, JEOL JSM-7600F). A reduction in the agglomeration state of the NPs was observed. For example, as shown in Figure 10(a), agglomerates (about 500 nm diameter) can be seen after 1.5 h of dynamic biaxial deformations. On the other hand, only aggregates in the range of 100 nm can be

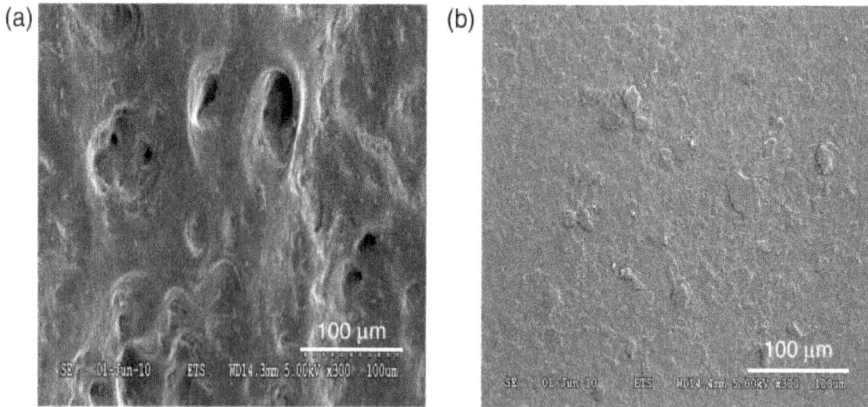

Figure 8.   (a) SEM images of the nitrile rubber and (b) butyl rubber external glove surface after 7 h of dynamic biaxial deformations.

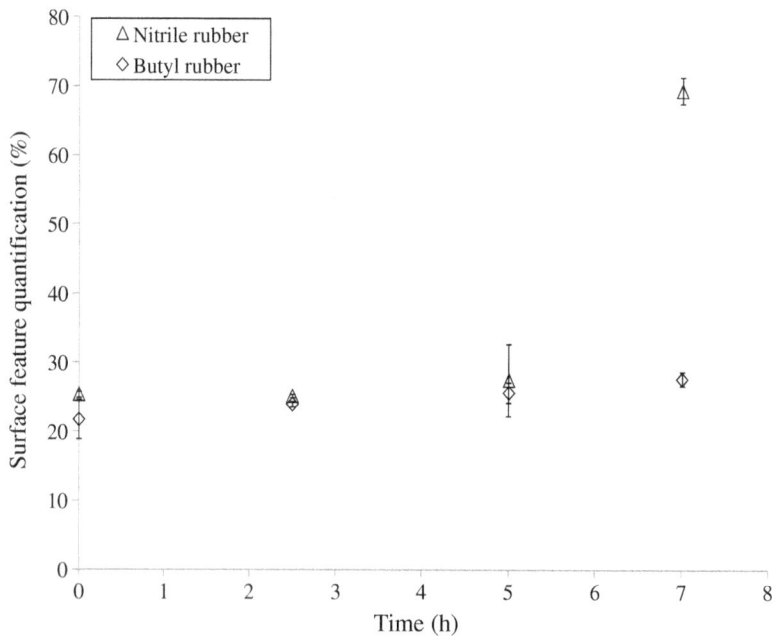

Figure 9.   Variation of the surface features as a function of the duration of dynamic biaxial deformations for the nitrile and butyl rubber (outer surface).

observed after 7 h of deformations (Figure 10(b)). This reduction in the agglomeration state can be attributed to shearing induced by the probe on the sample and may potentially contribute to increased penetration of the $nTiO_2$ particles through the protective gloves.

4.4.   *Protective glove materials and* $nTiO_2$ *in colloidal solutions*

Tests were also carried out by exposing the glove samples to a solution of $nTiO_2$ for up to 7 h of 50% dynamic biaxial deformations. As above, the sampling solutions were analysed by ICP–MS (Figure 11) and as above, no differences were observed for measurements made with and without the NPs for the samples composed of the butyl rubber. On the other hand, for the nitrile gloves, a

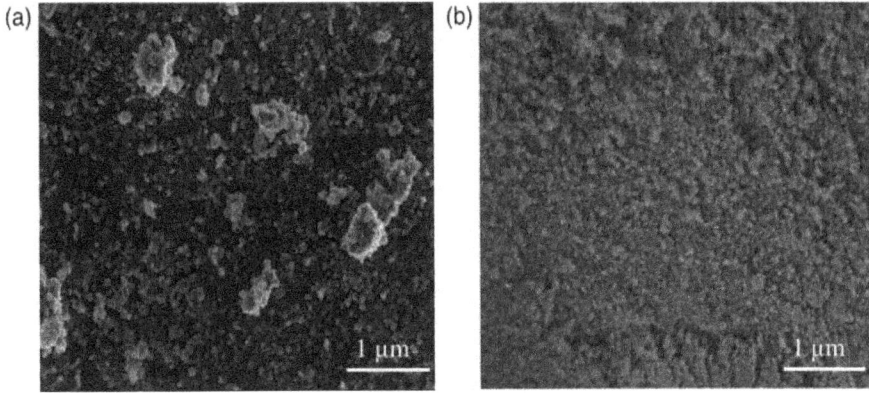

Figure 10. (a) FEG–SEM images of the nitrile rubber glove surface after 1.5 h of dynamic biaxial deformations and (b) after 7 h of dynamic biaxial deformations.

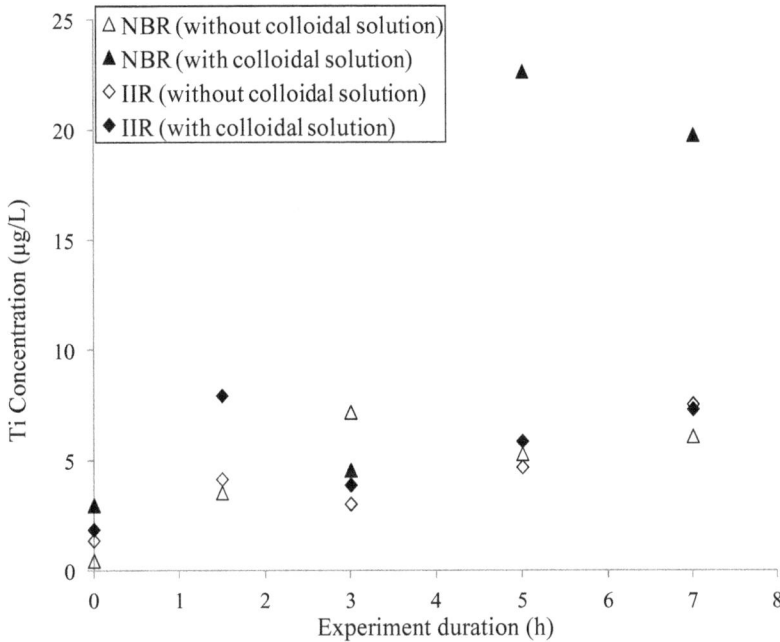

Figure 11. Variation in the concentration of titanium in the sampling solutions as a function of the duration of 50% biaxial deformations for nitrile (NBR) and butyl (IIR) rubber samples exposed to colloidal solutions of nTiO$_2$.

clear increase in the concentration of titanium was observed in the sampling solutions after 5 and 7 h of deformations nTiO$_2$. These results suggest the penetration of the nTiO$_2$ solutions nTiO$_2$ through the nitrile gloves when they were subjected to 50% dynamic biaxial deformations for periods of 5 h or more.

In order to investigate the cause of the apparent NP penetration through the nitrile gloves, the glove surface morphology was characterized following the mechanical deformations and contact with the solutions of nTiO$_2$. Dynamic biaxial deformations were applied to the glove samples over 7 h and in combination with the nTiO$_2$ solvents. The surface morphology of the glove samples following exposure to the solvents, to the mechanical deformations or to a combination of the two was imaged by SEM and analysed by Image J. Contact with 1,2-propanediol alone did not appear to cause any modification to the nitrile rubber surface while a small effect was produced on the butyl rubber (Figure 12). In contrast, when combined with mechanical deformations, the

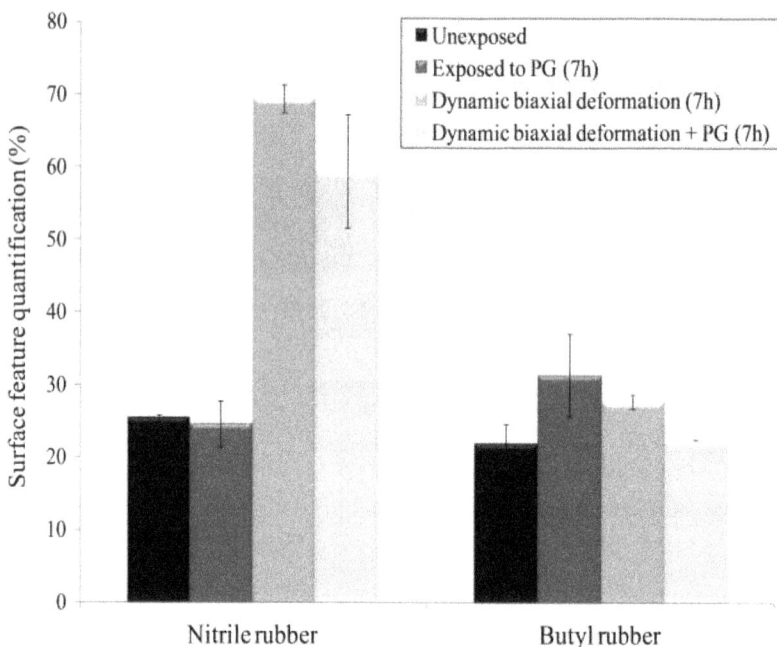

Figure 12. Effect of 7 h of exposure to 1,2-propanediol (propylene glycol (PG)), dynamic biaxial deformations and combination of both on the material-specific glove surface features for the nitrile and butyl rubber.

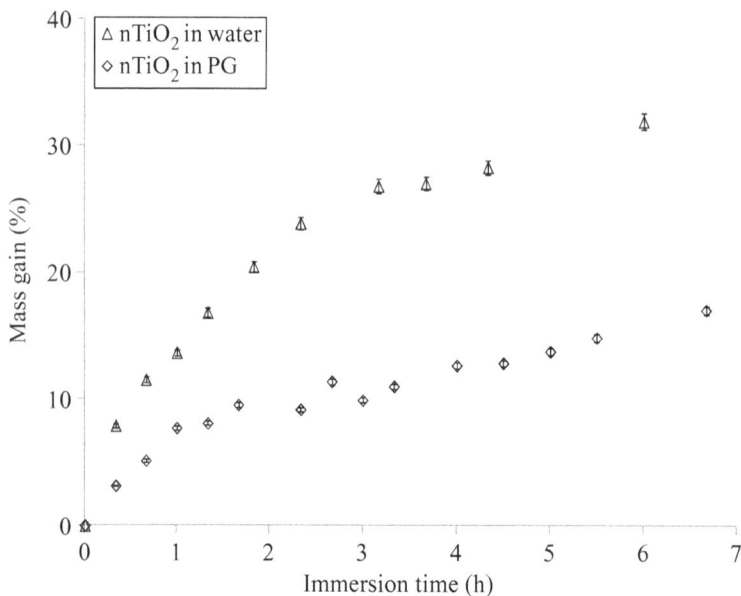

Figure 13. Mass gain of the nitrile rubber gloves as a function of immersion time in the $nTiO_2$ solution in water and 1,2-propanediol (PG).

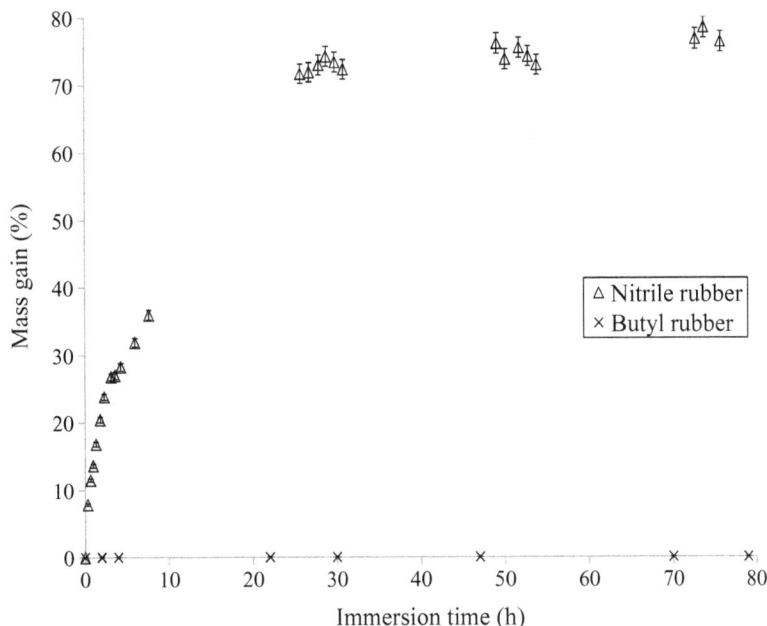

Figure 14. Mass gain as a function of immersion time in nTiO$_2$ in water for nitrile and butyl rubber samples.

presence of 1,2-propanediol reduced the number of glove deformations, probably due to its role as a lubricating agent. These data did not provide any insight into the nTiO$_2$ large increase in titanium that was observed in the nitrile rubber sampling solutions after 5 and 7 h.

Swelling measurements were performed by recording the mass of the samples after immersion in the nTiO$_2$ solutions. The results obtained for the nitrile rubber immersed in the solutions of nTiO$_2$ are presented in Figure 13. A gradual weight increase is observed for the two colloidal solutions, indicating significant penetration of both solutions of nTiO$_2$. Nonetheless, a larger swelling ratio was recorded for the aqueous solution of nTiO$_2$. For example, for an immersion time of 8 h, the swelling ratio for the nitrile rubber was 35% for the nTiO$_2$ in water and 15% when they were in nTiO$_2$ 1,2-propanediol. It should be noted that maximum was not attained for either of the solutions of nTiO$_2$ after 8 h of immersion.

A comparison of the swelling behaviour of the two glove materials nTiO$_2$ was therefore performed over a longer period (Figure 14). A large difference in the swelling behaviour was observed for the two elastomers. A 79% gain in mass was observed for the nitrile rubber after three days of immersion with a plateau after about 40 h of immersion. For the butyl rubber, no significant weight gain was recorded. This difference in behaviour between the two elastomers can be attributed to their respective affinity to the liquid carriers. In particular, the butyl rubber displays an excellent resistance to most polar solvents [26]. The difference in swelling behaviour between the nitrile and butyl rubber may well be one of the primary reasons that the nTiO$_2$ was able to more easily penetrate the nitrile rubber gloves, especially when the NPs were in water.

4.5. *Protective clothing materials and* nTiO$_2$ *in powder*

Some preliminary measurements were carried out to investigate the penetration of nTiO$_2$ powder through protective clothing following a slight adjustment of the test conditions. In particular, since the maximum deformation of the PE–PP non-woven is 10% and the cotton/polyester plain

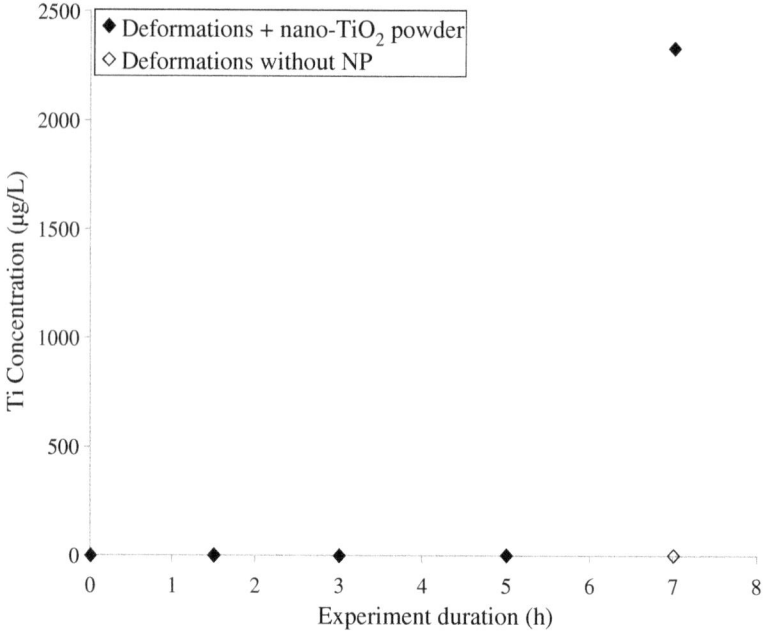

Figure 15.   Titanium concentration penetrating a non-woven coverall material (PE-PP) exposed to nTiO$_2$ and 6% biaxial deformations.

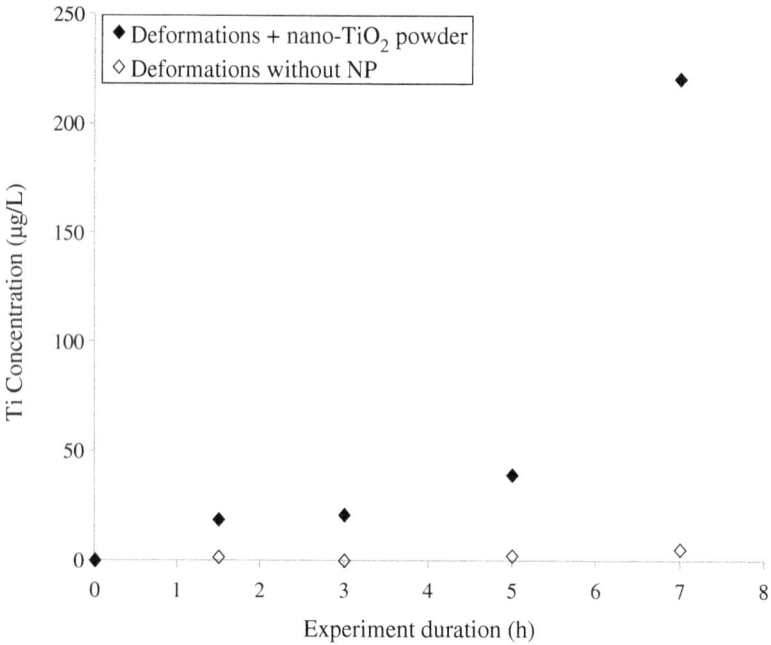

Figure 16.   Titanium concentration penetrating a cotton/polyester woven fabric exposed to nTiO$_2$ and 6% biaxial deformations.

woven fabric is 25%, and since these materials are usually only subjected to small deformations when they are used in protective clothing, dynamic biaxial deformations were limited to 6%.

As illustrated in Figure 15, the presence of NPs in the sampling solutions was detected after 7 h of deformations for the non-woven coverall material. In this case, the concentration of titanium that was detected reached 2400 μg/L. For the laboratory coat fabric, a gradual increase in the concentration of titanium was observed with a large jump after 5 h and a maximum concentration in titanium in the sampling solution was 220 μg/L (Figure 16). By comparison, no titanium was detected in the sampling solution when NPs were not present in the exposure chamber. These results suggest the passage of $nTiO_2$ powder through the protective clothing textiles in the presence of dynamic deformations. The higher level of protection that was measured for the non-woven material as compared with the woven fabric is in agreement with what has been reported by Golanski et al. [10] when studying graphite nanoaerosols. The sharp increase in the concentration of titanium in the sampling solutions that was observed for both materials after 7 h of deformations might have been caused by the mechanical failure of the membranes at the micron or submicron scale. In this case, no observations of the sample surface were performed and more work is clearly required in order to identify the underlying phenomena responsible for the loss of integrity of the protective materials.

5. Conclusion

This paper has investigated the penetration of $nTiO_2$ powder and two commercially obtained $nTiO_2$ solutions through protective clothing and gloves that were subjected to dynamic biaxial deformations, corresponding to conditions typical of occupational use. In the case of the protective glove materials, no penetration of NPs was measured through the butyl rubber, even after 7 h of deformations. On the other hand, penetration of $nTiO_2$ was recorded for the disposable nitrile gloves when exposed to both $nTiO_2$ powder and colloidal solutions of $nTiO_2$ (after 7 and 5 h, respectively). Similarly, penetration of the $nTiO_2$ powder was detected for two typical fabrics of laboratory coats, when mechanical deformations simulating occupational use were included in the simulation.

This penetration of the NPs through the protective materials was attributed in part to a degradation of their physical and mechanical properties following repetitive mechanical deformations and contact with $nTiO_2$ liquid carriers. A possible reduction in the agglomeration state of the $nTiO_2$ powder as a result of probe shearing was also observed. These results show that great care must be exercised when selecting protective clothing and gloves for the handling of NPs. In particular, these results suggest that exposure may be reduced through the frequent replacement of disposable gloves, especially when exposure of the NPs occurs in the liquid phase. In addition, while non-woven coveralls appear to be preferable to woven fabric laboratory coats, their prolonged use or reuse should be avoided.

Acknowledgements

This project was conducted within the context of the Research Chair in Protective Materials and Equipment for Occupational Safety and Health (École de technologie supérieure). It has been supported by the Institut Robert-Sauvé en santé et en sécurité du travail (IRSST), NanoQuébec, the Natural Sciences and Engineering Research Council of Canada (NSERC), the Agence nationale de sécurité sanitaire de l'alimentation, de l'environnement et du travail (ANSES) and the Agence de l'Environnement et de la Maîtrise de l'Énergie (ADEME). The authors would like to acknowledge the contribution of S. Mahé, A. Jambou and F. Deltombe (École de technologie supérieure) to the project as well as the collaboration of P. Plamondon and G. L'Espérance (École Polytechnique de Montréal).

References

[1] Roco MC, Mirkin CA, Hersam MC. Nanotechnology research directions for societal needs in 2020 – retrospective and outlook. Berlin and Boston: WTEC Study on Nanotechnology Research Directions; 2010.

[2] Hervé-Bazin B. Les nanoparticules: un enjeu majeur pour la santé au travail? Les Ulis Cedex A, France: EDP Science 2007. p. 19–36.

[3] Robichaud CO, Uyar AE, Darby MR, Zucker LG, Wiesner MR. Estimates of upper bounds and trends in nano-TiO2 production as a basis for exposure assessment. Environ Sci Technol. 2009;43:4227–4233.

[4] CCHST. Information de base sur le dioxyde de titane. Centre canadien d'hygiène et de sécurité au travail; 2007. Available from: http://www.cchst.ca/

[5] Höhr D, Steinfartz Y, Schins RPF, Knaapen AM, Martra G, Fubini B, Borm PJA. The surface area rather than the surface coating determines the acute inflammatory response after instillation of fine and ultrafine TiO_2 in the rat. Int J Hyg Environ Health. 2002;205(3):239–244. doi: 10.1078/1438-4639-00123

[6] Warheit DB, Webb TR, Reed KL, Frerichs S, Sayes CM. Pulmonary toxicity study in rats with three forms of ultrafine-TiO2 particles: differential responses related to surface properties. Toxicology. 2007;230(1):90–104. doi:10.1016/j.tox.2006.11.002

[7] Warheit DB, Webb TR, Sayes CM, Colvin VL, Reed KL. Pulmonary instillation studies with nanoscale TiO_2 rods and dots in rats: toxicity is not dependent upon particle size and surface area. Toxicol Sci. 2006;91(1):227–236. doi:10.1093/toxsci/kfj140

[8] IARC. Monographs on the evaluation of carcinogenic risks to humans – carbon black, titanium dioxide and talc. Lyon: World health organization; 2010.

[9] OECD. Comparison of guidance on selection of ski protective equipment and respirators for use in the workplace: manufactured nanomaterials. Series on the safety of manufactured nanomaterials. 2009; 12. (Report number: JT03267010)

[10] Golanski L, Guiot A, Tardif F. Are conventional protective devices such as fibrous filter media, respirator cartridges, protective clothing and gloves also efficient for nanoaerosols? NanoSafe: European Strategy for nanosafety; 2008. Available from: http://www.nanosafe.org/home/liblocal/docs/Dissemination%20report/DR1_s.pdf

[11] Golanski L, Guiot A, Rouillon F, Pocachard J, Tardif F. Experimental evaluation of personal protection devices against graphite nanoaerosols: fibrous filter media, masks, protective clothing, and gloves. Hum Exp Toxicol. 2009;28(6–7):353–359. doi:10.1177/0960327109105157

[12] Golanski L, Guoit A, Tardif F. Experimental evaluation of individual protection devices against different types of nanoaerosols: graphite, TiO_2 and Pt. J Phys Conf Ser. 2009;170(1):012025. doi:10.1088/1742-6596/170/1/012025

[13] Hanley JT. (Vêtements de protection) Aerosol system and swatch testing of chemical protective garments. In: Elevated Wind Studies International Conference, Arlington, VA; 2006:151

[14] Huang S-H, Chen C-W, Chang C-P, Lai C-Y, Chen C-C. Penetration of 4.5 nm to 10 nm aerosol particles through fibrous filters. J Aerosol Sci. 2007;38(7):719–727. doi:10.1016/j.jaerosci.2007.05.007

[15] Hofacre KC. (Textile) Aerosol penetration of fabric swatches. In: Elevated Wind Studies International Conference, Arlington, VA; 2006:150.

[16] Ahn K, Ellenbecker MJ. Dermal and respiratory protection in handling nanomaterials at the center for high-rate nanomanufacturing (CHN). Paper presented at: The AIHce Conference, Chicago, IL, 2006 May 13–18.

[17] Nohile C, Dolez PI, Vu-Khanh T. Parameters controlling the swelling of butyl rubber by solvents. J Appl Polym Sci. 2008;110(6):3926–3933.

[18] Vinches L, Dolez P, Vu-Khanh T. Study on the penetration of TiO_2 nanoparticles through nitrile and butyl protective gloves. Paper presented at: The International Conference on Nanotechnology: Fundamentals and Applications, Ottawa; 2011 juillet 27–29.

[19] Li Y, De Kee D, Fong CFCM, Pintauro P, Burczyk A. Influence of external stress on the barrier properties of rubbers. J Appl Polym Sci. 1999;74(6):1584–1595. doi:10.1002

[20] Nicolay CW, Walker AL. Grip strength and endurance: influences of anthropometric variation, hand dominance, and gender. Int J Ind Ergonomics. 2005;35(7):605–618. doi:10.1016/j.ergon.2005.01.007

[21] Dolez P, Vinches L, Wilkinson K, Plamondon P, Vu-Khanh T. Development of a test method for protective gloves against nanoparticles in conditions simulating occupational use. J Phy Conf Ser. 2011;304(1):012066.

[22] Harrabi L, Dolez PI, Vu-Khanh T, Lara J. Evaluation of the flexibility of protective gloves. Int J Occup Saf Ergonomics. 2008;14(1):61–68.

[23] Kaegi R, Ulrich A, Sinnet B, Vonbank R, Wichser A, Zuleeg S, Simmler H, Brunner S, Vonmont H, Burkhardt M, Boller M. Synthetic TiO_2 nanoparticle emission from exterior facades into the aquatic environment. Environ Pollut. 2008;156:233–239.

[24] Zayed J, L'Espérance G, Truchon G, Cloutier Y, Kennedy G. Contribution de la nanoscopie à l'échantillonnage et caractérisation physico-chimique des nanoparticules. Montréal: Institut de recherche Robert-Sauvé en santé et sécurité du travail; 2009.

[25] Dolez P, Vinches L, Perron G, Vu-Khanh T, Plamondon P, L'Espérance G, Wilkinson KJ, Cloutier Y, Dion C, Truchon G. Développement d'une méthode de mesure de la pénétration des nanoparticules à travers les matériaux de gants de protection dans des conditions simulant l'utilisation en milieu de travail. Montréal: Institut de recherche Robert-Sauvé en Santé et Sécurité du Travail; 2012.

[26] Jin JZ, Nguyen V, Gu WQ, Lu XY, Elliott BJ, Gin DL. Cross-linked lyotropic liquid crystal-butyl rubber composites: promising 'breathable' barrier materials for chemical protection applications. Chem Mater. 2005;17(2):224–226. doi:10.1021/cm040342u

9

Investigation of electroactive behaviour of polyaniline containing polyelectrolyte nanocomposite membranes

S.N. Soni, J. Bajpai and A.K. Bajpai*

Department of Chemistry, Bose Memorial Research Laboratory, Government Autonomous Science College, Jabalpur – 482 001, Madhya Pradesh, India

Novel electrically conducting composite hydrogel materials comprising of polyaniline (PANI) nanoparticles dispersed in a poly(vinyl alcohol) (PVA)–g–poly(2-acrylamido-2-methyl-1-propanesulfonic acid) matrix were prepared by *in situ* polymerisation of aniline (AN). The conversion yield of AN into PANI particles was determined gravimetrically while structural confirmation of the synthesised polymer was ascertained by Fourier transform infrared spectroscopy, UV–Visible analysis and X-ray diffraction techniques. The morphology and dimension of PANI particles embedded into the coloured optically semi-transparent polymer films were evaluated by scanning electron microscopy analysis and transmittance electron microscopy. The electrical conductivity of composite hydrogels containing different composition percentages was determined by LCR metre while the eletroactive behaviour of composite hydrogels swollen in electrolyte solution was investigated by effective bending angle measurements. The prepared ionic hydrogels of varying composites were also evaluated for their water-uptake capacity with their different composition ratio in distilled water.

Keywords: graft polymer; hydrogel composite; conducting polymer; characterization

1. Introduction

A gel containing ionic groups can be actuated isothermally by an electric field. When the gel is negatively charged, it swells near the anode and contracts near the cathode, the contraction rate being proportional to the external electric current [1]. Such polymer gels can be said to be intelligent or smart materials, that is they can be made to respond to changes in their environment such as pH, temperature, voltage and so on by changing their shape, solubility or their degree of ionisation. A thermodynamic system capable of transforming chemical energy directly into mechanical work is known as a chemomechanical system, such a gel bar made of eletroactive polymers (EAPs) or composite materials can bend towards electrode by the application of an electrical field.

Such EAPs or composite materials have received much attention in recent years because they have variety of applications in different fields such as sensor and biosensor materials [2–4], actuators and electromechanical devices [5], biomedical implant materials [6,7], electronic and optical materials [8], etc. These hydrogels do not dissolve in water at a physiological temperature and pH, but swell considerably in an aqueous medium. These polymers can change their volume significantly in response to small alterations of certain environmental parameters. Thus, the development of manifold stimuli-sensitive hydrogels with sensitivities across temperature [9],

*Corresponding author. Email: akbmrl@yahoo.co.in; akbajpailab@yahoo.co.in

electrical field [10], light [11], pH, solvent composition and specific ions [12–15] have opened up a new dimension of diversified technical applications [16]. Stimuli-responsive polymers constitute a versatile class of materials finding wide spectrum of applications in biomedical, pharmaceutical and industrial fields [17–19].

Poly(vinyl alcohol) (PVA) is a unique non-ionic synthetic polymer and credits wide spectrum of biomedical applications such as soft contact lenses, dialysis membranes, implants and artificial organs [20–22]. This is because of their inherent non-toxicity, non-carcinogenicity, good biocompatibility and desirable physical properties such as rubbery or elastic nature and high degree of swelling in aqueous solution. The chemical resistance and physical properties of PVA have led to its broad industrial usage. Chemical crosslinking of this linear polymer provides feasible routes for the improvement of the mechanical properties and thermal stability. Chemically cross-linked PVA hydrogels have received increasing attention in biomedical and biochemical applications because of their permeability, biocompatibility and biodegradability [23].

Poly(2-acrylamido-2-methyl-1-propanesulfonic acid) (PAMPS) is a well-known ionic, hydrophilic, polymer and its co-polymers have been widely used as carriers in controlled drug release technology because of their multifunctional nature, pH-dependent properties and good biocompatibility [24–26].

Polyaniline (PANI) is an electrically conducting polymer having a spatially extended π bonding system, which accounts for their intrinsic semi-conducting nature. PANI is one of the most promising conducting polymers due to a good combination of properties, stability, price and ease of synthesis by different routes and uncountable application [27–30]. PANI is difficult to process because it is soluble only in a limited number of organic solvents [31–34].

Thus, in this study, highly swelling grafted hydrogel composed of PAMPS and PVA have been prepared and PANI nanoparticles are dispersed in it by '*in situ*' polymerisation of AN. This study mainly focuses on the synthesis and characterisation of conducting nanocomposites and dependence of their swelling behaviour on chemical composition.

2.   Experimental

2.1.   *Materials*

PVA (98.6% hydrolysed, mol.wt. *ca.* 70,000 Da and hot processed) was obtained from Research Lab, Mumbai, India, and used as-received. AMPS was purchased from Aldrich (Germany). Other chemicals such as AN, hydrochloric acid, potassium per sulphate (KPS), ammonium per sulphate (APS) and N, N'-methylene bis acrylamide (MBA) were also of analytical grade and used without any further purification.

2.2.   *Methods*

2.2.1.   *Preparation of gel*

A polymer matrix composed of PVA–g–PAMPS was prepared using MBA as crosslinker and KPS as polymerising initiator by following a method reported in the literature [35–37]. The structure of graft co-polymer may be displayed as shown in Figure 1.

In a typical experiment, 1 g PVA was dissolved in 25 mL hot double-distilled water and to this solution, pre-calculated amounts of AMPS(4.825 mM), MBA (19.45×10^{-2} mM) and KPS (11.10×10^{-2} mM) were added. The whole reaction mixture was homogenised and kept in a Petri dish (corning glass, 2.5″ diameter) maintained at 35 ± 0.2°C for 24 h. After the reaction is over, the whole mass was converted into a semi-transparent film and was purified by equilibrating it in double-distilled water for a week. The swollen gel was dried at room temperature, cut into rectangular-shaped piece and stored in air-tight plastic bags.

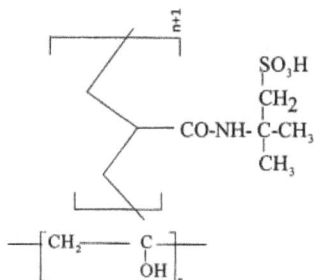

Figure 1.   The structure of graft co polymer of PAMPS onto PVA.

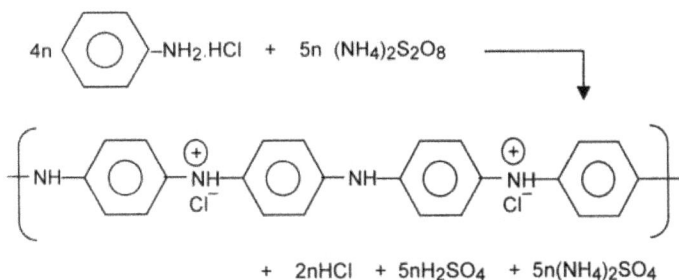

Figure 2.   Reaction scheme showing the polymerisation of AN.

2.2.2.   *PANI-impregnation*

Required quantity of AN (10.74 mM) was dissolved in 0.5 N HCl (50 mL) and the gel prepared as above was allowed to soak in the AN solution for 24 h. The AN containing swollen gel was dried and then again left in a 0.3 M APS bath to soak the required quantity of APS in solution (0.5 N HCl). As soon as the gel swells in APS solution, the entrapped APS initiates polymerisation of AN via the mechanism shown in Figure 2. As the polymerisation proceeds, the semi-transparent gel turns into black. The PANI impregnated gel is repeatedly washed with distilled water and allowed to dry at $30 \pm 0.2°C$ for 72 h.

The percentage impregnation of PANI into the gel was calculated using the following equation:

$$\% \text{ Impregnation of PANI} = \left(\frac{W_{PANI} - W_{Dry}}{W_{Dry}} \right) \times 100 \qquad (1)$$

where W_{PANI} is weight of the dry PANI impregnated gel and W_{Dry} is the initial weight of polymer gel.

In order to achieve the objectives undertaken in this study, hydrogels of different composition were prepared by varying the amounts of PVA, AMPS, MBA, KPS, AN and APS in the feed mixture of the hydrogel.

2.3.   *Characterization*

2.3.1.   *Fourier transforms infrared spectra*

To gain insights into the structural features of the prepared PANI-impregnated gel, the Fourier transform infrared (FTIR) spectra of PANI powder, polymer hydrogel and PANI-impregnated matrix were recorded on a FTIR spectrophotometer (FTIR–8400S, Shimadzu). For recording FTIR spectra of native and PANI-impregnated films, quite thin and transparent films of respective samples were prepared by solution cast method and the prepared films were directly mounted on spectrophotometer.

2.3.2. *UV–Visible analysis*

UV–Visible (UV-Vis) analysis was also carried out on a double beam UV–Vis spectrophotometer (Systronics, 2201, Ahmedabad, India). For scanning UV-spectra, thin films of samples (native and PANI impregnated) were prepared of sizes $3 \times 1 \times 0.05 \, cm^3$ by solution cast method and put into the quartz corvette in vertical orientation.

2.3.3. *X-ray diffraction analysis*

In order to gain insights into the crystalline nature of the prepared native and PANIimpregnated gel, the X-ray diffraction (XRD) spectra were recorded using a Philips (Holland) automated X-ray powder diffractometer. The dried gels were placed on the glass-slide specimen holder and exposed to X-rays in a vertical goniometer assembly. The scan was taken between $10°$ and $90°$ with a scanning speed of $2.4° \, min^{-1}$. The operating target voltage was $35 \, kV$, tube current was $20 \, mA$ and radiation used was Fe-Kα ($\lambda = 0.193 \, nm$).

2.3.4. *Microscopic analysis*

The scanning electron microscopic (SEM) analysis and transmission electron microscopy (TEM) analysis of native and PANI-impregnated films were performed on an SEM (LEO 435 VP variable pressure SEM) and TEM (Morgagni 268 D, Fei Company, The Netherlands), respectively.

2.3.5. *Electrical conductivity*

The electrical conductivity of the prepared gels were measured by Four probe LCR meter (Masstech Digital M/M No. MAS 830 L) with the help of silver electrode pressed on both sides of the gel. The electrical conductivity of gels of different compositions were also measured.

2.3.6. *Swelling measurements*

In order to evaluate water uptake potential of native and impregnated gel, a gravimetric procedure was followed [38]. In brief, a preweighed piece of gel was immersed in water at room temperature ($25°C$) and taken out after a predetermined time period. The swollen gel was gently pressed in between the filter papers to remove excess water and weighed on a sensitive balance. The degree of water sorption was calculated using the following equation:

$$\text{Swelling ratio} = \frac{W_{\text{Swollen}}}{W_{\text{Dry}}} \qquad (2)$$

where W_{Swollen} = weight of swollen gel and W_{Dry} = weight of dry gel.

2.3.7. *Reproducibility of data*

All measurements were carried out at least thrice and the average value was utilised for presenting results. It was found that the experimental errors had never been greater than 3%.

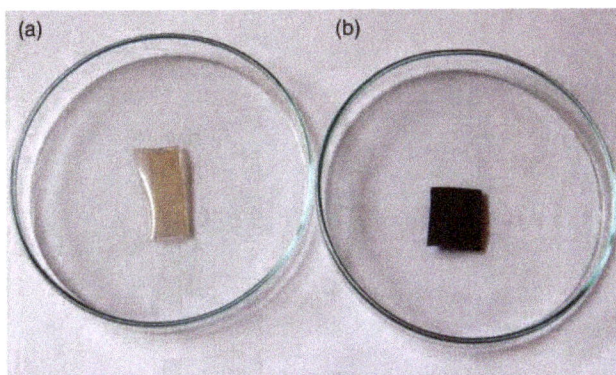

Figure 3.   Physical appearance of (a) PVA–g–PAMPS gel and (b) PANI-impregnated PVA–g–PAMPS gel.

3. Results and discussion

3.1. *Characterisation of gels*

3.1.1. *Physical appearance*

Figure 3(a) and (b) depicts the appearance of the native and PANI-impregnated gels which provide a clear evidence of impregnation of PANI into the polymer gel matrix. As evident from the photographs, the native gel is semi-transparent in colour while the impregnated gel appears dark-green, this could be attributed to the formation of PANI within the matrix.

3.1.2. *FTIR spectra*

Figure 4(a) and (b) represents the FTIR spectra of PANI powder and PANI-impregnated PVA–g–PAMPS gel film scanned in the range of 400–4000 cm^{-1}. The characteristic peaks appeared in Figure 4(a) at 824, 1144, 1312, 1505 and 1590 cm^{-1} indicate aromatic C–H, aromatic amide and aromatic C–C stretching vibrations, respectively. The spectra (b) of PANI-impregnated PVA–g–PAMPS gel film contain peaks at 866, 1126, 1270, 1510 and 1602 cm^{-1}, indicating the presence of aromatic C–H, aromatic amide and aromatic C–C stretching vibrations which confirm the impregnation of PANI into the gel (polymer matrix).

3.1.3. *UV–Vis spectral analysis*

The UV–Vis analysis of PVA–g–PAMPS film and PANI-impregnated PVA–g–PAMPS film was carried out in the range of 200–800 nm. It is observed that two characteristic peaks at 334 and at 632 in PANI-impregnated PVA–g–PAMPS film spectra (Figure 5) indicates the presence of PANI into the gel whereas these peaks were not visible in the spectra of PVA–g–PAMPS (not shown in this article). This obviously confirms the impregnation of PANI into the gel because the emeraldine form of PANI has two characteristic peaks at 334 and 632 nm originating from $\pi - \pi^*$ transition of benzenoid rings and the excitation absorption of the quinoid rings, respectively.

3.1.4. *XRD analysis*

The XRD patterns of the prepared native and PANI-impregnated gels are shown in Figure 6(a) and (b), respectively. Figure 6 (a) shows a prominent peak near 20.8° which corresponds to the

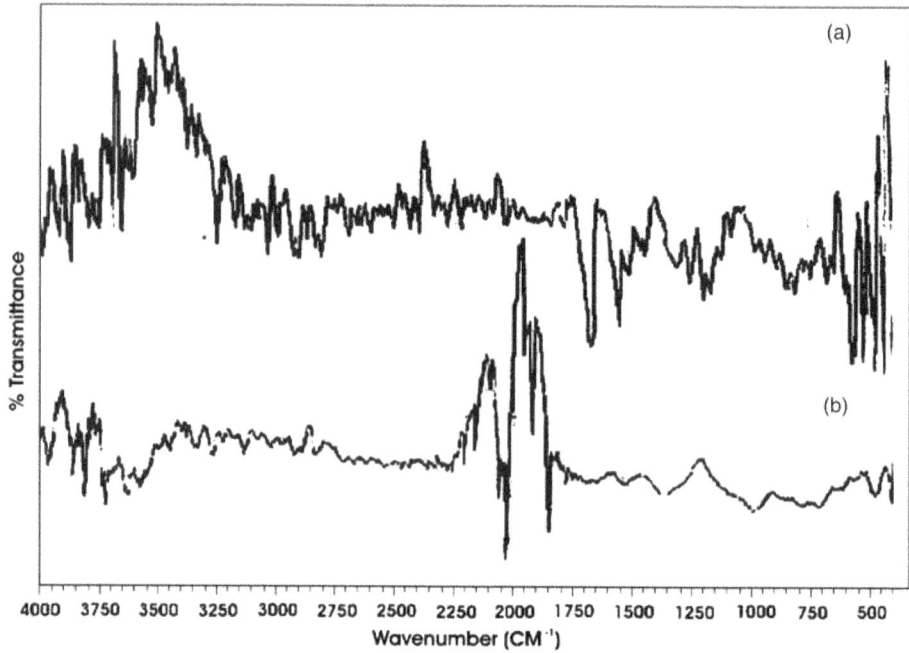

Figure 4.   FTIR spectra of (a) PANI powder and (b) PVA–g–PAMPS impregnated PANI gel.

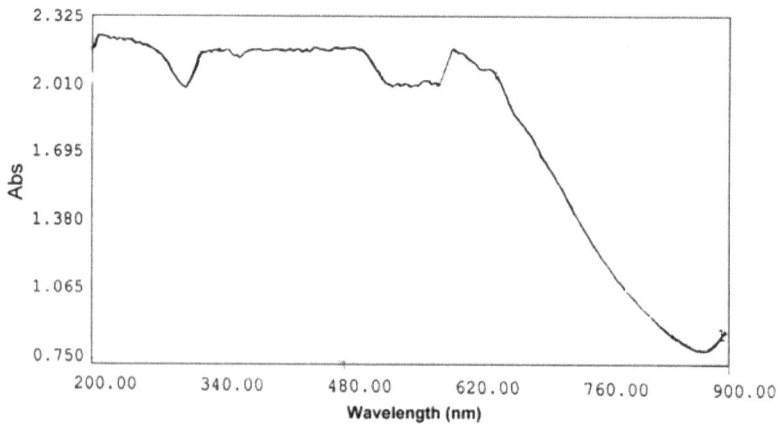

Figure 5.   UV–Vis spectra of PANI–impregnated gel.

(101) plane of the PVA crystal. Other minor peaks appeared around 33.6° and 41.7° could be attributed to crystallites of grafted PAMPS chains.

The diffraction patterns of PANI-impregnated gel are shown in Figure 6(b) which not only show a characteristic peak at 20.3° (due to PVA) but also depicts a prominent peak at 23.5°, indicative of the characteristic peak of PANI. Thus, the XRD patterns of impregnated gel provide an additional evidence of PANI formation within the polymer matrix.

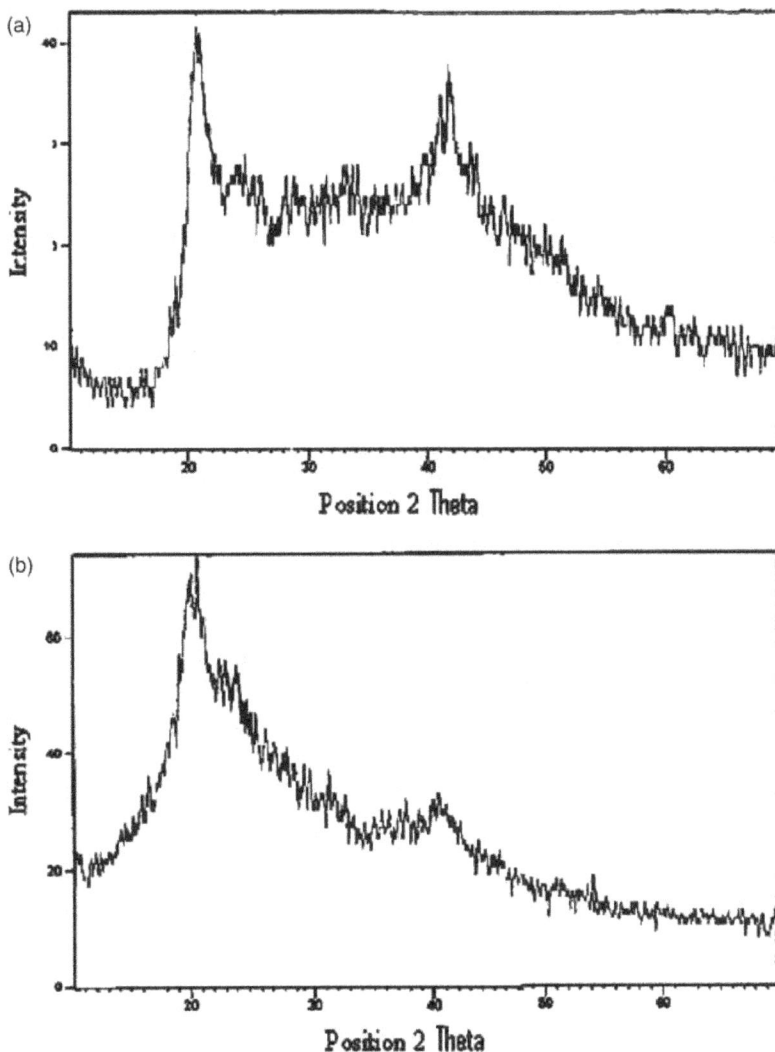

Figure 6.   XRD–spectra of (a) native (PVA–g–PAMPS) gel and (b) PANI-impregnated gel.

3.1.5.   *Scanning electron microscopy*

The morphological features of the prepared native and PANI-impregnated composite films have been studied by recording SEM images of the films, as shown in Figure 7(a) and (b), respectively. It is clear from the image (a) that the surface of the native gel is quite homogeneous and shows no cracks, voids or unevenness. This suggests that after grafting of PAMPS chains onto PVA backbone, the matrix remains homogeneous in composition. However, impregnation of PANI into the matrix develops heterogeneity in the matrix as evident from the SEM image (b). It is clear from the image (b) that impregnated PANI molecules form cluster-like morphology varying in the sizes 0.5–2 μm. The formation of PANI clusters within the polymer matrix could be attributed to hydrophobic nature of the PANI molecules, which may aggregate due to hydrophobic dispersion forces.

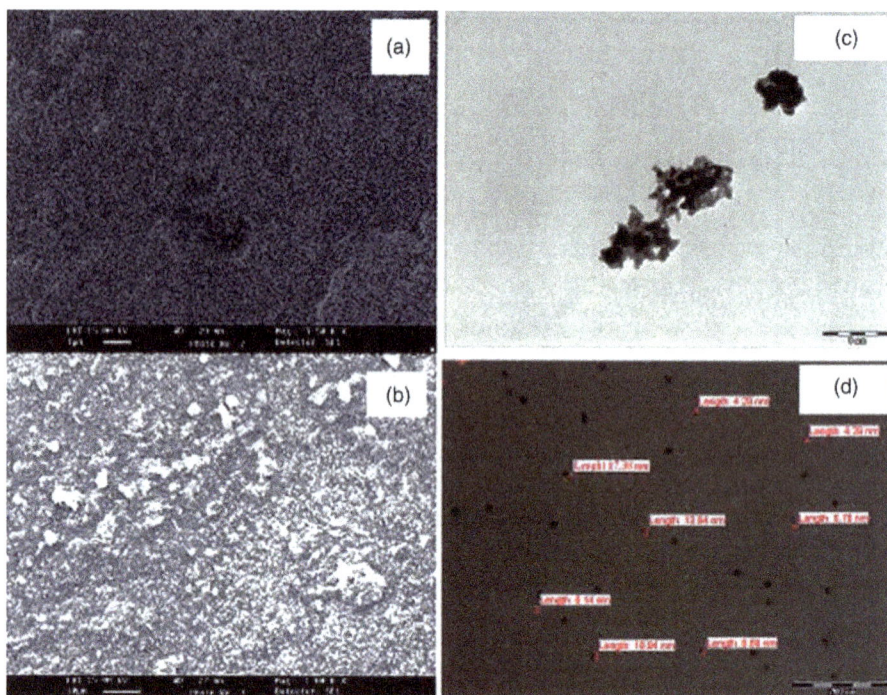

Figure 7.   SEM images of (a) native (PVA–g–PAMPS) gel, (b) impregnated gel, TEM image of PANI particles (c) cluster and particle distribution (d).

3.1.6.   *Transmission electron microscopy*

In order to investigate the size and morphology of the prepared PANI nanoparticles, TEM images were recorded as shown in Figure 7(c) and (d), respectively, which represent the images of nanoparticle clusters and single nanoparticles. It is clear from the image (c) that the formed clusters have a dimension up to $2\,\mu m$ while individual nanoparticles vary in their sizes in the range of 5–10 nm. The results obtained from TEM study are consistent to SEM investigations, and thus confirm the nanocomposite nature of the material.

3.2.   *PANI impregnation and effects of ingredients*

In this study, AN has been taken as a monomer and in order to get it polymerised within the polymer matrix, it is essential that AN molecules should go into the polymer gel. To achieve the desired objectives of AN polymerisation, the monomer AN was dissolved in HCl which formed AN hydrochloride, thus, yielding cationic specie of AN. Now, because of positive charge over the AN molecule, its diffusion into the matrix will be controlled by operative electrostatic forces as well as swelling nature of the hydrogel matrix. It is important to notice here that the polymer matrix is hydrophilic in nature while AN molecules are hydrophobic. In this way, one cannot expect hydrophobic/hydrophilic forces as majorly responsible for the diffusion of AN. It is, therefore, convincing to consider electrostatic forces as the main factor to cause diffusion of AN into the polymer matrix. Moreover, the presence of PAMPS segments in the matrix will also contribute to the electrostatic interaction between the entering AN molecules and the matrix itself. Impregnation of PANI into the polymer matrix is basically dependent on the extent of swelling of the polymer film in AN hydrochloride solution and its subsequent polymerisation

Table 1. Effect of concentration of constituents of the gel on the amount of PANI impregnation.

	PANI impregnation (%)	Other ingredients
PVA content (g)		
1.5	5.74	AMPS–2.412 mM
2	3.28	MBA–13.0×10^{-2} mM
3	2.32	KPS–7.39×10^{-2} mM
4	2.08	APS–0.3 M in 0.5 N HCl
		AN–10.74 mM in 0.5 N HCl
AMPS content (mM)		
2.412	10.58	PVA–1.5 g
4.825	15.38	MBA–13.0×10^{-2} mM
7.237	18.0	KPS–7.39×10^{-2} mM
9.650	27.27	APS–0.3 M in 0.5 N HCl
		AN–10.74 mM in 0.5 N HCl
MBA content (mM)		
6.5×10^{-2}	3.17	PVA–2 g
13.0×10^{-2}	7.66	AMPS–2.412 mM
19.45×10^{-2}	12.82	KPS–7.39×10^{-2} mM
25.9×10^{-2}	11.11	APS–0.3 M in 0.5 N HCl
		AN–10.74 mM in 0.5 N HCl
KPS content (mM)		
3.69×10^{-2}	16.30	PVA–1 g
7.39×10^{-2}	17.07	AMPS–2.412 mM
11.10×10^{-2}	16.96	MBA–19.45×10^{-2} mM
14.80×10^{-2}	24.13	APS–0.3 M in 0.5 N HCl
		AN–10.74 mM in 0.5 N HCl
AN content (mM)		
5.3	6.43	PVA–2 g
10.74	3.82	AMPS–2.412 mM
16.11	5.63	MBA–19.45×10^{-2} mM
21.48	6.45	KPS–11.10×10^{-2} mM
		APS–0.3 M in 0.5 N HCl
APS conc. (M)		
0.2	1.2	PVA–2 g
0.3	1.3	AMPS–2.412 mM
0.4	4.65	MBA–19.45×10^{-2} mM
0.5	1.5	KPS–14.80×10^{-2} mM
		AN–10.74 mM in 0.5 N HCl

within the matrix. Thus, the inclusion of AN into the polymer should definitely be a function of the chemical composition (and nature also) of the hydrogel, and this has been investigated further by varying the concentration of the constituents components of the gel as discussed in the following section.

3.2.1. *Effect of PVA*

To study the effect of PVA on PANI impregnation, PVA was varied in the range 1.5–4.0 g keeping quantities of other ingredients constant. The results summarised in Table 1 reveal that the amount of impregnated PANI constantly decreases. This is due to the fact that the increasing PVA content decreases the interionic repulsion between the PAMPS chain which allows lesser number of AN molecules to enter the polymer matrix that results in a decrease in PANI impregnation.

3.2.2. *Effect of AMPS*

The influence of AMPS variation on the extent of PANI impregnation has been investigated by varying the concentration of AMPS in the range of 2.412–9.650 mM while keeping the other constituents' concentration constant. The results are summarised in Table 1, which indicate an increase in the impregnated amount of PANI with increasing PAMPS content. The increase in PANI impregnation may be explained by the fact that due to ionic nature of PAMPS, its increasing content in the gel results in interionic repulsion between the PAMPS chains, which permits a large number of AN molecules to enter the polymer matrix. This results in an enhanced impregnation.

3.2.3. *Effect of MBA*

The effect of increasing concentration of MBA on the extent of PANI impregnation has been studied by varying its concentration in the range of 6.5×10^{-2}–25.9×10^{-2} mM while keeping the other concentration terms constant. The results are summarised in Table 1, which show that the amount of impregnated PANI increases initially with the increasing concentration of MBA in the range of 6.5×10^{-2}–19.45×10^{-2} mM while beyond 19.45×10^{-2} mM, a fall is noticed. The results may be interpreted by the fact that in the initial concentration range of MBA, the increase observed is due to an enhanced hydrophilicity of the matrix, which in turn, attracts a larger number of AN hydrochloride molecules to diffuse into the gel which upon subsequent polymerisation forms greater amount of PANI. However, beyond 19.45×10^{-2} mM of MBA, much greater crosslinked network becomes compact and restrain the mobility of both incoming AN molecules as well as the relaxation of polymer matrix chain. This clearly results in a less amount of PANI impregnation.

3.2.4. *Effect of KPS*

In this study, the effect of KPS on PANI impregnation has been investigated by varying the concentration of KPS in the range of 3.69×10^{-2}–14.8×10^{-2} mM. The results are summarised in Table 1, which clearly reveal that the amount of impregnated PANI increases with increasing KPS. The observed increase may be attributed to the reason that as the concentration of KPS increases, the molecular weight (and also size) of polymer chain decreases, which results in the formation of network with large number of pores, but of smaller size, which obviously intakes greater number of AN molecules, thus, giving rise to greater impregnation.

3.2.5. *Effect of AN*

The effect of AN (monomer) concentration on PANI impregnation within the polymer matrix has been studied by varying the concentration of AN in 0.5N HCl solution in the range of 5.3–21.48 mM while keeping the concentration of other constitution of the gel constant. The results are summarised in Table 1, which indicate that the extent of PANI impregnation shows an optimum impregnation at 2.412 mM of AN content while a fall is noticed, thereafter again increases with increasing AN concentration. The observed findings may be explained by the fact that as AN is a monomer of PANI, its lower concentration facilitate to penetrate compact network due to higher degree of swelling of native hydrogel due to the availability of large number of small pores in the matrix and its increasing concentration at 4.825 mM in the gel. A decrease observed in PANI impregnation could be due to a lower degree of swelling in AN hydrochloride solution, as much greater ionic concentration (because of AN hydrochloride ion) in the external solution may result in lower swelling which permits less number of AN molecules for polymerisation. However, thereafter the extent of impregnation increases with increasing AN concentration may be explained by the fact that with increasing concentration of AN solution, greater number of AN molecules will be available for polymerisation within the matrix and, as a consequence, the extent of impregnation also increases.

3.2.6. *Effect of APS concentration*

The influence of concentration of APS on PANI impregnation was studied by varying its concentration in the range of 0.2–0.5 M, keeping other concentration terms constant. The results summarised in Table 1 show that the amount of impregnated PANI increases with increasing APS concentration up to 0.4 M, thereafter a fall is noticed. The reason for the observed increase may be explained by the fact that increasing the concentration of APS causes greater polymerisation of AN monomer, which consequently results in higher impregnation of PANI. However, beyond 0.4 M of APS concentration a decrease in impregnation may be due to higher APS concentration, the PANI formed would be of lower molecular weight and, therefore, the possibility of leaching of low molecular weight, PANI could not be ruled out. This may consequently result in low impregnation.

3.3. *Electrical conductivity and effects of ingredients*

The electrical conductivity of polymer composite in different weight ratio of ingredients has been studied and it is found that the electrical conductivity of the gel is enhanced after impregnation of PANI. Moreover, the conductivity also varies with the constituents of the polymer matrix, as discussed below.

3.3.1. *Effect of PVA*

The effect of PVA content on the conductivity of the matrix has been studied by varying the amount of PVA in the range of 1.5–4 g. The results are depicted in Table 2, which reveal that the matrix shows an optimum conductivity at 1.5 g of PVA content while a drastic fall of about 10 times in decreasing order is noticed in conductivity when the concentration of PVA increases. The results may be explained as follows.

The conductivity is determined by both the extent of PANI impregnation as well as PAMPS content of the matrix. The former, in turn, is dependent on the hydrophilicity of the matrix. When the amount of PVA increases from 1.5 to 4.0 g, the weight fraction of ionic polymer (PAMPS) decreases in the matrix. This obviously results in fall in the conductivity of the matrix.

3.3.2. *Effect of PAMPS*

In order to study the influence of PAMPS on the conductivity of the matrix, the concentration of the AMPS was increased in the range of 2.412–9.650 mM and electrical conductivity was measured. The results are summarised in Table 2, which reveal that the conductivity gradually increases with increasing PAMPS content in the gel. The results are quite expected and may be explained by the fact that on increasing the concentration of PAMPS, the number of ionic groups increases along the macromolecular chain which add to the electrical conductivity of the matrix by facilitating conduction of electrons along the PANI chain.

Table 2.   Effect of concentration of constituents of the gel on conductivity.

	Conductivity native S/cm	Conductivity PANI impregnated S/cm	Other ingredients
PVA content (g)			
1.5	8.12×10^{-3}	8.38×10^{-3}	AMPS–2.412 mM
2	1.01×10^{-4}	1.93×10^{-3}	MBA–13.0×10^{-2} mM
3	1.64×10^{-4}	0.08×10^{-3}	KPS–7.39×10^{-2} mM
4	6.17×10^{-4}	0.09×10^{-3}	APS–0.3 M in 0.5 N HCl
			AN–10.74 mM in 0.5 N HCl
AMPS content (mM)			
2.412	9.4×10^{-3}	1.81×10^{-2}	PVA–1.5 g
4.825	2.02×10^{-2}	2.05×10^{-2}	MBA–13.0×10^{-2} mM
7.237	2.8×10^{-2}	3.3×10^{-2}	KPS–7.39×10^{-2} mM
9.650	3.4×10^{-2}	27.65×10^{-2}	APS–0.3 M in 0.5 N HCl
			AN–10.74 mM in 0.5N HCl
MBA content (mM)			
6.5×10^{-2}	4.95×10^{-4}	5.60×10^{-3}	PVA–2 g
13.0×10^{-2}	1.10×10^{-3}	11.32×10^{-3}	AMPS–2.412 mM
19.45×10^{-2}	0.21×10^{-3}	4.85×10^{-3}	KPS–7.39×10^{-2} mM
25.9×10^{-2}	0.11×10^{-3}	1.48×10^{-3}	APS–0.3 M in 0.5 N HCl
			AN–10.74 mM in 0.5 N HCl
AN content (mM)			
5.3	0.11×10^{-3}	1.49×10^{-2}	PVA–2 g
10.74	0.11×10^{-3}	4.69×10^{-2}	AMPS–2.412 mM
16.11	0.11×10^{-3}	4.77×10^{-2}	MBA–19.45×10^{-2} mM
21.48	0.11×10^{-3}	3.19×10^{-2}	KPS–11.10×10^{-2} mM
			APS–0.3 M in 0.5 N HCl

It is also necessary to mention that with increasing concentration of PAMPS, percent impregnation of PANI also increases in the studied range, which thus contributes to the observed increase in conductivity.

3.3.3. *Effect of MBA*

In the present investigation, the role of crosslinker has been investigated by varying the amount of crosslinker in the concentration range of 6.5×10^{-2}–25.9×10^{-2} mM and observing the change in electrical conductivity. The results are summarised in Table 2, which show that the conductivity sharply increases with increasing concentration of MBA in the range of 6.5×10^{-2}–13.0×10^{-2} mM while after 13.0×10^{-2} mM of crosslinker, the conductivity acquires a limiting value.

The observed initial steep increase in the conductivity of impregnated gel with increasing concentration of MBA may be attributed to the fact that with increasing crosslinker concentration, the network density increases which results in a compact network, thus, facilitating flow of electrons responsible for conductivity behaviour. However, beyond an optimum concentration of crosslinking agent, the gels acquire optimum compactness and, therefore, no further increase in conductivity is observed.

3.3.4. *Effect of AN*

The effect of PANI content on the conductivity of the matrix has been investigated by increasing the concentration of AN in the range of 5.3–21.48 mM in the feed mixture of the gel.

The results are given in Table 2, which indicate that the conductivity increases with increasing concentration of AN up to 16.11 mM concentration of AN, thereafter a fall in conductivity is noticed.

The mechanism of electrical conductivity in conducting polymers has been reported in the literature that the DC conductivity of conducting polymers depend on their morphology and certain other factors such as the type of monomer, doping level, degree of crystallinity, etc., [39,40]. It has been reported that the large increase in conductivity originates from the fact that doped (protonated) PANI is a polyelectrolyte, i.e. a macromolecule bearing a large number of ionisable groups. It is also reported that upon reaction with the vapour of appropriate substance, its molecular conformation change from compact coil structure to an expanded coil-like structure. The attainment of expanded molecular conformation acts to reduce π-conjugation defects in the polymer backbone and the opening up of coil tends to promote linear conformation necessary for crystallisation. Thus, an increase in the crystallinity of the polymer occurs with enhancement of the intermolecular component of the bulk conductivity.

Therefore, the results may be explained by the fact that with increasing concentration of AN solution, greater number of AN molecules will be available within the matrix for polymerisation and as a subsequence, the extent of impregnation also increases.

3.3.4.1. Bending behaviour of matrix.   Swollen polyelectrolyte gels when kept under an applied electric field normally exhibits a bending towards the electrodes. This is called electromechanochemical (EMC) behaviour and has been recognised as a promising phenomenon in designing smart materials. The effect of various factors on the bending behaviour may be discussed as follows.

(a) **Effect of PANI content.** Inclusion of PANI molecules within the polymer matrix is expected to enhance the EMC behaviour and, therefore, has been studied by varying the PANI content in the range of 0–25% at different applied voltages, as shown in Figure 8. The results clearly reveal that the effective bending angle (EBA) constantly increases with increasing PANI content. The observed results may be explained by the fact that greater the PANI content of the matrix, larger would be the conductivity which, in turn, will respond frequently to the applied voltage.

   Figure 8 also indicates that the magnitude of EBA also increase with increasing applied voltage across the electrodes. The increase in EBA may be attributed to the fact that with increase in applied voltage, the charged matrix are attracted more readily towards the electrodes and, thus, the bending angle increases.

(b) **Effect of voltage.** There is a lower critical voltage (LCV) below which no bending of matrix is observed. In this study, the LCV is found to be 4.5 V. Thus, the bending phenomenon is observed only after 4.5 V and, therefore, the effect of voltage on bending phenomenon has been investigated in the range of 6.0–12.0 V for a given PANI content. The influence of applied voltage on the EMC behaviour of the matrix has been studied by varying the voltage in the range of 6.0–12.0 V. The results are depicted in Figure 8(a)–(c), which indicate that the EMC behaviour of the matrix shows an increase in EBA with increasing voltage at 0.1, 0.2 and 0.3 N NaCl solution. The observed results are quite obvious as with increasing voltage, the charged matrix shows greater attraction towards electrodes and, therefore, bending angle increases.

(c) **Effect of electrolyte concentration.** The concentration of electrolyte solution is an important experimental parameter to exert a significant influence on the EBA of the polymer matrix. The effect of increasing concentration of NaCl solution on bending of the gel

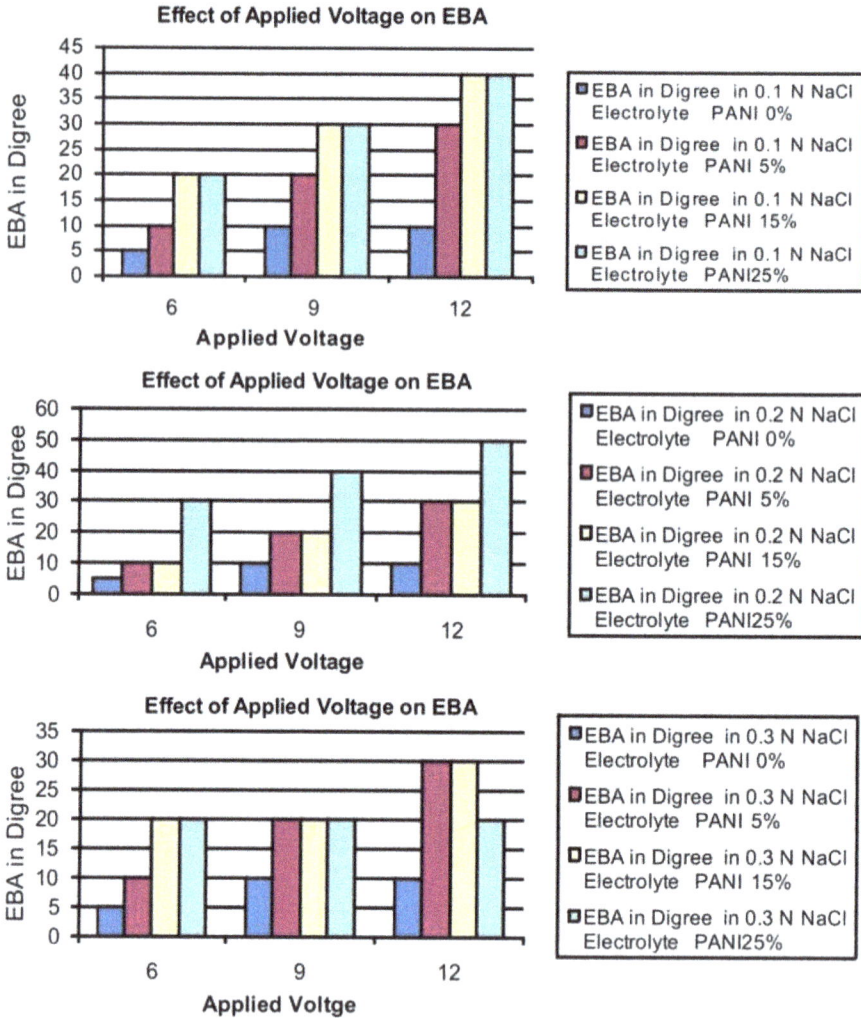

Figure 8.   Effects of applied voltage on EBA in (a) 0.1 N NaCl electrolyte solution, (b) 0.2 N NaCl electrolyte solution and (c) 0.3 N NaCl electrolyte solution.

has been studied by varying the concentration of NaCl in the range of 0.1–0.3 N. The results are shown in Figure 9(a)–(c), which reveal that the bending angle substantially increases up to 0.2 N NaCl solution and, thereafter, it becomes almost constant showing no further increase in bending. The results may be attributed to the fact that an increased concentration of NaCl solution facilitates bending of the polymer matrix by binding of salt ion to the polyelectrolyte matrix so as to make it more responsive to the applied field. However, beyond 0.2 N NaCl concentration when the charged centres present along the polyelectrolyte molecules are almost completely ionised (so it may be a critical concentration), a shielding effect of polyions caused by the other ions in the electrolyte solute occurs leading to the electrostatic repulsion of the polyions and a decrease in the degree of bending [41]. According to Flory's [42] theory, an increasing amount of ions could reduce the electrostatic repulsion of the polyions by the screening of fixed charges and bring about a decrease in the degree of bending. In such a situation, the network could

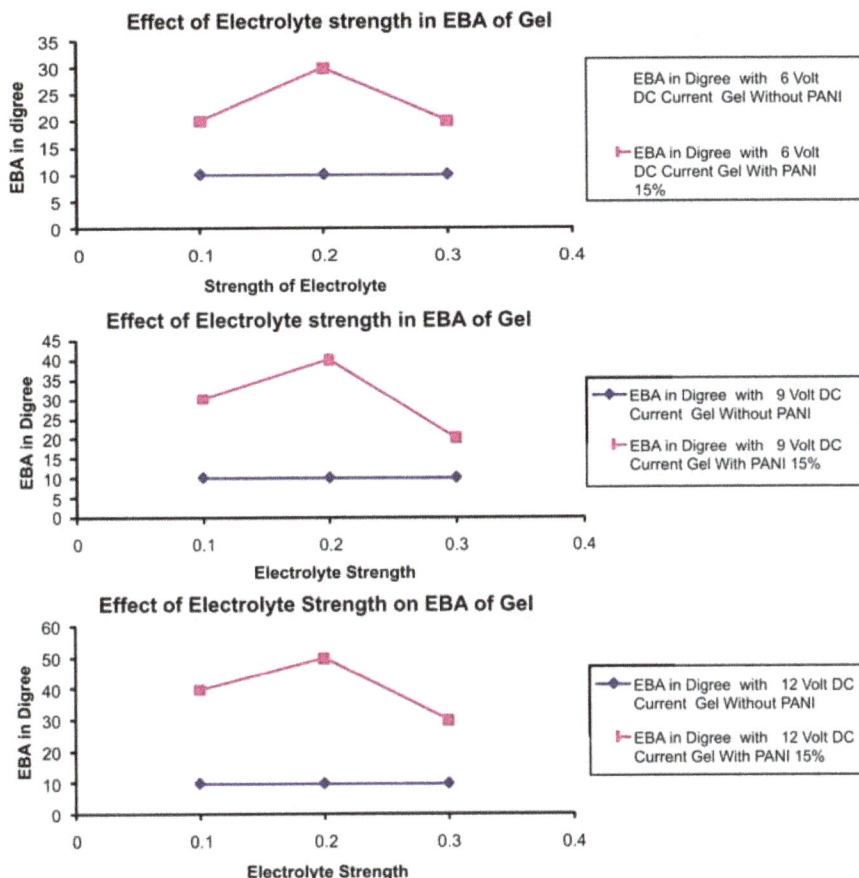

Figure 9.   Effect of the strength of electrolyte solution in (a) 6, (b) 9 and (c) 12 V.

shrink into a more compact structure, making the diffusion of ions more difficult through the gel. Because of this, the ionic gradient and osmotic pressure may not occur above the critical concentration of the electrolyte, and a decrease in EBA is observed above 0.2 N ionic strength of the electrolyte.

3.4. *Swelling study and effect of ingredients*

The swelling behaviour of the native and impregnated hydrogels with varying ingredients has been measured in distilled water. The results are summarised in Table 3 and may be explained as follows.

When the amount of PVA increases in the range from 1.5 to 4.0 g, the swelling ratio decreases – which may be attributed to the fact that with increase in polymer fraction in the hydrogel, the number of network chains increases in unit volume of the gel. This restrains both the inclusion of water molecules into the gel and subsequent relaxation of polymer chains. This obviously decreases the swelling capacity of the gel.

The influence of ionic monomer, AMPS, on the swelling ratio of the gel has been investigated by varying its concentration in the range of 2.412–9.650 mM. The results show that the swelling ratio constantly increases with increasing PAMPS content in the gel. The observed increase may

Table 3.   Effect of varying concentration of constituents of the gel on swelling ratio.

	SR (distilled water)	Other ingredients
PVA (g)		
1.5	6.82	AMPS–2.412 mM
2	6.69	MBA–13.0 $\times$ 10^{-2} mM
3	5.43	KPS–7.39 $\times$ 10^{-2} mM
4	4.21	APS–0.3 M in 0.5 N HCl
		AN–10.74 mM in 0.5 N HCl
AMPS (mM)		
2.412	3.3	PVA–1.5 g
4.825	5.67	MBA–13.0 $\times$ 10^{-2} mM
7.237	7.03	KPS–7.39 $\times$ 10^{-2} mM
9.650	12.14	APS–0.3 M in 0.5 N HCl
		AN–10.74 mM in 0.5 N HCl
MBA (mM)		
6.5 $\times$ 10^{-2}	9.03	PVA–2 g
13.0 $\times$ 10^{-2}	6.35	AMPS–2.412 mM
19.45 $\times$ 10^{-2}	5.05	KPS–7.39 $\times$ 10^{-2} mM
25.9 $\times$ 10^{-2}	6.16	APS–0.3 M in 0.5 N HCl
		AN–10.74 mM in 0.5 N HCl
KPS (mM)		
3.69 $\times$ 10^{-2}	4.70	PVA–1 g
7.39 $\times$ 10^{-2}	4.39	AMPS–2.412 mM
11.10 $\times$ 10^{-2}	4.26	MBA–19.45 $\times$ 10^{-2} mM
14.80 $\times$ 10^{-2}	4.52	APS–0.3 M in 0.5 N HCl
		AN–10.74 mM in 0.5 N HCl
AN (mM)		
5.3	2.57	PVA–2 g
10.74	2.42	AMPS–2.412 mM
16.11	2.43	MBA–19.45 $\times$ 10^{-2} mM
21.48	2.53	KPS–11 $\times$ 10^{-2} mM
		APS–0.3 M in 0.5 N HCl
APS (M)		
0.2	3.15	PVA–2 g
0.3	3.07	AMPS–2.412 mM
0.4	3.41	MBA–19.45 $\times$ 10^{-2} mM
0.5	3.67	KPS–14.80 $\times$ 10^{-2} mM
		AN–10.74 mM in 0.5 N HCl

be explained by the reason that with increase in the concentration of AMPS, the number of sulphonic groups present along the polymer chain also increases which is due to the enhanced repulsion results in a greater mobility of gel segments. This, in turn, brings about a rise in the swelling ratio.

When the concentration of crosslinking agent (MBA) increases in the range of 6.5–19.45 $\times$ 10^{-2} mM, the swelling ratio decreases while beyond 19.45 $\times$ 10^{-2} mM the swelling ratio increases up to 25.9 $\times$ 10^{-2} mM. The results may be explained by the fact that since MBA is a bifunctional monomer, its increasing amount in the gel caused compactness in gel which decreases water sorption capacity – while beyond a certain concentration, the PANI impregnation increases. The impregnated PANI not only occupied more space in the gel, but also its hydrophobic nature restricts the water sorption capacity of the gel. The effect of PANI content in the hydrogel on its swelling ratio has been studied by increasing the concentration of AN in the range of 5.3–21.48 mM. The results indicate that the swelling ratio slightly decreases with increasing concentration of AN. The observed decrease in the swelling ratio may be explained by the fact that due to the increase in PANI impregnation in matrix, the water sorption capacity of the gel was restricted and controlled.

4. Conclusions

Impregnation of PANI into PVA–g–PAMPS results in a composite hydrogel which shows fair electro-conductive and electro-active behaviours. The FTIR spectra of PANI impregnation hydrogel show characteristic peaks of PANI and other functional groups of constituent polymers, i.e. PVA and PAMPS. The impregnation of PANI into polymer matrix is further confirmed by UV-spectral analysis. The impregnation of PANI within the polymer matrix brings about a loss in crystallinity, as confirmed by the XRD spectra. The hydrogel composite shows cluster-like morphology varying in size between 0.5 and 2.0 μm. The PANI particles undergo aggregation and show a wide variation in their sizes ranging from 1 to 100 μm.

The extent of PANI impregnation depends on chemical composition of the composite hydrogel. It is noticed that with increasing AMPS, MBA and KPS concentrations, the extent of PANI impregnation increases up to a certain concentration range, and thereafter decreases. On the other hand, an increase in impregnation results with increasing APS and AN content, a fall is observed with increasing PVA concentration.

The electrical conductivity also varies with varying composition of the composite. The conductivity constantly increases with increasing concentration of AMPS, APS and AN. In the case of MBA, however, a decrease in conductivity is observed, and in the case of PVA the conductivity constantly decreases.

The PANI impregnated matrix shows bending behaviour when a fixed voltage is applied in the presence of an electrolyte solution (NaCl). The gels show an enhanced bending behaviour with increasing PANI content and applied voltage. Moreover, a greater bending is noticed with increasing concentration of electrolyte solution. The native and PANI-impregnated matrices also show a display swelling properties in distilled water.

References

[1] K. Kajiwara and S.B. Ross-Murphy, *Synthetic gels on the move*, Nature 355 (1992), pp. 242–244.
[2] H. Bai and G. Shi, *Gas sensors based on conducting polymers*, Sensors 7 (2007), pp. 267–307.
[3] M. Baibarac and P. Gomez-Romero, *Nano composites based on conducting polymers and carbon nanotubes from fancy materials to functional applications*, J. Nanosci. Nanotechnol. 6 (2006), pp. 289–302.
[4] D. Wei and A. Ivaska, *Electrochemical biosensors based on polyaniline*, Chem. Anal. (Warsaw) 51 (2006), pp. 839–852.
[5] H. Suzuki, *Stimulus-responsive gels: Promising materials for the construction of micro actuators and sensors*, J. Intell. Mater. Syst. Struct. 17 (2006), pp. 1091–1096.
[6] M.J. Carlos and C.M. Teresa, *Artificial muscles based on conducting polymers*, e-polymers 41 (2003), pp. 1–42.
[7] Y. Chun-ting, H. Yao-Xiong, Z. Hai-yah, C. Hong-hui, and P. Yan-hao, *Development of polyvinyl alcohol–collagen composite and its biocompatibility as tissue substitute*, J. Clin. Rehabil. Tiss. Eng. Res. 12 (2008), p. 153.
[8] C. Ponce de Leon, S.A. Campbell, J.R. Smith, and F.C. Walsh, *Conducting polymer coatings in electrochemical technology. Part 2 – Application areas*, Trans. Inst. Metal Finish. 86 (2008), pp. 34–40.
[9] T. Tanaka, *Collapse of gels and the critical endpoint*, Phys. Rev. Lett. 40 (1978), pp. 820–823.
[10] T. Tanaka, I. Nishio, S.T. Sun, and S. Ueno-Nishio, *Collapse of gels in an electric-field*, Science 218 (1982), pp. 467–469.
[11] A. Suzuki and T. Tanaka, *Phase-transition in polymer gels induced by visible-light*, Nature 346 (1990), pp. 345–347.
[12] Y. Zhang, H.F. Ji, G.M. Brown, and T. Thundat, *Detection of CrO_4^{2-} using a hydrogel swelling microcantilever sensor*, Anal. Chem. 75 (2003), pp. 4773–4777.
[13] M. Irie, Y. Misumi, and T. Tanaka, *Stimuli-responsive polymers: Chemical-induced reversible phase separation of an aqueous solution of poly(N-isopropylacrylamide) with pendent crown ether groups*, Polymer 34 (1993), pp. 4531–4535.

[14] T. Tanaka, C. Wang, V. Pande, A.Y. Grosberg, A. English, S. Masamune, H. Gold, R. Levy, and K. King, *Polymer gels that can recognize and recover molecules*, Faraday Discuss. 101 (1995), pp. 201–206.

[15] K. Dong-Hwan, M. Abidian, and D.C. Martin, *Conducting polymers grown in hydrogel scaffolds coated on neutral prosthetic devices*, Biomed. Mater. Res. 71A (2004), pp. 577–585.

[16] W. Li, H. Zhao, P.R. Teasdale, R. John, and S. Zhang, *Synthesis and characterization of a polyacrylamidepoly-acrylic acid copolymer hydrogel for environmental analysis of Cu and Cd*, React. Funct. Polym. 52 (2002), pp. 31–41.

[17] A.K. Bajpai and R. Saini, *Preparation and characterization of spongy cryogels of poly(vinyl alcohol)-casein system: Water sorption and blood compatibility study*, Polym. Int. 54 (2005), pp. 796–806.

[18] A.K. Bajpai and R. Saini, *Preparation and characterization of novel biocompatible cryogels of poly(vinyl alcohol) and egg-albumin and their water sorption study*, J. Mater. Sci. Mater. Med. 17 (2006), pp. 49–61.

[19] A.K. Bajpai and M. Sharma, *Preparation and characterization of novel pH-sensitive binary grafted polymeric blends of gelatin and poly(vinyl alcohol): Water sorption and blood compatibility study*, J. Appl. Polym. Sci. 100 (2006), pp. 599–617.

[20] P. Raul, *Electrical Characterization of Polyaniline Synthesized Using One Phase Emulsion Polymerization Process*, Proceeding of the National Conference on Undergraduate Research (NCUR) 2004, Indiana University, Purdue University, Indianapolis, Indiana, 2004, pp. 15–17.

[21] A. Naser, K. Mitra, and A.E. Ali, *Synthesis and characterization of novel-N-substituted polyaniline by Triton X–100*, Iran. Polym. J. 12(3) (2003), pp. 237–242.

[22] P.D. Gaikwad, D.J. Shirale, V.K. Gade, P.A. Savale, H.J. Kharat, K.P. Kakde, S.S. Hussaini, N.R. Dhumane, and M.D. Shirsat, *Synthesis of H_2SO_4 doped polyaniline film by potentiometric method*, Bull. Mater. Sci. 29(2) (2006), pp. 169–172.

[23] K. Mallick, M.J. Witcomb, A. Binsmore, and S. Scurrell, *Polymerization of aniline by cupric sulfate: A facile synthetic route for producing polyaniline*, J Polym. Res. 13 (2006), pp. 397–401.

[24] G.D. Nestorovic, K.B. Jeremic, and S.M. Jovanovic, *Kinetics of aniline polymerization initiated with iron III chloride*, J. Serb. Chem. Soc. 71(8–9) (2006), pp. 895–904.

[25] V. Gupta and N. Miura, *Electrochemically deposited polyanilines network: A high performance electrode materials for redox super capacitors*, Electrochem. Solid State Lett. 8(12) (2005), pp. A630–A632.

[26] M. Mazur, M. Tagowska, B. Palys, and K. Jackowska, *Template synthesis of poly aniline and poly(2–methoxy aniline) nanotubes: Comparison of the formation mechanism*, Electrochem. Commun. 5 (2003), pp. 403–407.

[27] C. Aispenza, C.L. Presti, C. Belfione, G. Spadaro, and S. Piazza, *Electrically conductive hydrogel composites made of polyaniline nanoparticles and poly(N-vinyl-2-pyrrolidone)*, Polymer 47 (2006), pp. 961–971.

[28] P. Savitha, P. Swapna Rao, and D.N. Sathyanarayana, *Highly conductive new aniline copolymers: Poly(aniline-coamino acetophenone)s*, Polym. Int. 54 (2005), pp. 1243–1250.

[29] H.C. Pant, M.K. Patra, S.C. Negi, A. Bhatia, S.R. Vadera, and N. Kumar, *Studies on conductivity and dielectric properties of polyaniline–zinc sulphide composites*, Bull. Mater. Sci. 29(4) (2006), pp. 379–384.

[30] S.D. Patil, S.C. Raghavendra, M. Revansiddappa, P. Narsimha, and M.V.N. Ambika Prasad, *Synthesis, transport and dielectric properties of polyaniline/Co_3O_4 composites*, Bull. Mater. Sci. 30(2) (2007), pp. 89–92.

[31] K.S. Lee, G.B. Blanchet, F. Gao, and L. Yueh-Lin, *Direct patterning of conductive water-soluble polyaniline for thin film organic electrodes*, Appl. Phys. Lett. 86 (2005), pp. 076102/1–076102/3.

[32] A. Pud, N. Ogurtsov, A. Korzhenko, and G. Shapoval, *Some aspects of preparation methods and properties of polyaniline blends and composites with organic polymers*, Progr. Polym. Sci. 28 (2003), pp. 1701–1753.

[33] S. Shukla and A.K. Bajpai, *Preparation and characterization of highly swelling smart grafted polymer networks of poly(vinyl alcohol) and poly(acrylic acid-co-acrylamide)*, J. Appl. Polym. Sci. 102 (2006), pp. 84–95.

[34] A.K. Bajpai. and A. Mishra, *Preparation and characterization of tetracycline-loaded interpenetrating polymer networks of carboxymethyl cellulose and poly(acrylic acid): Water sorption and drug release study*, Polym. Int. 54 (2005), pp. 1347–1356.

[35] A.K. Bajpai, J. Bajpai, and S.N. Soni, *Electroactive actuation and conductive behavior of polyaniline-impregnated blood compatible nanocomposites*, J. Compos. Mater. 45(5) (2010), pp. 485–497.

[36] A.K. Bajpai, J. Bajpai, and S.N. Soni, *Preparation and characterization of electrically conductive composites of poly(vinyl alcohol)-g-poly(acrylic acid) hydrogels impregnated with polyaniline (PANI)*, Exp. Polym. Lett. 2(1) (2008), pp. 26–39.

[37] A.K. Bajpai, J. Bajpai, and S.N. Soni, *Designing polyaniline (PANI) and polyvinyl alcohol (PVA) based electrically conductive nanocomposites: Preparation, characterization and blood compatible study*, J. Macro. Sci. Part A Pure Appl. Chem. 46(8) (2009), pp. 774–782.

[38] Y. Murali Mohan, P. Dickson Joseph, and E. Geckeler Kurt, *Swelling and diffusion characteristics of novel semi-interpenetrating network hydrogels composed of poly[(acrylamide)-co-(sodium acrylate)] and poly[(vinylsulfonic acid), sodium salt]*, Polym. Int. 56 (2007), pp. 175–185.

[39] H.C. Pant, M.K. Patra, S.C. Negi, A. Bhatia, S.R. Vadera, and N. Kumar, *Studies on conductivity and dielectric properties of polyaniline–zinc sulphide composites*, Bull. Mater. Sci. 29(4) (2006), pp. 379–384.

[40] S.C. Raghavenrra, S. Khasim, M. Revanasiddappa, M.V.N. Ambika Prasad, and A.B. Kulkarni, *Synthesis, characterization and low frequency AC conduction of polyaniline/fly ash composites*, Bull. Mater. Sci. 26(7) (2003), pp. 733–739.
[41] S.J. Kim, S.G. Yoon, Y.H. Lee, and S.I. Kim, *Bending behavior of hydrogels composed of poly(methacrylic acid) and alginate by electrical stimulus*, Polym. Int. 53 (2004), pp. 1456–1460.
[42] P.J. Flory, *Principle of Polymer Chemistry*, Cornell University Press, Ithaca, NY, 1953.

10

Magnetic field induced assembly of polyvinylpyrrolidone stabilised cobalt ferrite nanoparticles in different dispersion medium

Devasish Chowdhury*

Material Science Division, Polymer Unit, Institute of Advanced Study in Science and Technology, Paschim Boragaon, Garchuk, Guwahati 781035, Assam, India

In this article, a report on the assembly of hydrophilic polyvinylpyrrolidone (PVP) stabilised $CoFe_2O_4$ nanoparticles ($PVP@CoFe_2O_4$) in different dispersion media under external magnetic field is presented. The assembly process in water, ethanol and toluene as solvent was studied. It was observed that water dispersion of $PVP@CoFe_2O_4$ gives longer chain-like assembly on applying unidirectional magnetic field as compared to ethanol which gives shorter chain assembly. On the contrary, in toluene dispersion, there is no such assembly on applying magnetic field. The effect of magnetic field on the assembly process was also studied. A possible explanation accounting for the observations is also presented.

Keywords: magnetic field; magnetisation; ferromagnetic; nanoparticles; assembly; nanostructure

1. Introduction

The synthesis and characterisation of nanostructure materials have gained considerable interest in the past decade or so, due to prospective applications in nanoelectronic devices [1,2], data storage application [3] miniaturised chemical and biological sensors [4–7] and synthesis of nanocomposites for the improvement of properties [8,9]. The challenges are to device methods that will control the crystalline structure, magnetic properties and inter-particle ordering of nanoparticles (NPs) for their use as building blocks for new devices [10–12]. The development of methods for controlling the organisation of nanometre-sized object relative to the one another and to larger macro object is crucial for the overall advancement of using such building blocks for devices [13]. The unique electronic, magnetic and photonic properties of NPs will be best exploited if they can be integrated into large devices with the development of self-assembly of NP into layer structures [14]. Magnetic nanostructures are being used for data storage applications to push the limit of existing technologies and device new techniques for better capabilities. Magnetic NPs find application in various biomedical and biological applications, such as magnetic resonance

*Email: devasishc@gmail.com

imaging (MRI) contrast enhancement [15], hyperthermia [16], gene and drug delivery [17]. Metal–oxide NPs have been the subject of much interest because of their unusual optical, electronic and magnetic properties, which often differ from the bulk. The most important ferromagnetic oxide materials are ferrites [18–20]. Cobalt ferrite ($CoFe_2O_4$) is a well known hard magnetic material with high coercivity and moderate magnetisation [21,22]. $CoFe_2O_4$ NPs also find their use in biological and clinical applications. It is already a material in choice for magnetic fluid hyperthermia (MFH) [23]. Thus, in order to use them for *in vivo* applications, the additional requirement is that $CoFe_2O_4$ NPs have to be hydrophilic and smaller or comparable in size than to those of a cell (10–100 μm), a virus (20–450 nm), a protein (5–50 nm) or a gene (2 nm wide and 10–100 nm long), which will facilitate such magnetic NPs to get closer to a biological entity of interest. $CoFe_2O_4$ NPs can be made hydrophilic/lypophilic with a hydrophilic/lypophilic coating, for example with polyvinylpyrrolidone (PVP) [24], with mono/difunctional phosphonic and hydro-xamic acid [25] or with starch [26].

Magnetic NPs dispersed in a liquid medium are interesting systems in view of the interplay of various forces operating like the magnetic dipole forces, particle–particle van der Waals forces and the particle substrate interaction. Extending the self-assembly beyond the molecular scale to mesoscale and macroscale levels will not only ensure ordered nanostructure but will also realise the functional system and sub-system at the scale of structural organisation of matter. The assemblies of magnetic NPs are quite different from the other systems, as there is an additional magnetic force of interaction which favours the formation of ordered chains. These studies will help in understanding more complex systems and will provide further avenues for manipulating assemblies of NPs with greater precision. As a general method for control over NP self-assembly, variation in solvent provides the first step in assembling more complex systems. The interaction of solvent with the solvophobic regions and the relative volume of the hydrophilic and hydrophobic regions and the thermodynamics of the interaction of the two domains with one another are important in determining the size and shape of the resultant nanostructures.

In the past, magnetic assembly of magnetic NPs has been demonstrated. Niu et al. [27] have shown growth and assembly of cobalt magnetic nanocrystallites under external magnetic field to fabricate a large array of uniform wires of magnetic nanomaterial. Platt et al. [28] have shown that magnetic NPs functionalised microtubes could be aligned and deposited onto kinesin-coated glass surfaces by the application of a magnetic field. $CoFe_2O_4$ nanowires were fabricated using $CoFe_2O_4$ NPs in porous anodic alumina template assisted by the magnetic field of a permanent magnet placed under the substrate [21]. Magnetically directed assembly of polystyrene spheres with coating of 12 nm Fe_3O_4 nanocrystal was also demonstrated [29]. The separation and the length of the individual chain were tuned by magnetic field and the concentration of the particle solution. But there is no report on the study of the effect of dispersion solvent on the assembling of magnetic NP under external magnetic field.

In this article, we aimed to study the role of different dispersion solvents on the assembling of hydrophilic magnetic NPs under external magnetic field. In order to conduct the study, PVP-stabilised cobalt ferrite ($PVP@CoFe_2O_4$) NPs, less than 20 nm in size having a hydrophilic surface coating were synthesised. These hydrophilic ferrite NPs are aligned and assembled in the form of wires and junctions on surfaces by applying a magnetic field. The effect of strength of the magnetic field on the assembly process in different dispersion solvents was also discussed.

2. Materials and methods

2.1. *Materials*

FeCl$_3$ · 6H$_2$O (Merck India), CoCl$_3$ · 6H$_2$O (Merck India), NaOH (Merck India), oleic acid (Aldrich), ethanol (Tedia Company Inc., Canada) and PVP (Aldrich) were used as received.

2.2. *Preparation of CoFe$_2$O$_4$ NPs and PVP@CoFe$_2$O$_4$ NPs*

CoFe$_2$O$_4$ NPs were prepared by a small modification of an earlier reported method [30]; typically, 0.4 M solution of FeCl$_3$ · 6H$_2$O and 0.2 M solution of CoCl$_3$ · 6H$_2$O were prepared in Milli-Q water followed by the addition of 3 M solution of NaOH in a dropwise manner till the pH of the solution was around 11–12. A small amount of oleic acid was added as surfactant and the liquid mixture was brought to reaction temperature at 80°C and stirred for 1 h. The resultant black precipitate formed after the reaction was cooled to room temperature. The precipitate was washed with Milli-Q water to remove Na$^+$ and Cl$^-$ and with ethanol to remove excess surfactant. In order to render the CoFe$_2$O$_4$ surface hydrophilic, the CoFe$_2$O$_4$ precipitate was mixed in 1% PVP w/w and heated together at 50°C for 1 h. The colour of the precipitate changed from black to dark brown.

2.3. *Assembly of PVP@CoFe$_2$O$_4$ NPs*

The assembly of PVP@CoFe$_2$O$_4$ in the presence of external magnetic field was studied. A drop of dispersion made in a suitable solvent was placed on a substrate (grid, glass or mica as used for different studies) and put under external magnetic field. Magnetic field was applied in the horizontal direction generated by an electromagnet (direct current; DC electromagnet, Model EMU-50V, Scientific Equipment & Services, Roorkee, India) till the solvent got evaporated to obtain the alignment of the magnetic NPs. The magnetic field generated was measured with gauss meter. The magnetic field in and around the drop was uniform and the variation was in the range of ±5 guass (G). After evaporation of the solvent, PVP@CoFe$_2$O$_4$ NPs dispersion was dropped at the same place with applying magnetic field perpendicular to the first one. This will generate a junction of PVP@CoFe$_2$O$_4$ chain-like assembly aligned perpendicular to each other. The assembly process in different dispersion media, i.e. water, ethanol and toluene was studied by repeating the same process with different solvents. The effect of magnetic field on the assembling process in different dispersion media was also investigated by placing the dispersion of magnetic NPs under varied magnetic field. The magnetic field in the electromagnet was varied by changing current flow into the coil. The one-dimensional (1-D) chain-like assembly process was investigated under different magnetic field, e.g. 1560, 750, 305 and 75 G. The 1-D assembly of magnetic NPs thus formed was viewed under different microscopes (transmission electron microscope (TEM), scanning electron microscope (SEM), atomic force microscope (AFM) and optical microscope).

2.4. Characterisation methods

SEM images were obtained using a Leo 1430vp microscope operating at 15 kV. Powder X-ray diffractometer (XRD) analysis was performed using Bruker AXS D8. Magnetic properties of PVP@CoFe$_2$O$_4$ were measured using variable sample magnetron, model no. 7410, Lakeshore Cryotronics Inc., USA. Magnetic force microscopy was done using a digital multi-mode scanning probe microscope (SPM) from Veeco with Nanoscope IVa controller imaged at ambient condition. TEM analysis was done in JEOL JEM 2100. Optical microscopy was done using Metzer Megavision (Metzer Biomedical, Mumbai, India).

3. Results and discussion

3.1. Characterisation of PVP@CoFe$_2$O$_4$ magnetic NPs

First, CoFe$_2$O$_4$ NP was prepared with little modification of earlier reported method of Claesson et al. [30], details are given in experimental section. In the next step, CoFe$_2$O$_4$ was rendered hydrophilic by mixing with PVP. Figure 1(a) depicts the TEM image of CoFe$_2$O$_4$ synthesised. The image shows that the NPs are somewhat aggregated; although looking inside the thinner regions shows that the particle size of these NPs are ~15 nm. TEM image of the PVP@CoFe$_2$O$_4$ is shown in Figure 1(b). PVP@CoFe$_2$O$_4$ NPs are formed without visible core shell-type structure after coating the polymeric chains on the surface of CoFe$_2$O$_4$. The reason can be attributed to the exchange of surface coating of oleic acid with PVP. There is no visible change in the size of the particle either after coating of CoFe$_2$O$_4$ with PVP. The sizes of PVP@CoFe$_2$O$_4$ NPs were similar and were ~15 nm. The hydrophilic coating of PVP on the CoFe$_2$O$_4$ is confirmed by the stable dispersion PVP@CoFe$_2$O$_4$ in water without any noticeable settlement of particles even after 1 h. The inset of Figure 1(b) shows the selective area electron diffraction (SAED) pattern with several rings characteristic of the different diffraction peaks of PVP@CoFe$_2$O$_4$ suggesting

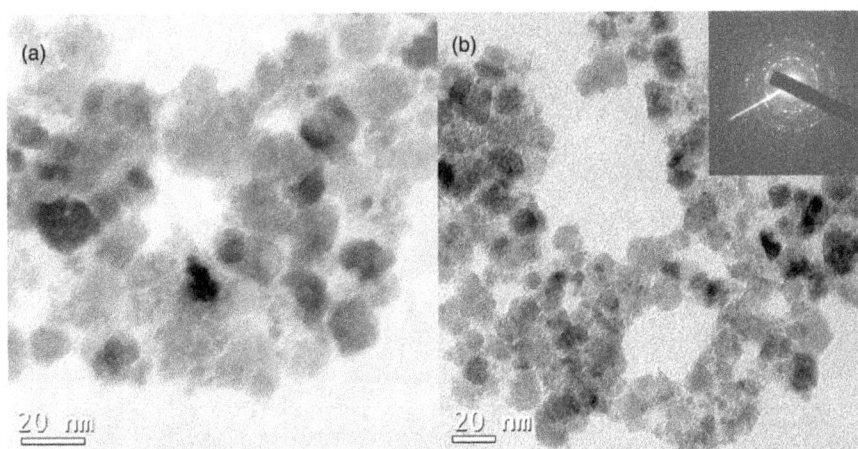

Figure 1. TEM images of (a) CoFe$_2$O$_4$ NPs and (b) PVP@ CoFe$_2$O$_4$ NPs (inset) SAED pattern of PVP@ CoFe$_2$O$_4$ NPs. The respective dispersion solution was prepared in hexane.

crystalline nature of the NPs. Nanocrystal's XRD studies were carried out to characterise the PVP@CoFe$_2$O$_4$ NPs with powder XRD as shown in Figure 2. The spectrum matches one of the spinal phase characteristics of CoFe$_2$O$_4$ with peaks at 2θ values: 30.20, 35.59, 43.07, 53.19, 57.07 and 62.86. The broader natures of the peaks are indicative of the nanocrystalline feature of the PVP@CoFe$_2$O$_4$. The average crystallite size was estimated from the XRD line broadening measurement by using the Scherrer formula [31]. The analysis of the maximum intensity (3 1 1) peak gave a mean diameter of 10 nm of the PVP @ CoFe$_2$O$_4$ particles.

Magnetic properties of both CoFe$_2$O$_4$ and PVP@ CoFe$_2$O$_4$ were measured using variable sample magnetron at room temperature (Figure 3). The magnetisation curve shows low hysteresis for the CoFe$_2$O$_4$ NPs as well as for the PVP@CoFe$_2$O$_4$ NPs. The synthesised ferrite NPs are not very hard magnetic materials because of their low coercivity. The coercivity recorded was 383 Oe for CoFe$_2$O$_4$ and 350 Oe for PVP@CoFe$_2$O$_4$. The coercivity is lower for PVP@CoFe$_2$O$_4$ because of the polymer coating on the CoFe$_2$O$_4$ NPs. The saturation magnetisation and remanence magnetisation of CoFe$_2$O$_4$ were determined to be 36.0685 and 8.187 emu/g, respectively, while PVP@CoFe$_2$O$_4$ NPs have saturation magnetisation and remanence magnetisation of 6.87 and 1.87 emu/g, respectively, as obtained from Figure 3.

Under the influence of external magnetic field, the PVP@CoFe$_2$O$_4$ can be aligned in a chain-like assembly in the direction of the field. Figure 4(a) and (b) shows the TEM images of one such assembly of PVP@ CoFe$_2$O$_4$ under external magnetic field of 1560 G in two different solvents: water and ethanol, respectively. It is evident from the images that the magnetic NPs align in the direction of the field. There is also a visible assembly

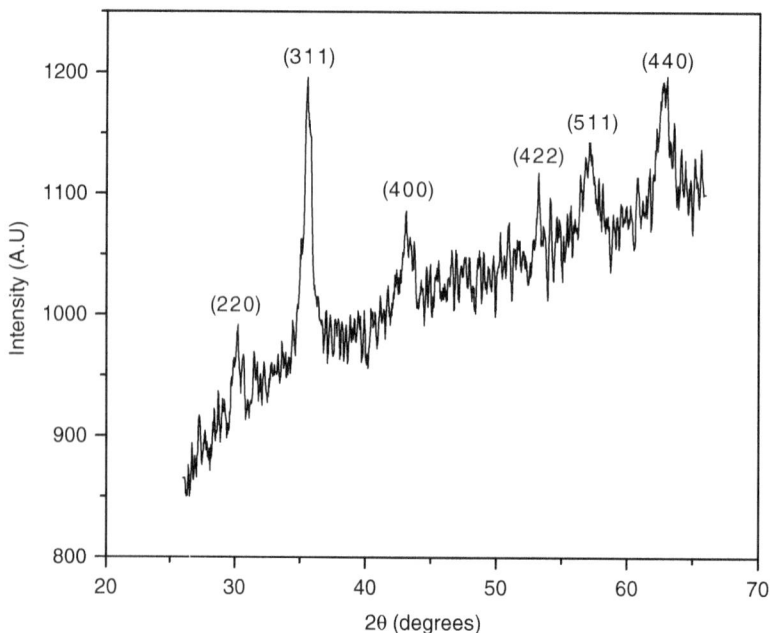

Figure 2. X-ray diffraction pattern of PVP@CoFe$_2$O$_4$ NPs.

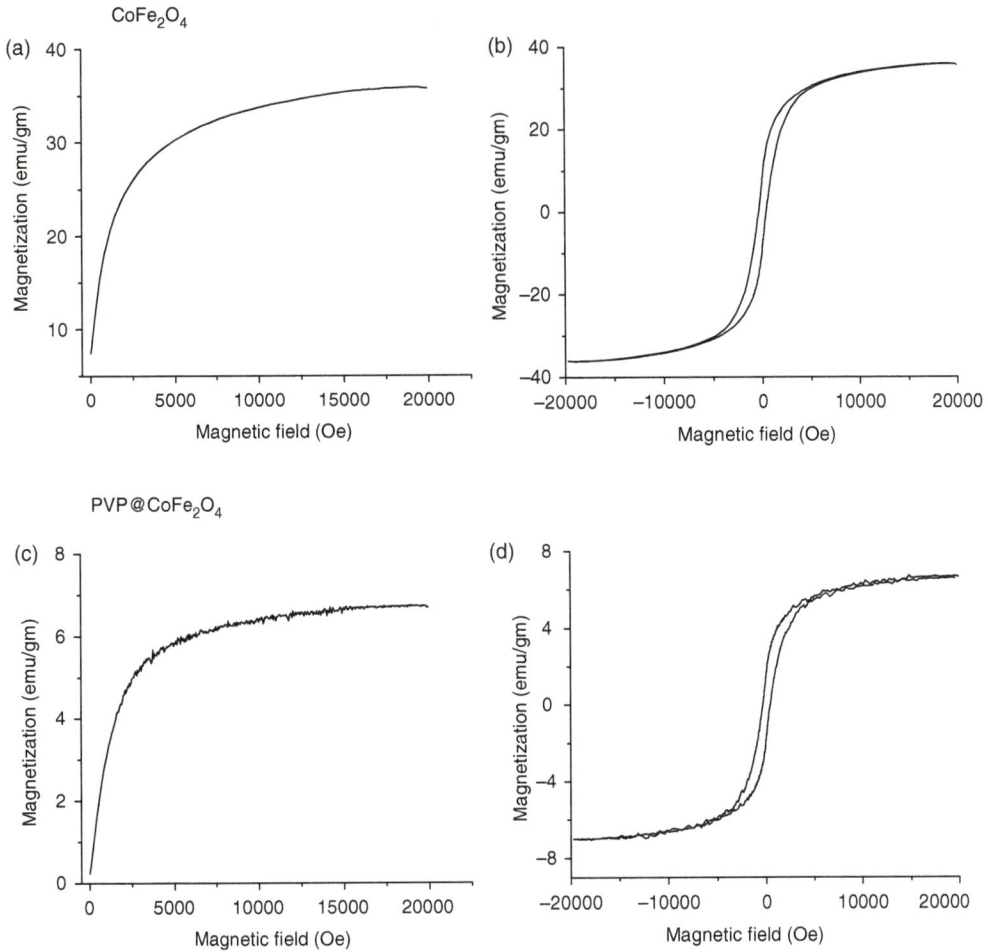

Figure 3. Magnetisation curve with applied magnetic field and the corresponding hysteresis loop at room temperature (300°K) of (a), (b) $CoFe_2O_4$ NPs and (c), (d) PVP@$CoFe_2O_4$ NPs.

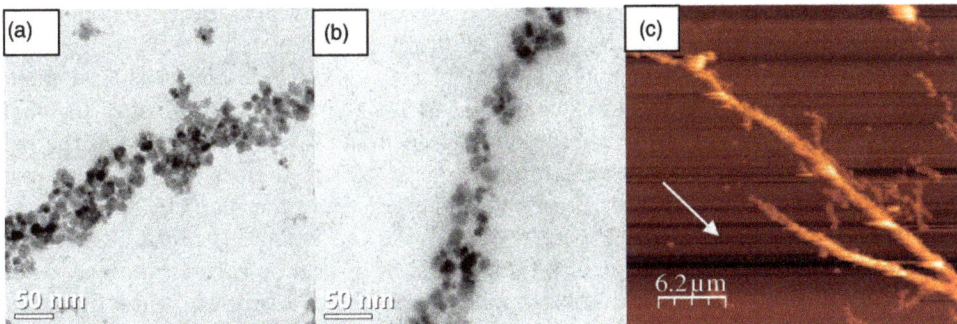

Figure 4. TEM images of a section of PVP@$CoFe_2O_4$ NPs chain assembly under external magnetic field of 1560 G dispersed in (a) water, (b) ethanol and (c) AFM image of the chain assembly.

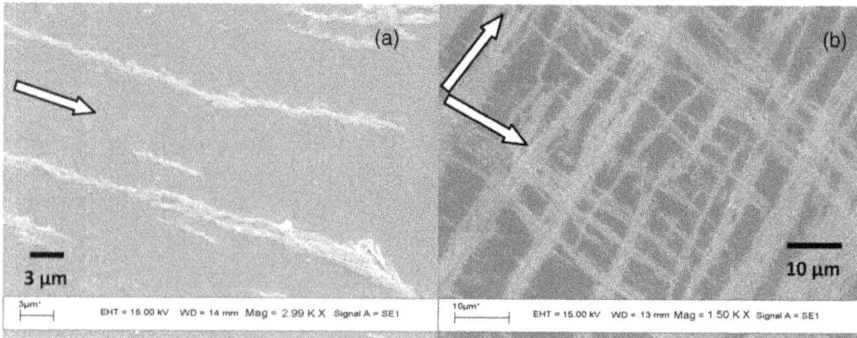

Figure 5. SEM image of (a) line pattern assembly and (b) cross pattern assembly of PVP@CoFe$_2$O$_4$ obtained by applying magnetic field (1540 G). The direction of the field applied in each case is shown by arrows.

of a few particles perpendicular to the field direction. This is explained by the fact that chain–chain interaction between fluctuating parallel dipole chains aligned by magnetic field can give rise to attraction perpendicular to the field direction [32], although the particles in our case are less than 20 nm in size. Representative AFM image of one such assembly of PVP@CoFe$_2$O$_4$ NPs in the form of wires is presented in Figure 4(c).

The aligned nanowires were also studied using SEM. Figure 5(a) and (b) shows the representative SEM images of the unidirectional magnetic field induced chains of PVP@CoFe$_2$O$_4$ NPs formed along the field direction and junctions assemblies formed by changing the field direction (perpendicular to initial field direction). Thus, the magnetic NP assembly can be manipulated with external magnetic field.

3.2. *Dispersion of PVP@CoFe$_2$O$_4$ in different solvents*

The assembly of PVP@CoFe$_2$O$_4$ in different dispersion media was investigated. For this study, a dispersion taking equal amount of PVP@CoFe$_2$O$_4$ was prepared in different solvents, namely, water, ethanol and toluene. It was observed that under the influence of external magnetic field, the magnetic NPs dispersed in water form 1-D chain-like assembly which are longer and can have length up to millimetre. On the other hand, dispersion of magnetic NPs in ethanol when placed under same magnetic field makes chains which are shorter in length as compared with water dispersion. Most of the chain assemblies in this case are 50–400 µm in length depending upon the strength of the field applied. On the contrary, the dispersion of magnetic NPs in toluene when placed under magnetic field shows no assembly at low field but only very few short chain assemblies at high fields (discussed later). Figure 6 depicts the representative optical microscope images of the assembly of PVP@CoFe$_2$O$_4$ in different dispersion solvents, namely water, ethanol and toluene. The external field applied in all the cases shown in the figure was 305 G. Contrary to the reports by Buttler et al. [33] in their system of Fe ferrofluids, we did not observe any sort of directional assembly of PVP@CoFe$_2$O$_4$ in the absence of magnetic field.

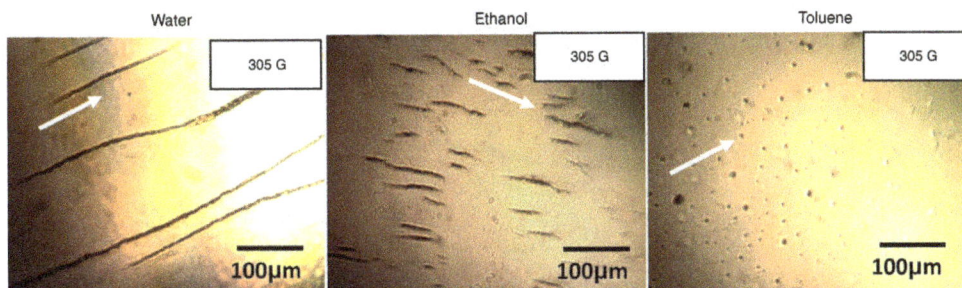

Figure 6. Optical microscope images of chain assembly of PVP@CoFe$_2$O$_4$ NPs in different dispersion media: water, ethanol and toluene under magnetic field of 305 G. The direction of the field in each case is shown by arrows.

3.3. *Effect of strength of external magnetic field on assembly of PVP@CoFe$_2$O$_4$*

The effect of magnetic field on the length of the chain-like assembly formed in different dispersion media was studied. Various specimens of dispersion of PVP@CoFe$_2$O$_4$ in different solvent media were prepared by taking equal amount of PVP@CoFe$_2$O$_4$ (0.1 mg/mL) in each case. Assembly process at external fields 1560, 775, 305 and 75 G in different dispersion media was observed. In water dispersion, application of external magnetic field yielded chain assembly along the direction of field which was several hundred micrometres in length even up to 305 G of magnetic field. Thus in water dispersion, the magnetic NPs form a similar type of assembly in high field (1560 G) as well as comparatively low field (305 G). Only when the external field was 75 G, there was no assembly observed (Figure 7 W1–W4). In the case of ethanol dispersion, shorter chain assembly was observed. As visible from the optical microscope, as the field is reduced from 1560 G (Figure 7 E1) to 305 G (Figure 7 E3) the chain assembly got shorter. Similar to the case as in water dispersion, there was no assembly observed under the influence of 75 G field. In the case of toluene dispersion, only a few magnetic NPs assemblies are seen in high field of 1560 and 775 G (marked with arrows, Figure 7, T1 and T2). However, unlike in water and ethanol dispersions, there was no assembly at lower fields: 305 and 75 G (Figure 7, T3 and T4). It can be added here that there was no assembly observed when a drop of magnetic NPs dispersed in solvents was allowed to evaporate and then magnetic field applied.

3.4. *Why PVP@CoFe$_2$O$_4$ in water makes longer chain assembly*

The structure, dynamics and thermodynamics of many physical and chemical systems are controlled by the interplay of short-range isotropic van der Waals forces and long-range anisotropic magnetic dipole forces. There are reports of measurement of these colloidal forces in ferrofluid droplet [34] and in the presence of polyelectrolyte coating [35] by Bragg's diffraction. Surface-force measurement apparatus can also be used to measure the interaction of polymer-bearing surfaces [36]. Magnetic NPs dispersed in solvents are ideal model systems to study the interplay between the different types of forces. Magnetic NPs dispersed in solutions are subjected to electrostatic repulsion between the electric double layers of the particles, van der Waals attractions between the particles and also the

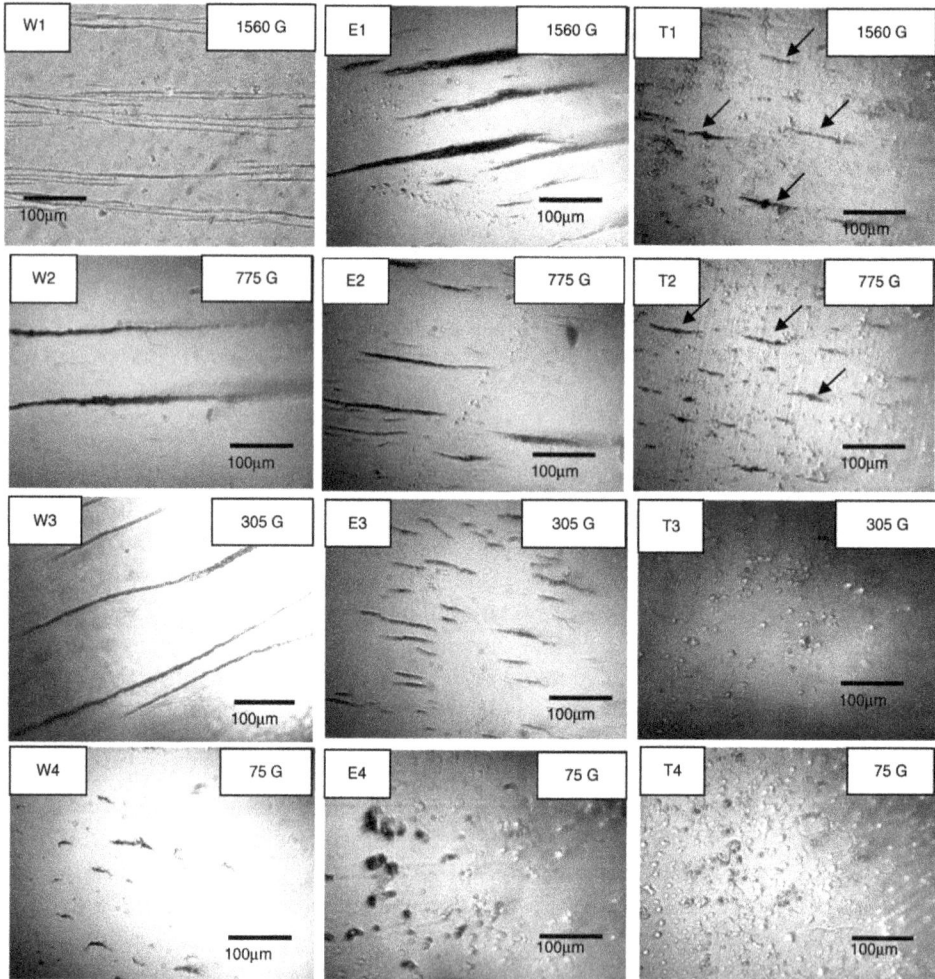

Figure 7. Optical microscope images of chain assembly formed with various magnetic field strengths in different dispersion media. Specimens were coded according to the solvent used: W, water; E, ethanol and T, toulene.

magnetic interaction between them. The electrostatic repulsion between electric double layers of the particles is given by the expression [37]

$$V_{EL} = 2\pi\varepsilon_r\varepsilon_0 a\zeta^2 \ln[1 + e^{-\kappa s}],$$

where it is assumed that the particle has a constant and moderate surface potential, ζ^2 is the zeta potential, ε_r the relative dielectric constant of the liquid, ε_0 the permittivity of the vacuum and κ the reciprocal Debye length.

The van der Waals attraction between the particles is given by the expression [38]

$$V_{LW} = (-A/6)[2a^2/s(4a+s) + 2a^2/(2a+s)^2 + \ln s(4a+s)/(2a+s)^2],$$

where A is the Hamaker constant.

The magnetic interaction energy between the particles is given by [39]

$$V_M = -8\pi\mu_0 M^2 a^3 / 9(s/a + 2)^3,$$

where M is the magnetisation of the material, a the radius of the particles, s the distance between the surfaces of two interacting particles and μ_0 the permeability of the vacuum. The magnetic dipole interactions, which are long-range forces, come into play when the magnetic particles are in direct contact with each other making clusters. In such a scenario, their electron wave function overlaps. If the forces of interaction are strong, then the whole array of NP will be ferromagnetic. In such a case, there will be coupling of the moments of neighbouring NPs, thus producing long-range magnetic order. On the contrary, if the magnetic moments of neighbouring particles interact weakly, the magnetic order breaks down and the magnetic moment of each NP fluctuates independently. In any ferromagnetic system with the increase in particle size from few to hundreds of nanometres, the system will pass through three distinct states namely superparamagnetic state, single domain ferromagnetic state and multi-domain ferromagnetic state. The last one is thermodynamically advantageous as it balances the local magnetic field created by the neighbouring magnetic domains. The aggregation of single domain can also create multi-domain NPs. Under external magnetic field, an NP acquires a magnetic dipole given by $m = \mu_0 \chi VH$, where μ_0 is the magnetic permeability of vacuum, χ the magnetic susceptibility and V the volume of the NP. This induces a magnetic dipole–dipole interaction (u) between two NPs separated by a distance r with an angle θ between r and m: $u\,(r, h) = -(m^2/4\pi\mu_0 r^3)(3\cos^2\theta - 1)$. Two NPs separated by a distance r_1 and $\theta = 0$ interact with each other attractively and is be given by ($u_1 = 2m^2/4\pi\mu_0 r_1^3$). In contrast, two NPs at a distance of r_2 and $\theta = \pi/2$ repel each other due to a repulsive interaction, $u_2 = m^2/4\pi\mu_0 r_2^3$.

The stability of NPs in the dispersion medium will determine the assembly of the NPs (chain length) under external magnetic field. PVP@CoFe$_2$O$_4$ dispersion in high dielectric constant solvent (polar), as water ($\varepsilon_r = 80$ at 68°F) will have strong hydrophilic/lypophilic attraction, which is consistent with the observed stability of the dispersion system (no sedimentation) and will facilitate the formation of clusters. When external magnetic field is applied, the direction of moments inside different clusters may not be synchronous with the external field. If the moments inside neighbouring clusters are anti-parallel, then the magnetic force between two moments will be attractive and this attractive force will bring the two clusters together. At the initial stage of NP assembly, such dipolar interaction is very weak. However, as the solvent evaporates, the dipolar interaction energy becomes significant. Due to attractive dipolar interaction between NPs aligned through the external field direction and repulsive interactions along the direction perpendicular to the field axis, the formation of the assembly of NPs along the external magnetic field direction is facilitated. On the other hand, PVP@CoFe$_2$O$_4$ dispersion in ethanol ($\varepsilon_r = 24.3$ at 77°F) will have weaker hydrophilic/lypophilic attraction than in water dispersion but still the attraction is good enough to facilitate the formation of clusters, a prerequisite for the magnetic dipole interaction to come into play. In the case of PVP@CoFe$_2$O$_4$ dispersion in toluene having very low dielectric constant ($\varepsilon_r = 2.4$ at 68°F), the phase separation is very fast in this system which is consistent with the observed instability of the dispersion with visible sedimentation thus preventing formation of clusters. Although Gao et al. [40] have shown that magnetic dipole interaction inherently associated with magnetic NP is sufficient to maintain 1D assemblies with hollow CoSe$_2$

nanocrystal, it will not lead to longer wire assembly as in our case. Therefore, stability of the NPs in dispersion medium has a role in determining the length of chain assembly.

4. Conclusion

In this article, the assembly of PVP@$CoFe_2O_4$ NPs in the form of chains and junctions on surfaces by applying a magnetic field is demonstrated. The assemblies of magnetic NPs depend on the dispersion solvent. The use of solvent which can stabilise the magnetic NP, such as water dispersion of PVP@$CoFe_2O_4$ will give longer chain assembly on application of unidirectional magnetic field. On the contrary, with toluene dispersion of PVP@$CoFe_2O_4$ NPs, there are few assemblies/no assembly on applying magnetic field. The role of van der Waals forces, magnetic dipole interaction and the electrostatic factors determine the chain length of assembly. The assembly process also depends on the magnetic field and the minimum field required to ensure chain assemblies varies with different dispersion media.

Acknowledgements

I thank the Department of Science and Technology, GoI for the project no. SR/FTP/CS-45/2007 under SERC Fast Track Scheme and project no. SR/S5/NM-108/2006 & 2/2/2005-S.F. Special thanks to I. Talukdar and K. Acharyya. The support provided by Central Instrumentation Facility of IIT Guwahati and Centre for Nanotechnology, IIT Guwahati is also acknowledged. I also thank Prof. Arun Chattopadhyay, Dr Neelotpal Sen Sarma and Dr Gitanjali Majumdar for helpful discussions.

References

[1] D.L. Feldheim and C.D. Keating, *Self-assembly of single electron transistors and related devices*, Chem. Soc. Rev. 27 (1998), pp. 1–12.

[2] J.H. Fendler, *Chemical self-assembly for electronic applications*, Chem. Mater. 13 (2001), pp. 3196–3210.

[3] V. Skumryev, S. Stoyanov, Y. Zhang, G. Hadjipanayis, D. Givord, and J. Nogués, *Beating the superparamagnetic limit with exchange bias*, Nature 423 (2003), pp. 850–853.

[4] Z.-M. Liu, H.-F. Yang, Y.-F. Li, Y.-L. Liu, G.-L. Shen, and R.-Q. Yu, *Core–shell magnetic nanoparticles applied for immobilization of antibody on carbon paste electrode and amperometric immunosensing*, Sensors Actuators, B 113 (2006), pp. 956–962.

[5] G. Li, V. Joshi, R.L. White, S.X. Wang, J.T. Kemp, C. Webb, R.W. Davis, and S.J. Sun, *Detection of single micron-sized magnetic bead and magnetic nanoparticles using spin valve sensors for biological applications*, J. Appl. Phys. 93 (2003), pp. 7557–7559.

[6] J.M. Perez, F.J. Simeone, Y. Saeki, L. Josephson, and R. Weissleder, *Viral-induced self-assembly of magnetic nanoparticles allows the detection of viral particles in biological media*, J. Am. Chem. Soc. 125 (2003), pp. 10192–10193.

[7] J.M. Perez, T. O'Loughin, F.J. Simeone, R. Weissleder, and L. Josephson, *DNA-based magnetic nanoparticle assembly acts as a magnetic relaxation nanoswitch allowing screening of DNA-cleaving agents*, J. Am. Chem. Soc. 124 (2002), pp. 2856–2857.

[8] Y. Long, K. Huang, J. Yuan, D. Han, L. Niu, Z. Chen, C. Gu, A. Jin, and J.L. Duvail, *Electrical conductivity of a single Au/polyaniline microfiber*, Appl. Phys. Lett. 88 (2006), pp. 162113-1–162113-3.

[9] U. Jeong, J.X. Teng, Y. Wang, H. Yang, and Y. Xia, *Superparamagnetic colloids: controlled synthesis and niche application*, Adv. Mater. 19 (2007), pp. 33–60.

[10] S. Sun, H. Zeng, D.B. Robinson, S. Raoux, P.M. Rice, S.X. Wang, and G. Li, *Monodisperse MFe_2O_4 (M=Fe, Co, Mn) nanoparticles*, J. Am. Chem. Soc. 126 (2004), pp. 273–279.

[11] C. Liu, B. Zou, A.J. Rondinone, and Z.J. Zhang, *Chemical control of superparamagnetic properties of magnesium and cobalt spinel ferrite nanoparticles through atomic level magnetic couplings*, J. Am. Chem. Soc. 122 (2000), pp. 6263–6267.

[12] T. Hyeon, *Chemical synthesis of magnetic nanoparticles*, Chem. Commun. 8 (2003), pp. 927–934.

[13] S.L. Tripp, S.V Pusztay, A.E. Ribbe, and A. Wei, *Self-assembly of cobalt nanoparticle rings*, J. Am. Chem. Soc. 124 (2002), pp. 7914–7915.

[14] H.S. Kim, B.H. Sohn, W. Lee, J.-K. Lee, S.J. Choi, and S.J. Kwon, *Multifunctional layer-by-layer self-assembly of conducting polymers and magnetic nanoparticles*, Thin Solid Films 419 (2002), pp. 173–177.

[15] R. Lawaczeck, H. Bauer, T. Frenzel, M. Hasegawa, Y. Ito, K. Kito, N. Miwa, H. Tsutsui, H. Vogler, and H. Weinmann, *Magnetic iron oxide particles coated with carboxydextran for parenteral administration and liver contrasting*, J. Acta Radiol. 38 (1997), pp. 584–597.

[16] A. Jordan, R. Scholz, P. Wust, H. Fahling, and R. Felix, *Magnetic fluid hyperthermia (MFH): cancer treatment with AC magnetic field induced excitation of biocompatible superparamagnetic nanoparticles*, J. Magn. Magn. Mater. 201 (1999), pp. 413–419.

[17] A.S. Lübbe, C. Bergemann, J. Brock, and D.G. McClure, *Physiological aspects in magnetic drug-targeting*, J. Magn. Magn. Mater. 194 (1999), pp. 149–155.

[18] W. Jiang, H.C. Yang, S.Y. Yang, H.E. Horng, J.C. Hung, Y.C. Chen, and C.-Y. Hong, *Preparation and properties of superparamagnetic nanoparticles with narrow size distribution and biocompatible*, J. Magn. Magn. Mater. 283 (2004), pp. 210–214.

[19] T. Meron, Y. Rosenberg, Y. Lereah, and G. Markovich, *Synthesis and assembly of high-quality cobalt ferrite nanocrystals prepared by a modified sol–gel technique*, J. Magn. Magn. Mater. 292 (2005), pp. 11–16.

[20] D.K. Kim, Y. Zhang, W. Voit, K.V. Rao, and M. Muhammed, *Synthesis and characterization of surfactant-coated superparamagnetic monodispersed iron oxide nanoparticles*, J. Magn. Magn. Mater. 225 (2001), pp. 30–36.

[21] J.-S. Jung, J.-H. Lim, K.-H. Choi, S.-L. Oh, Y.-R. Kim, S.-H. Lee, D.A. Smith, K.L. Stokes, L. Malkinski, and C.J. O'Connor, *$CoFe_2O_4$ nanostructures with high coercivity*, J. Appl. Phys. 97 (2005), pp. 10F306/1–10F306/3.

[22] Y. Liu, Y. Zhang, J.D. Feng, C.F. Li, J. Shi, and R. Xiong, *Dependence of magnetic properties on crystallite size of $CoFe_2O_4$ nanoparticles synthesised by auto combustion method*, J. Exp. Nanosci. 4 (2009), pp. 159–168.

[23] Q.A. Pankhurst, J. Connolly, S.K. Jones, and J. Dobson, *Applications of magnetic nanoparticles in biomedicine*, J. Phys. D: Appl. Phys. 36 (2003), pp. R167–R181.

[24] C. Mateo-Mateo, C. Vázquez-Vázquez, M.C. Buján-Núñez, M.A. López-Quintela, D. Serantes, D. Baldomir, and J. Rivas, *Electrolyte influence on the anodic synthesis of TiO2 nanotube arrays*, J. Non-Cryst. Solids 354 (2008), pp. 5236–5237.

[25] G. Baldi, D. Bonacchi, M.C. Franchini, D. Gentili, G. Lorenzi, A. Ricci, and C. Ravagli, *Synthesis and coating of cobalt ferrite nanoparticles: a first step toward the obtainment of new magnetic nanocarriers*, Langmuir 23 (2007), pp. 4026–4028.

[26] T.T. Dung, T.M. Danh, L.T.M. Hoa, D.M. Chien, and N.H. Duc, *Structural and magnetic properties of starch-coated magnetite nanoparticles*, J. Exp. Nanosci. 4 (2009), pp. 259–267.

[27] H. Niu, Q. Chen, H. Zhu, Y. Lin, and X. Zhang, *Magnetic field-induced growth and self-assembly of cobalt nanocrystallites*, J. Mater. Chem. 13 (2003), pp. 1803–1805.

[28] M. Platt, G. Muthukrishnan, W.O. Hancock, and M.E. Williams, *Millimeter scale alignment of magnetic nanoparticle functionalized microtubules in magnetic fields*, J. Am. Chem. Soc. 127 (2005), pp. 15686–15687.

[29] E.L. Bizdoaca, M. Spasova, M. Farle, M. Hilgendorff, and F. Carusco, *Magnetically directed self-assembly of submicron spheres with a Fe3O4 nanoparticle shell*, J. Magn. Magn. Mater. 240 (2002), pp. 44–46.

[30] E.M. Claesson and A.P. Philipse, *Monodisperse magnetizable composite silica spheres with tunable dipolar interactions*, Langmuir 21 (2005), pp. 9412–9419.

[31] A.L. Patterson, *The Scherrer formula for X-ray particle size determination*, Phys. Rev. 56 (1939), pp. 978–982.

[32] T.C. Halsey, *Electrorheological fluids*, Science 258 (1992), pp. 761–766.

[33] K. Buttler, P.H.H. Bomans, P.M. Frederik, G.J. Vroege, and A.P Philipse, *Direct observation of dipolar chains in iron ferrofluids by cryogenic electron microscopy*, Nature Mater. 2 (2003), pp. 88–91.

[34] F.L. Calderon, T. Stora, O.M. Monval, P. Poulin, and J. Bibette, *Direct measurement of colloidal forces*, Phys. Rev. Lett. 72 (1994), pp. 2959–2962.

[35] J. Philip, O. Mondain-Monval, F.L. Calderon, and J. Bibette, *Colloidal force measurements in the presence of a polyelectrolyte*, J. Phys. D: Appl. Phys. 30 (1997), pp. 2798–2803.

[36] P.F. Luckham and J. Klien, *Forces between mica surfaces bearing adsorbed polyelectrolyte, poly-L-lysine, in aqueous media*, J. Chem. Soc. Faraday Trans. I 80 (1984), pp. 865–878.

[37] R.J. Hunter, *Foundations of Colloid Science*, Vol. 1, Clarendon Press, Oxford, UK, 1987.

[38] J.K. Marshall and J.A. Kitchener, *The deposition of colloidal particles on smooth solids*, J. Colloid Interface Sci. 22 (1966), pp. 342–351.

[39] R.E. Rosensweig, *Ferrohydrodynamics*, Cambridge University Press, Cambridge, UK, 1985.

[40] J. Gao, B. Zhang, X. Zhang, and B. Xu, *Magnetic-dipolar-interaction-induced self-assembly affords wires of hollow nanocrystals of cobalt selenide*, Angew. Chem. Int. Ed. 45 (2006), pp. 1220–1223.

11

Preparation and characterisation of acephate nano-encapsulated complex

Samrat Roy Choudhury*, Saheli Pradhan and Arunava Goswami

Agricultural and Ecological Research Unit, Biological Sciences Division, Indian Statistical Institute, 203 Barrackpore Trunk Road, Kolkata 700108, India

Acephate is the most widely used organophosphorous insecticide in Indian subcontinent. In order to prepare ecofriendly and hydrophilic formulation of acephate, nanotechnology-based research is necessary. Here, we report a novel method of preparation and subsequent characterisation of the acephate nano-encapsulation complex. Nanoparticles of the encapsulation complex were characterised by dynamic light scattering (DLS), scanning electron microscopy (SEM), transmission electron microscopy (TEM) and Fourier transform infrared (FTIR) spectroscopy. The TEM image reveals the size of particles in the range 90–120 nm. This method would be very useful for industries for making farmer-friendly pesticide formulations.

Keywords: nanoscience; insecticide; acephate; nano-encapsulation complex

1. Introduction

Acephate (R,S)-N-[methoxy(methylthio)phosphinoyl]acetamide is an organophosphate systemic insecticide which has long been used in agricultural sectors to control a wide range of chewing and sucking insect pests like aphids, thrips, sawflies, leafminers, leafhoppers, cut worm of cotton, paddy, soybean, sugarcane, chilies, maize, tobacco, etc. This entomotoxic compound is metabolically converted into methamiodophos and subsequently inhibits the activity of acetylcholinesterase (AChE) [1,2]. In India, 9000 metric tons of acephate are being used per year. Such a large-scale use of acephate is not only expensive but also confers enhanced toxicity in agricultural fields [3]. Moreover, indiscriminate use of this insecticide has also induced resistance in the target species. Research is absolutely necessary towards considerable reduction in the usage amount, so that the cost is cut down, toxic effect of acephate can be lowered and at the same time, the buildup of resistance in the target species can be minimised. Nanotechnology-based application could be a suitable solution to the aforesaid problems.

Nanostructured materials like nanocapsules and nanospheres have been extensively studied in the recent years and well-established as potential drug carriers owing to their quantum size properties, high surface-to-volume ratio [4] and high intracellular uptake potential. Surface modification of these nano-encapsulated complexes with hydrophilic

*Corresponding author. Email: samratroychoudhury@gmail.com

polymer is a common practice in agriculture and medicine in order to generate biocompatible pesticide/drug formulation and to lower the risk of opsonisation [5,6]. This study has been designed to prepare a nano-encapsulation complex of an agriculturally important neurotoxic insecticide acephate as core component and a hydrophilic polymer polyethylene glycol-400, (PEG-400) [7,8] as a surface stabiliser. It is hypothesised that acephate contains phosphate group as well as amide linkage where the NH proton is capable of forming a –H bond with the carbonyl group, which might be a major factor for the stability of this encapsulation complex. PEG-400 is a neutral ligand (with high HLB ratio) that hydrophilises the surface and induces a steric barrier by anchoring long, mobile PEG chains on the surface of the core component and exerts protective action (stealth properties®) [9,10], subcellular size, sustained release properties and biocompatibility of the encapsulation complex would enhance the overall activity of acephate.

2. Materials and methods

2.1. *Materials*

Acephate powder (technical grade 97.8%, Bayer Crop Sciences, India), dichloromethane (GR, MERCK) and PEG-400 (for synthesis, MERCK) were used. The water used throughout this study was distilled water (18 ′Ω; Sartorius-arium@611VF).

2.2. *Preparation of acephate nano-encapsulation*

Amorphous acephate powder was fully ground in a mortar at an ambient temperature to form particles of around 1 µm; 200 mL of PEG-400 solumer were prepared by mixing PEG-400 and distilled water in the ratio of 9 : 1 under continuous stirring. Two hundred millilitres of 1% acephate solution in dichloromethane were dripped into the reaction flask, under continuous stirring of 2500 rpm and heating at 45°C for around 240 min. It is expected that this procedure would help PEG-400 to encapsulate a few acephate molecules. The reaction mixture was then subjected to rotary vacuum evaporator (EYELA N-N series) for the removal of residual organic solvent.

2.3. *Characterisation*

Particle size distribution was obtained by dynamic light scattering (DLS) (MALVERN Nano S (red badge), Ze-1600) from 25°C to 55°C. Surface topology was determined through scanning electron microscope (SEM; FEI Quantum-200 MK-2) at 10 kV [Spot: 2.5]. The sizes and shapes of particles were observed through transmission electron microscope (TEM) [JEOL 2010F]. Fourier transform infrared (FTIR) (Shimadzu) spectroscopy was used to characterise and identify the organic molecule.

3. Results and discussion

3.1. *DLS analysis*

In this study, 5 mL of acephate nano-encapsulation complex was suspended in 15 mL of ddH$_2$O for analysis. The hydrodynamic diameter of the colloid thus obtained was analysed

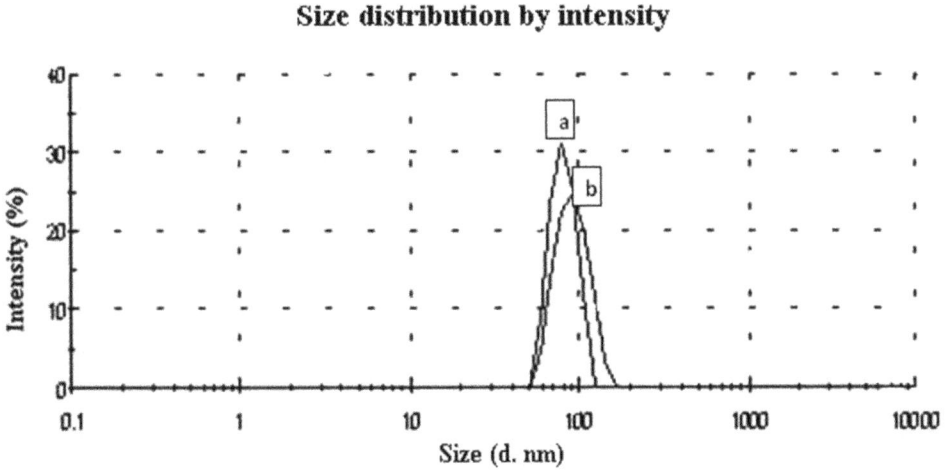

Figure 1. Size distributions of acephate nano-encapsulation complex at (a) 25°C and (b) 55°C.

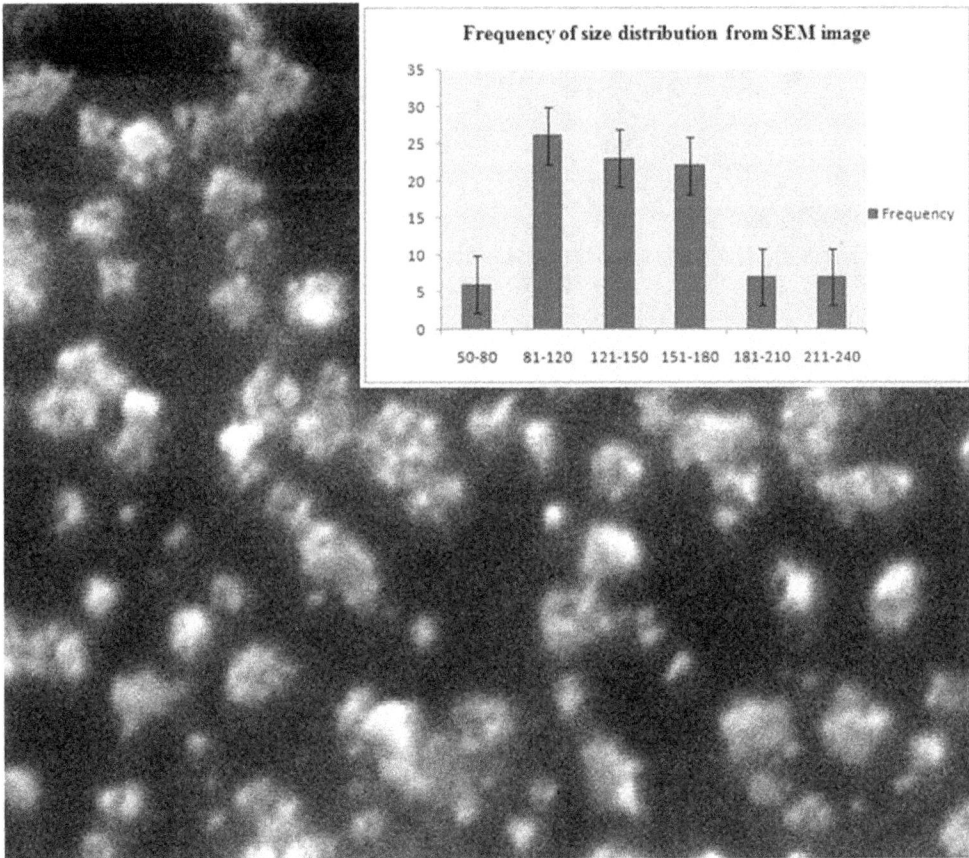

Figure 2. SEM image for acephate nano-encapsulations.

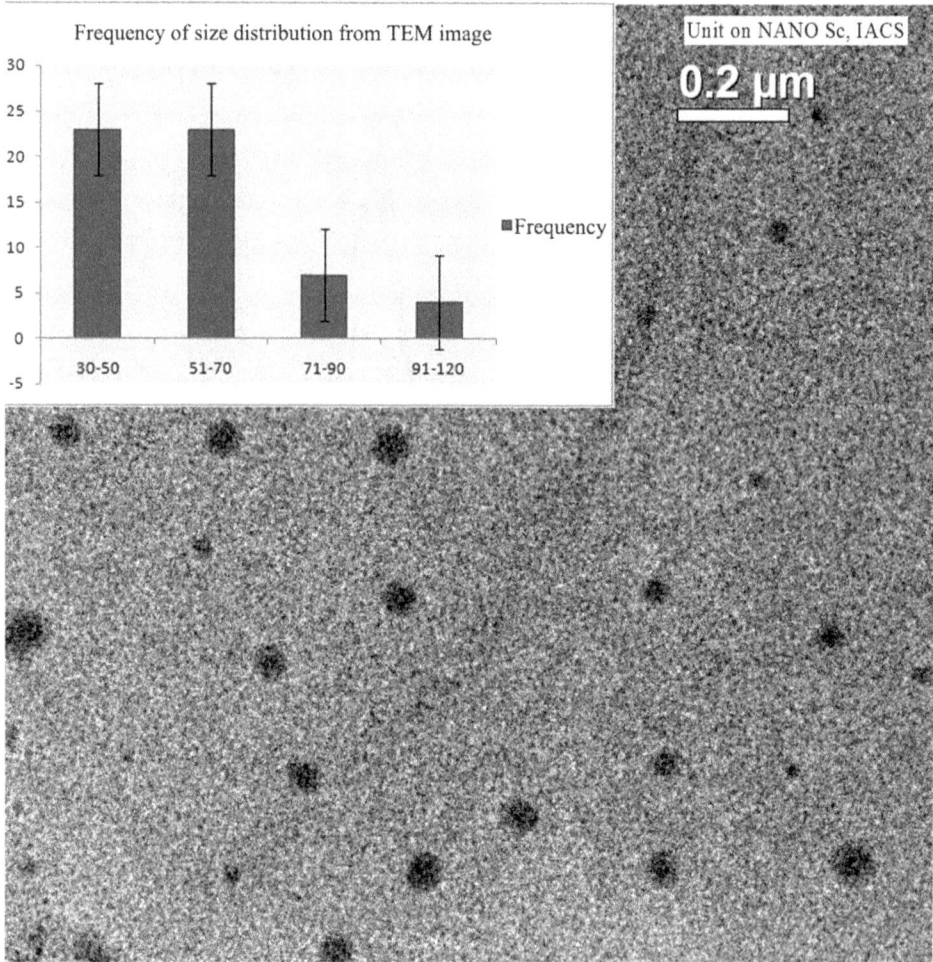

Figure 3. TEM image for acephate nano-encapsulatons.

with DLS from 25°C to 55°C. The particle size distribution was around 90–120 nm (Figure 1) and found to be stable under the aforesaid temperature range.

3.2. *Analysis of SEM*

The surface topology of the acephate nano-encapsulations, as shown in Figure 2, was analysed using SEM at 50,000 × magnification and at 10 kV vacuum. SEM sample was prepared by dispersing 100 µL of the acephate nano-encapsulation solution in 5 mL of absolute ethanol and ultrasonically dispersed prior to the imaging. SEM micrograph reveals irregular surface of the particles, though the homogeneity in shape was persistent among different scan fields.

Polyethylene glycol-400

Commercial acephate

Acephate nanoencapsulation

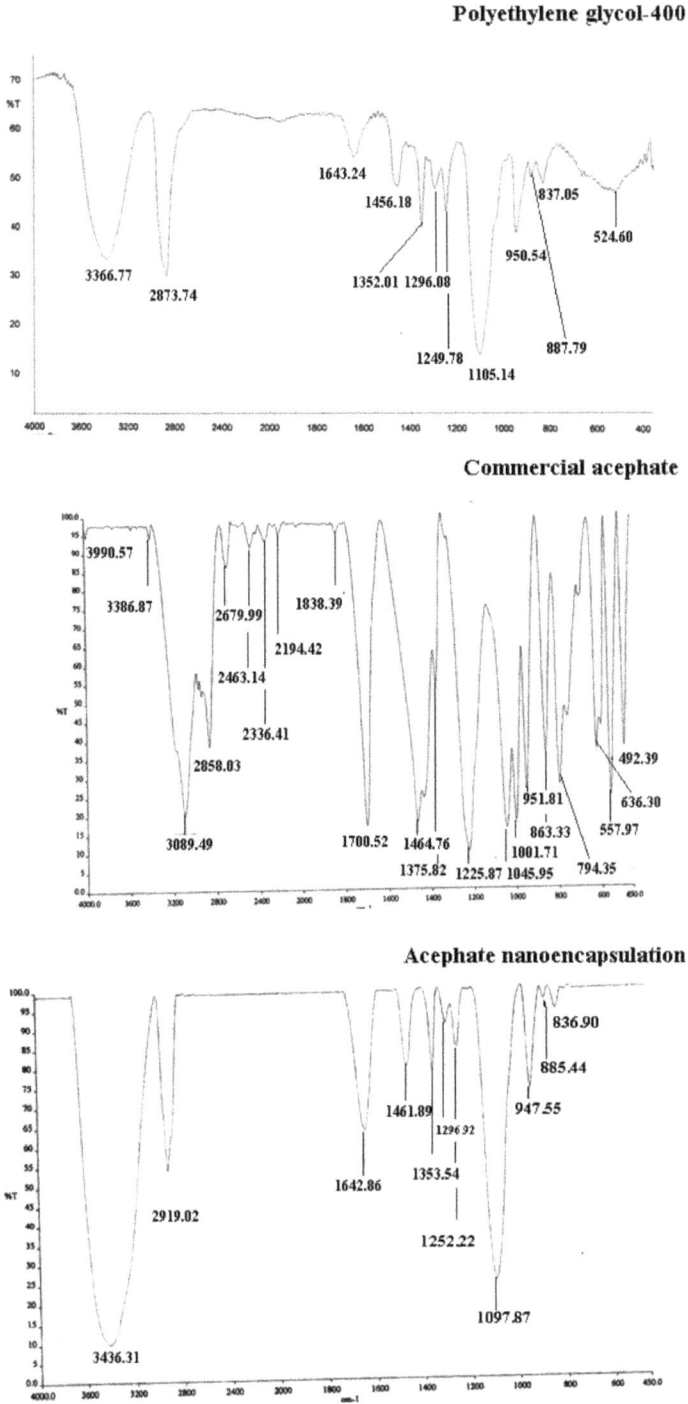

Figure 4. FTIR spectra for PEG-400, Commercial acephate and acephate nano-encapsulation.

3.3. *Ananlysis of TEM*

The actual size of synthesised acephate nano-encapsulation was analysed using TEM. The sample for TEM measurement was prepared by the deposition of acephate nano-encapsulation complex in ethyl alcohol on a carbon-coated copper grid. Particle size distribution was in the range 30–120 nm. Figure 3 represents the TEM image and frequency of size distribution of the acephate nano-encapsulation complexes.

3.4. *Analysis of FTIR analysis*

FTIR analysis has been made with commercial acephate, PEG and PEG-coated acephate nano-encapsulation (Figure 4). FTIR spectrum generated after encapsulation by PEG-400 on being compared with FTIR spectrum of acephate only shows that acephate was successfully encapsulated by PEG-400.

4. Conclusions

The nano-encapsulation method is promising due to its low cost and simplified formulation of acephate insecticide. Moreover, PEG-400 coating on the surface of acephate makes it water soluble and active at a lower dose, and also reduces acephate toxicity in the agricultural fields and enhances the overall stability of the complex.

Acknowledgements

We thank Professor Ratan Lal Brahmachary for his critical reading of the manuscript. The research was funded by NAIP-ICAR-World Bank (Comp-4/C3004/2008-09) and Department of Biotechnology (DBT), Government of India (BT/PR9050/NNT/28/21/2007 & BT/PR8931/NNT/28/07/2007) and ISI plan project for 2008–2011. We thank for their generous financial support.

References

[1] M. Mahajna, G.B. Quistad, and J.E. Casida, *Acephate insecticide toxicity: Safety conferred by inhibition of the bioactivating carboxyamidase by the metabolite methamidophos*, Chem. Res. Toxicol. 10(1) (1997), pp. 64–69.
[2] J. Yen, K. Lin, and Y. Wang, *Potential of the insecticides acephate and methamidophos to contaminate groundwater*, Ecotoxicol. Environ. Saf. 45(1) (2000), pp. 79–86.
[3] A. Navrotsky, *Nanomaterials in the environment, agriculture, and technology (NEAT)*, J. Nanopart. Res. 2 (2000), pp. 321–323.
[4] M. Daniel and D. Astruc, *Gold nanoparticles: Assembly, supramolecular chemistry, quantum-size-related properties, and applications toward biology, catalysis and nanotechnology*, Chem. Rev. 104 (2004), pp. 293–346.
[5] C.P. Reis, R.J. Neufeld, A.J. Ribeiro, and F. Veiga, *Nanoencapsulation I. Methods for preparation of drug-loaded polymeric nanoparticles*, Nanomedicine: NBM, 2 (2006), pp. 8–21.
[6] C.P. Reis, R. Neufeld, A. Ribeiro, and F. Veiga, *Nanoencapsulation II. Biomedical applications and current status of peptide and protein nanoparticulate delivery systems*, Nanomedicine: NBM, 2(2) (2006), pp. 53–65.

[7] R. Jimmie, A. Baran Jr Michael, and J.J.D. Laperre, *Polymer blends including surface-modified nanoparticles and methods of making the same origin*, 10 August 2006, ST. PAUL, MN US, IPC8 Class: AC08K700FI. USPC Class: 523206.

[8] R. Goldshtein, O. Sklyarsky, I. Zelkind, M. Kopylov, B. Tulbovich, and V. Goldshtein, *Inclusion complexes of active compounds in acrylate (co)polymers and methods for their production*, 13 October 2005, Pub. no: US 2005/0226934 A1. Pub, US, 2005.

[9] V.P. Torchilin, *Polymer coated long circulating microparticulate pharmaceuticals*, J. Microencapsul. 15(1) (1998), pp. 1–19.

[10] S. Zalipsky, *Functionalized poly(ethylene glycol)-modified gelatin nanoparticles for intracellular delivery*, Pharm. Res. 19(7) (2002), pp. 1062–1068.

12

The effect of microwave-assisted synthesis on the physico-chemical properties of pamoate-intercalated layered double hydroxide

Z. Jubri[a]*, M.Z. Hussein[b], A. Yahaya[b] and Z. Zainal[b]

[a]Department of Engineering Sciences and Mathematics, College of Engineering, Universiti Tenaga Nasional, 43000 Kajang, Selangor, Malaysia; [b]Department of Chemistry, Universiti Putra Malaysia, 43400 Serdang, Selangor, Malaysia

Layered organic-inorganic hybrid nanocomposite was prepared by inserting the pamoate anion (PA) between Zn and Al layered double hydroxide (LDH) using co-precipitation method. The ageing process was done by two different methods; conventional oil bath and microwave-assisted method at various ageing times, from 15 to 60 min. As a result of successful intercalation of PA anion into Zn–Al LDH lamella, the expansion of the interlayer spacing from 8.9 to 18.1 Å was observed in the powder X-ray diffractogram of the (ZAP) nanocomposite. Percentage of PA intercalated between the LDH was higher in the nanocomposite material aged using microwave irradiation (ZAPM) compared to conventional oil bath method (ZAP). The BET surface areas of ZAPM15 and ZAPM30 were in the range of 100–106 m^2/g, which were higher compared to ZAP, 90 m^2/g. In general, the microwave-assisted method has improved the physico-chemical properties of the nanocomposite material with shorter ageing time of 30 min compared to 18 h ageing time by conventional oil bath method.

Keywords: pamoate anion; microwave-assisted; layered double hydroxide

1. Introduction

At present, the interest in the chemistry of the layered double hydroxides (LDHs or the so-called anionic clays) still attracts attention. The LDHs are layered compounds with well-defined structures consisting of positively charge layers alternately interspersed with change balancing sheets of anions. The positively charged layers of M^{II} and M^{III} cations are octahedrally co-ordinated by six oxygen anions as hydroxides [1]. Various organic anions can be intercalated into the layer of double hydroxide to form a hybrid of organic-inorganic nanocomposite materials. This material has attracted considerable attention recently, owing to its anion exchange ability as well as its application as selective sorbents and potential catalysts [2–7].

Conventionally, this type of material can be directly synthesised by co-precipitation technique in which the guest species is included in the reaction solution, followed by ageing process at elevated temperature for 18 h to form nanocomposite. Alternatively, microwave-assisted synthesis can be used to speed up the ageing process from 18 h for conventional method to 15–60 min. Microwave method offered several advantages such as shorter reaction time and new physico-chemical properties of the resulting materials. For example, microwave-assisted synthesis of Al-intercalated clays showed a higher surface area and shorter synthesis time compared to the one synthesised by conventional method [8].

*Corresponding author. Email: Zaemah@uniten.edu.my

During microwave processing, microwave energy penetrates through the material. Some of the energy is absorbed by the material and converted to heat, which in turn raises the temperature of the material such that the interior parts of the material are hotter than its surface as the surface loses more heat to the surroundings. This characteristic has the potential to uniformly heat large sections of the material. The reverse thermal in microwave heating does provide some advantages. These include rapid heating of materials without overheating the surface; removal of gases from porous materials without cracking; improvement in product quality and yield and synthesis of new materials and composites [9]. Faster crystallisation rate was achieved when chemicals were synthesised using microwave irradiation. Microwave heating has also been applied to the synthesis of LDH, more precisely to the ageing of the slurry obtained by precipitation [10–12].

In this article, we discuss our work on the microwave-assisted synthesis of pamoate, i.e. the anion of pamoic acid, 1, 1'-metheylene-bis-[2-hydroxy-3-naphthoic acid] into the layered Zn–Al LDH for the formation of hybrid organic-inorganic nanocomposite materials. The materials obtained were also compared with a conventional ageing process for 18 h in oil bath shaker [13]. The effect of microwave irradiation time on the crystallinity and composition of the resulting nanocomposites is studied and will be discussed here.

2. Experimental

All the chemicals used in this synthesis were obtained from various chemical suppliers and used without any further purification. All solutions were prepared using deionised water. The synthesis of the nanocomposite, ZAP, was done by the spontaneous self-assembly method. A mother liquor containing Zn^{2+} and Al^{3+} cations with Zn/Al initial molar ratio ($R_i = 4$) and pamoate anion (PA) was prepared and the pH was adjusted to about pH 7. The concentration of PA used was 0.02 M and the reaction was carried out with stirring under nitrogen atmosphere. The solution was aged for 18 h in an oil bath shaker at 70°C. The resulting precipitate was centrifuged, thoroughly washed and dried in an oven at 70°C for 3 days and kept in a sample bottle for further use and characterisations. A similar method was adopted for the preparation of Zn–Al LDH with nitrate as the intergallery anion (ZAL) by omitting the addition of PA solution in the mother liquor. A similar method was repeated using 0.02 M PA, but the ageing process was done in microwave digestion system at various times; 15, 30, 45 and 60 min and the nanocomposites formed were called ZAPM15, ZAPM30, ZAPM45 and ZAPM60, respectively.

The ageing process of the nanocomposite was done in the Microwave Digestion System, Model MDS-2100 with a pressure of 50 bar and 100% power of 950 W. The microwave was also equipped with removable 12-position carousel. The slurry formed was first poured into a standard advanced composite vessel of 100 mL volume, with the ageing temperature of 70°C. Powder X-ray diffraction (PXRD) patterns of the samples were obtained by a Shimadzu Diffractometer XRD-6000, using filtered Cu-Kα radiation ($\lambda = 1.54$ Å), at 40 kV and 30 mA. FTIR spectra were recorded by a Perkin Elmer 1750 Spectrophotometer. KBr pellet containing 1% sample was used to obtain the FTIR spectra. The surface morphologies of the samples were observed by a scanning electron microscope (SEM), using JOEL JSM-6400. CHNS analyser, model EA 1108 of Finons Instruments was used for CHNS analyses. UV-visible technique was used to determine the percent composition of PA in ZAP nanocomposite using a Perkin Elmer UV-visible Spectrophotometer model Lambda 20.

The Zn/Al molar ratio of the resulting ZAP materials were determined by an inductively coupled plasma emission spectrometry, using a Labtest Equipment Model 710 Plasmascan sequential emission spectrometer. Surface characterisation of the materials was carried out by nitrogen gas adsorption–desorption at 77 K using a micromeritics ASAP 2000. Samples were degassed in an evacuated-heated chamber at 120°C, overnight. The thermal analysis study was done using Perkin Elmer TGA 700 thermal analyser, and were recorded starting from 35°C to 1000°C at heating rate of 10°C/ min in a flow of nitrogen gas. The thermogravimetric analyser measures the weight changes in materials with regard to temperature.

3. Results and discussion

3.1. *Powder X-ray diffraction*

The PXRD patterns for ZAL and ZAP prepared by microwave-assisted and conventional oil bath methods, under various times (15–60 min), are shown in Figure 1. The intercalation was confirmed by the enhancement of the interlayer spacing to be around 18 Å when PA was intercalated into the lamella and produced a sharp, symmetric and intense peaks, especially for the (003) reflection, indicating well-ordered nanolayered structure.

Figure 1 shows that the basal spacing of ZAP synthesised using conventional oil bath method is 18.1 Å. When microwave method was used for the ageing process in the synthesis of ZAPMs, the basal spacing of the nanolayered materials obtained was in the range of 18.0 to 18.3 Å. ZAPM15 gave the highest basal spacing of 18.3 Å, and the basal spacing decreased slightly to 18.2, 18.1 and 18.0 Å as the ageing times using microwave radiation were increased to 30, 45 and 60 min, respectively. Comparison of Figure 1 also shows that not much difference could be observed between the ZAPs in terms of crystallinity whether they were prepared using microwave-assisted or conventional oil bath method.

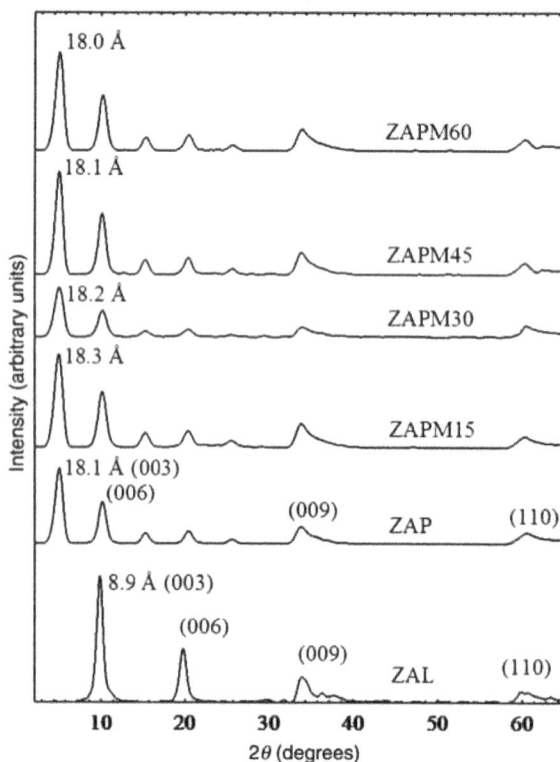

Figure 1.   PXRD patterns of ZAL, ZAP, ZAPM15, ZAPM30, ZAPM45 and ZAPM60.

Table 1.   XRD structure and parameters of ZAL, nanocomposites prepared by conventional oil bath (ZAP) and microwave-assisted methods (ZAPM15, ZAPM30, ZAPM45 and ZAPM60).

Structural parameters	ZAL	ZAP	ZAPM15	ZAPM30	ZAPM45	ZAPM60
d_{003} (nm)	0.8984	1.8225	1.8239	1.819	1.811	1.8016
d_{006} (nm)	0.4484	0.8779	0.8754	0.8761	0.877	0.8729
d_{009} (nm)	0.2656	0.2632	0.2634	0.2641	0.2638	0.2653
d_{110} (nm)	0.1529	0.1465	0.1536	0.1536	0.1529	0.1533
FWHM for $(003)(°)$	0.6739	1.026	1.0783	1.0520	0.9356	1.0239
FWHM for (110) (°)	0.840	1.5200	1.7700	1.6990	1.6333	1.6000
Lattice parameter a (nm)	0.3058	0.293	0.3072	0.3072	0.3058	0.3066
Lattice parameter c (nm)	2.6953	5.4675	5.4717	5.457	5.433	5.4048
Crystallite size in direction a (nm)	11.43	6.41	6.06	6.15	7.06	6.76
Crystallite size in direction c (nm)	12.35	8.08	7.69	7.89	8.88	8.11

Key XRD data and structural parameters of the materials are listed in Table 1. Assuming a 3R stacking of the layers and from the positions of the (110) and $(003/006)$ reflections, the lattice parameters a and c may be calculated [2]. The unit cell parameter a is the average distance between two metal ions in the layers and c is three times the distance from the centre of one layer to the next. The value of $a(= 2d_{110})$ is a function of the average radii of the metal cations while the value of $c(= 3d_{003})$ is a function of the average charge of the metal cations, the nature of the interlayer anion and the water content [14]. The lattice parameter c increases from 2.69 nm in ZAL to 5.47 nm in the nanocomposite materials of ZAPs synthesised using both conventional oil bath and microwave-assisted methods. It shows that the intercalation of PA has taken place in the layered structure of metal double hydroxide.

The average crystallite size in the a and c directions may be estimated from the values of full-width at half-maximum (FWHM) of the (110) and $(003)/(006)$ diffraction peaks, respectively, by means of the Scherrer equation $[L = 0.89\lambda/\beta(\theta) \cdot \cos\theta]$. Where L is the crystallite size λ is the wavelength of the radiation used, θ is the Bragg diffraction angle and $\beta(\theta)$ is the FWHM [14]. The values given in Table 1 confirmed that the crystal grain size of the samples is nanoscale. The nanocomposite synthesised using microwave-assisted method with the ageing times of 15 and 30 min gave slightly smaller crystallite size values compared to the nanocomposite materials obtained using conventional oil bath method. The advantage of using microwave-assisted method is that the time for ageing process has been shorten to only 15 min to achieve the cell ordered nanocomposite material compared to 18 h to age the sample if the conventional oil bath method was used to synthesise the nanocomposite.

3.2.   FTIR spectroscopy

Figure 2 shows the FTIR spectra for PA, ZAL and ZAPs synthesised using both conventional oil bath and microwave-assisted methods at various ageing times As a result of the intercalation of PA into the interlayer spacing of Zn–Al LDH, the spectra of ZAP and ZAPMs resembles a mixture of both the spectra of PA and ZAL.

The FTIR spectrum for ZAP and ZAPMs shows a broad absorption band centre around $3463\,cm^{-1}$, which is due to the presence of OH stretching of the hydroxyl groups of layers and/or physically adsorbed water molecule. The band at $1636\,cm^{-1}$ is due to ν_{H-O-H} bending vibrations. A sharp and very intense band located at approximately $1385\,cm^{-1}$ is attributed to the $\nu_3(NO_3^-)$ vibrations, which are not totally replaced by PA [5]. Strong bands at 1553 and $1385\,cm^{-1}$ are due to the antisymmetric and symmetric stretching of $-COO$, respectively [15]. The band at $1513\,cm^{-1}$ is attributed to the stretching of aromatic rings, $C=C$ and another sharp intense band

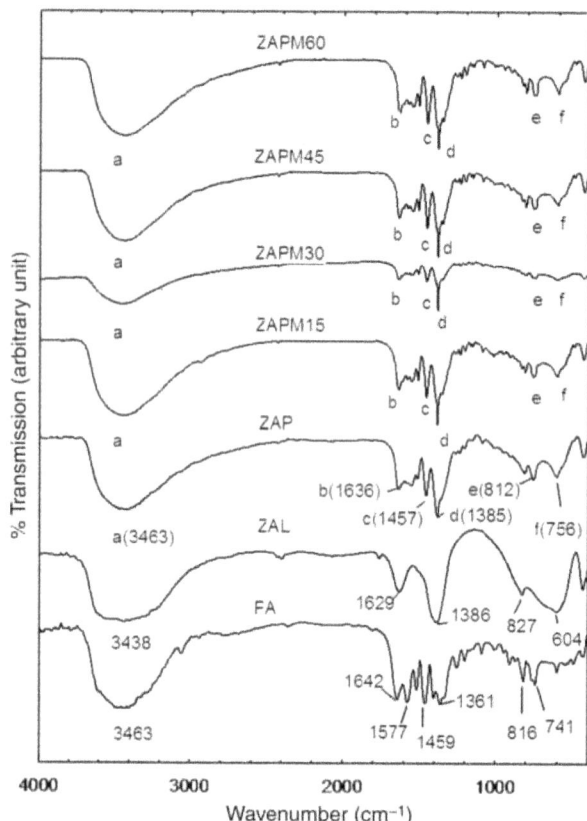

Figure 2. FTIR spectra of ZAl, PA and Zn–Al LDH–pamoate nanocomposite aged by conventional oil bath (ZAP) and microwave-assisted methods with various ageing times (ZAPMs) $a = 3463, b = 1636, c = 1457, d = 1385, e = 812$ and $f = 756$.

at $1457 \, cm^{-1}$ is due to CH_2 scissoring. Strong bands near 741–$816 \, cm^{-1}$ can be attributed to the presence of phenyl ring substitution band [16].

Not much difference can be seen from the FTIR spectrum of nanocomposite prepared using either conventional oil bath or microwave-assisted method. Both spectrums exhibited the same absorption band showing the same functional groups present in the materials.

3.3. Organic–inorganic composition

Table 2 compares the organic and inorganic composition of ZAPs synthesised by conventional oil bath and microwave-assisted methods. As given in the table, the molar ratio of Zn/Al present (R_{form}) in ZAP and ZAPMs are in the range of 3.2–3.4, compared to 4.0 for the initial ratio in the mother liquor, a slightly lower R_{form} value than $R_{initial}$ indicates that not all the Zn^{2+} ion in the mother liquor was used for the formation of the positively charged layers of the inorganic hydroxide during the formation of ZAP and ZAPMs ZAPM30 gives the best R_f value of 3.4 compared to other R_f values for other nanocomposite materials of ZAPM.

The CHNS analysis shows that ZAP contained 24.1% of carbon, which indicated that PA was successfully intercalated inside the interlamella of LDH, Table 2 also shows that the percentage of carbon for ZAPMs nanocomposites prepared using microwave-assisted method are slightly

Table 2. The organic and inorganic contents and surface properties of ZAL, nanocomposites prepared by conventional oil bath (ZAP) and by microwave-assisted methods (ZAPM15, ZAPM30, ZAPM45 and ZAPM60).

Parameters	ZAL	ZAP	ZAPM15	ZAPM30	ZAPM45	ZAPM60
Basal spacing (Å)	8.9	18.1	18.3	18.2	18.1	18.0
R_f (Zn/Al)	3.0	3.3	3.3	3.4	3.2	3.2
N (wt%) CHNS	3.8	0.8	0.5	0.8	0.8	0.7
C (wt%) CHNS	–	24.1	26.1	25.4	25.9	26.5
PA (wt%) CHNS	–	37.8	40.9	39.8	40.6	41.5
PA (wt%) UV	–	44.2	44.4	51.3	50.4	36.7
BET surface area (m^2/g)	5.8	90.1	100.7	106.9	76.7	62.1
Pore volume (cm^3/g)	0.028	0.342	0.364	0.404	0.277	0.206
Average pore diameter (Å)	87.5	115.1	110.1	116.2	108.4	87.2

Note: R_f = final ratio.

higher than prepared by conventional oil bath method. The presence of less that 1% of nitrogen from CHNS analysis in all samples of ZAP and ZAPMs shows that the nitrate anion still remains and not totally replaced by PA anion during the formation of nanocomposite. This is in agreement with the presence of strong and sharp bands at about 1385 cm^{-1} in the FTIR spectrum in Figure 2.

The weight percent of PA in the nanocomposite was determined from UV-Visible analysis measured at λ_{max} = 364.9 nm given in Table 2. Except for ZAPM60, the percentage of PA intercalated into the interlayer is higher for ZAPMs nanocomposite synthesised by microwave- assisted method compared to conventional oil bath method. The percentage of PA intercalated in the ZAPMs nanocomposite aged by microwave irradiation for 30 min is 51.3%, which shows the highest amount of PA intercalated compared to the other ZAPMs nanocomposites.

3.4. *Isotherm, surface area and pore size distribution*

The surface area and pore size distributions are measured using nitrogen gas adsorption–desorption technique at 77 K. Figure 3 shows the adsorption–desorption isotherms for ZAL, ZAP and ZAPMs. As shown in the figure, the adsorption–desorption isotherm for ZAP and ZAPMs are of Type IV, indicating mesopore-type material (20–500 Å). The adsorption increased slowly at low relative pressure in a range of 0.0–0.6, further increase of relative pressure to >0.6 resulted in the rapid adsorption of the adsorbent, reaching a maximum at more than 220 cm^3/g at STP for ZAP.

The maximum volumes of the nitrogen gas adsorbed by ZAPM15 and ZAPM30 are 260 and 240 cm^3/g, respectively, which are greater than those adsorbed by ZAP. As the ageing times for the microwave irradiation increased to 45 and 60 min, the maximum volumes adsorbed by the materials are 180 and 140 cm^3/g, respectively, which are lower than those adsorbed by nanocomposites prepared by the conventional oil bath method. The desorption branch of the hysteresis loop for ZAP and ZAPMs are much narrower compared to ZAL, indicating different pore textures of the resulting material as a result of successful intercalation of PA into the layered of Zn–Al LDH for the formation of nanocomposite.

The BET surface area and BET average pore diameter for ZAPs prepared using conventional oil bath and microwave-assisted methods at various ageing times are summarised in Table 1. As given in the table, the BET surface areas for ZAPM15 (100.7 m^2/g) and ZAPM30 (106.9 m^2/g) are higher than ZAP (90.1 m^2/g). But, as the ageing times were increased to 45 and 60 min in microwave-assisted method, the BET surface areas for ZAPM45 and ZAPM60 were decreased to

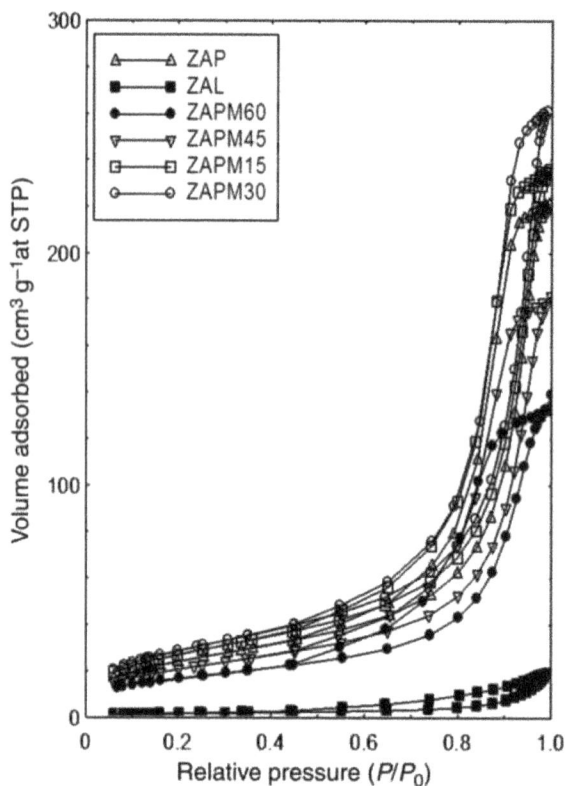

Figure 3. Adsorption–desorption isotherms of nitrogen gas on ZAL (■), ZAP (Δ), ZAPM15 (□), ZAPM30 (○), ZAPM45 (∇) and ZAPM60 (●).

76.7 and 62.1 m^2/g, respectively. This means that for the microwave-assisted synthesis, the ageing times needed to form a well-ordered nanocomposite structure are between 15 and 30 min. The data in Table 1 show that the ageing time for microwave radiation is also an important factor to be considered in order to produce materials of better crystallinity and physico-chemical properties.

The results obtained were in agreement with previous results reported by Kannan et al. [12] and Fetter et al. [17] using microwave radiation as a heating energy source, a significant increase in the S_{BET} values was observed when the slurry was heated under microwave field for short period of times, which correspond to 10 min. After this increase in S_{BET}, a reduction of the values was observed after submitting the sample to longer treatment periods. These results confirmed that the main factor that determined the development of specific surface area is the ageing process. The increase of the specific areas upon microwave treatment can be related to the simultaneous nucleation of many nuclei giving rise to smaller particles with a higher (surface area/volume unit) ratio. The decrease observed when the irradiation time was further increased could be explained on the basis of crystal growth and defect removal during the sintering process [18].

The BJH pore size distributions for ZAL and ZAPs prepared by conventional oil bath and microwave-assisted methods are shown in Figure 4. As shown in the figure, all materials show mesoporous property, in agreement with Type IV adsorption isotherm. BJH pore size distribution for ZAL shows a broad peak at around 100 Å, while for ZAP shows an intense peak centred at around 145 Å, indicating modification of pore texture in agreement with the formation of new nanocomposite phase with a basal spacing of 18.1 Å. The isotherm of ZAPMs is similar to the isotherm of ZAP prepared by the conventional oil bath method. The intensities of pore size

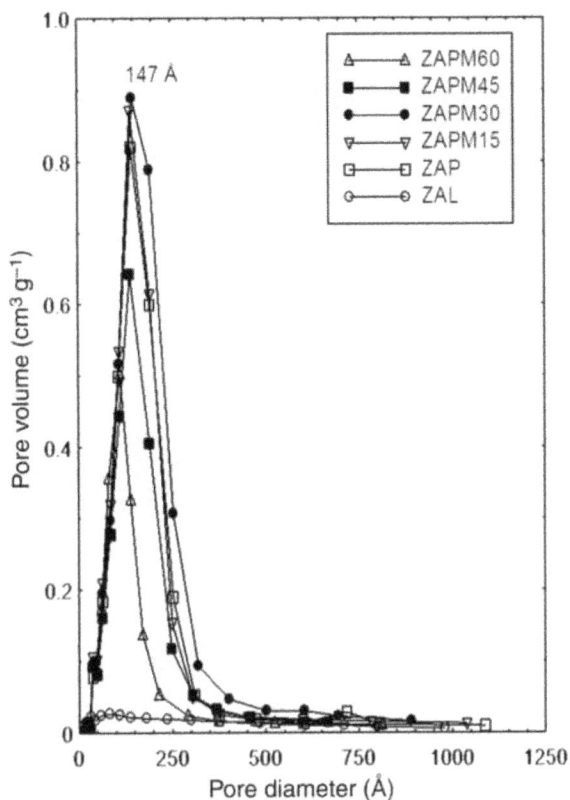

Figure 4.   BJH desorption pore size distributions for ZAL ($\bigcirc$), ZAP ($\square$), ZAPM15 (∇), ZAPM30 ($\bullet$), ZAPM45 ($\blacksquare$) and ZAPM60 ($\triangle$).

distributions of ZAPM15 and ZAPM30 are higher than those of ZAP, while the intensities of pore size distributions of ZAPM45 and ZAPM60 are lower than those of ZAP. As discussed earlier, the increase of the irradiation time leads to a decrease in the particle size and a homogeneous particle size distribution of small particle is achieved [17]. This indicated that the most suitable ageing time for microwave irradiation to synthesise the nanocomposite sample is between 15 and 30 min.

3.5.   *Thermal analysis*

The thermogravimetry analysis (TGA) and differential TGA (DTG) of ZAP prepared by both conventional oil bath and microwave-assisted methods are shown in Figure 5. The DTG curves of ZAP show three major stages of weight loss. The first weight loss is due to the removal of water physisorbed on the external surface of the powder particles. The second one should be due to the simultaneous loss of carbonate and dehydroxylation of the brucite-like layer and the third one, about 800°C, is due to the decomposition of the non-vapourisable of the organic anion present in the interlayer lamella of the nanohybrid material [19]. The percent weight losses are 16%, 19% and 29% at 213°C, 417°C and 877°C, respectively.

The asymmetric shape of the DTG curve for the first weight loss might be an indication of the existence of two types of interlamellar water molecules: free ones and those solvating the anionic species. These observations suggest that the interlamellar water molecules are held differently in

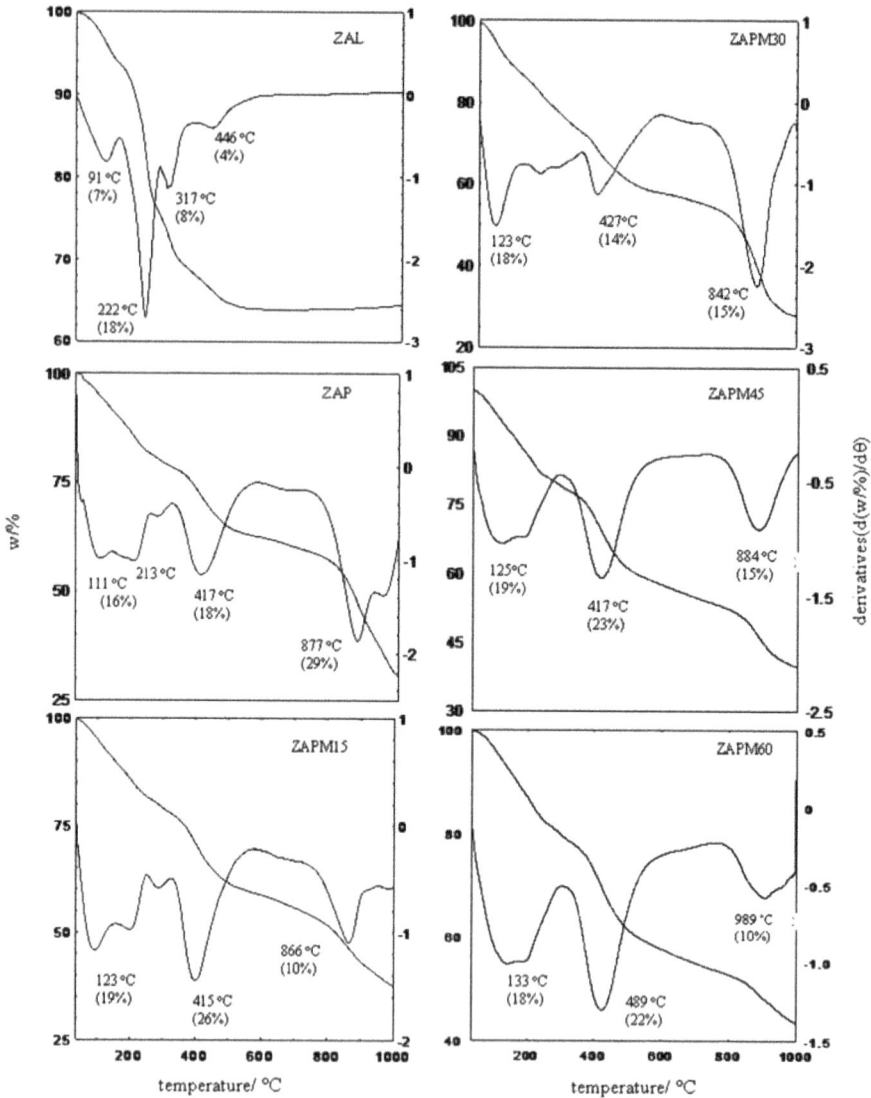

Figure 5.   TGA and DTG thermograms of ZAL ZAP, ZAPM15, ZAPM30, ZAPM45 and ZAPM60.

the materials considered, according to the local environment. They are involved in the solvation of the oxoanions through hydrogen bonding. These hydrogen bonds are expected to be stronger with oxoanions than with other anions [20].

Both ZAP and ZAPMs show similar patterns of TGA–DTG curves with three major weight losses. The first weight loss of interlayer water in ZAPM15 is about 19%, it is slightly higher than that in ZAP (16%) with temperature maxima around 123°C. The higher amount of interlamellar water and the improved order in the interlamellar space could be a consequence of microwave coupling with the species located in the interlayer domain. It is well known that water molecules and carbonate anions are linked with each other and to the host structure by weak hydrogen bonds. Due to the reorientation of water molecules and to the displacement of carbonate anions in response to microwave field, it is possible that hydrogen bonds are finally broken and re-formed

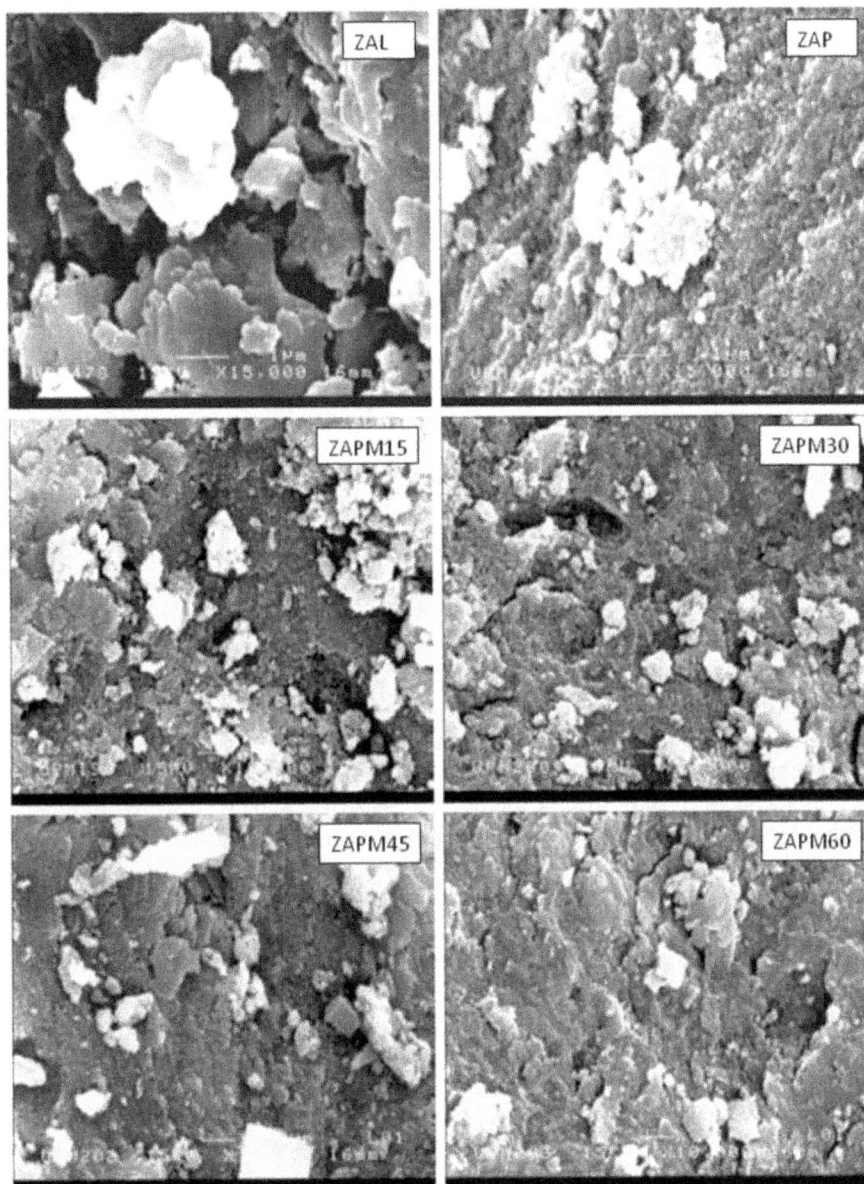

Figure 6.   Surface morphologies of ZAL and ZAP at $15,000\times$ magnification and ZAPM15, ZAPM30, ZAPM45 and ZAPM60 at $10,000\times$ magnification, respectively.

resulting in a better ordering of water molecules and carbonate anions in the interlayer space, such as enhanced ordering, would permit the accommodation of additional water molecules [11].

   The second weight loss of ZAPM15 due to the dehydroxylation of the brucite-like layer is about 26%, also slightly greater than the weight loss in ZAP (19%). The last TG step recorded at around 800°C has been attributed to the heat absorbed during the formation of the spinel phase, $ZnAl_2O_4$, which must be probably accompanied by the release of some residual carbonate anions adsorbed onto the zinc oxide surface [17]. As the ageing time for the microwave radiation increased to 60 min, the temperature maxima of the three weight losses in ZAPM60 shifted to higher temperatures at $133°C, 489°C$ and $989°C$ and percent weight losses of 18%, 22% and 10%, respectively. This shows that ZAPM60 is thermally more stable than the other three ZAPMs samples.

3.6. *Surface morphology*

Figure 6 shows the morphology of ZAL, ZAP and ZAPMs obtained by SEM. The obtained surface morphology shows agglomerates of compact and nonporous structure. As shown in the figure, there is no significant difference in the morphology of the nanocomposite samples synthesised using two different ageing methods, i.e. conventional oil bath for 18 h and microwave-assisted method with irradiation time 15–60 min. They are very similar to the morphologies of other nanocomposites, such as Zn–Al–ibuprofen LDH [15].

4. Conclusions

Hybrid organic–inorganic of Zn–Al–pamoate nanocomposites were prepared by direct co-precipitation method. The samples were aged using two different methods; conventional oil bath and microwave-assisted methods. The advantage of using microwave-assisted method is that the time for ageing process was shorter, about 15–30 min to achieve a well-ordered nanocomposite material compared to 18 h to age the sample, if conventional oil bath method was used to synthesise the nanocomposite. The BET surface areas for nanocomposites synthesised by microwave-assisted method, ZAPM15 ($100.7\,m^2/g$) and ZAPM30 ($106.9\,m^2/g$), were higher than those synthesised by conventional oil bath method, ZAP ($90.1\,m^2/g$). Percentage of PA intercalated between the LDH was higher in ZAPMs than in ZAP nanocomposite materials. Generally, the results obtained show that the microwave-assisted method with shorter irradiation time gave nanocomposite materials with improved physico-chemical properties.

Acknowledgements

The authors are grateful to the Malaysian Government (MOSTI) for providing the grant under IRPA no. 09-02-0897-EA001.

References

[1] S. Carlino and M J. Hudson, *A thermal decomposition study on the intercalation of tris-(oxalato) ferrate(III) trihydrate into a layered (Mg/Al) double hydroxide*, Solid State Ionics 110(1–2) (1998), pp. 153–161.

[2] F. Cavani, F. Trifiro, and A. Vaccari, *Hydrotalcite-type anionic clays: Preparation, properties and applications*, Catal Today, 11(2) (1991), pp. 175–301.

[3] W.T. Reichle, *Catalytic reactions by thermally activated, synthetic, anionic clay minerals*, J Catal 94(2) (1985), pp. 547–557.

[4] M.Z. Hussein, Z. Zainal, and T.C. Beng, *The use of Mg/Al layered double hydroxide for color removal of textile wastewater*, J. Environ. Sci. Health A 36 (2001), pp. 565–573.

[5] A. Legrouri, A. Badreddine, A. Barroug, J.P. Roy, and P. Besse, *Influence of pH on the synthesis of the Zn-Al-nitrate layered double hydroxide and the exchange of nitrate by phosphate ions*, J. Mater. Sci. Lett. 18 (1999), pp. 1077–1079.

[6] J. Orthman, H.Y. Zhu, and G.Q. Lu, *Use of anion clay hydrotalcite to remove coloured organics from aqueous solutions*, Sep. Purif. Technol. 31(1) (2003), pp. 53–59.

[7] Y. You, G.F. Vance, and H. Zhao, *Selenium adsorption on Mg–Al and Zn–Al layered double hydroxides*, Appl. Clay Sci. 20(1–2) (2001), pp. 13–25.

[8] G. Fetter, G. Heredia, A.M. Maubert, and P. Bosch, *Synthesis of Al-intercalated montmorillonites using microwave irradiation*, J. Mater. Chem. 6 (1996), pp. 1857–1858.

[9] H.S. Ku, E. Siores, A. Taube, and J.A.R. Ball, *Productivity improvement through the use of industrial microwave technologies*, Comput Ind. Eng. 42 (2002), pp. 281–290.

[10] S. Komarneni, Q.H. Li, and R. Roy, *Microwave-hydrothermal processing of layered anion exchangers*, J. Mater. Res. 11 (1996), pp. 1866–1869.

[11] M.Z. Hussein, Z. Zainal, and C.Y. Ming, *Microwave-assisted synthesis of Zn-Al layered double hydroxide-sodium dodecyl sulfate nanocomposite*, J. Mater. Sci. Lett. 19 (2000), pp. 879–883.

[12] S. Kannan and R.V. Jasra, *Microwave assisted rapid crystallization of Mg–M(III) hydrotalcite where M(III) = Al, Fe or Cr*, J. Mater. Chem. 10 (2000), pp. 2311–2314.

[13] M.Z. Hussein, Z.B. Jubri, Z. Zainal, and A.H. Yahaya, *Pamoate intercalated Zn-Al layered double hydroxide for the formation of layered organic-inorganic intercalate*, Mater. Sci. – Poland 22 (2004), pp. 57–67.

[14] L. Yan-Jun, L. Dian-Qing, G. David, and D. Xue, *Modulating effect of Mg–Al–CO3 layered double hydroxides on the thermal stability of PVC resin*, Polym. Degrad. Stab. 88(2) (2005), pp. 286–293.

[15] V. Ambrogi, G. Fardella, G. Grandolini, and L. Perioli, *Intercalation compounds of hydrotalcite-like anionic clays with antiinflammatory agents — I. Intercalation and in vitro release of ibuprofen*, Int. J. Pharm. 220 (2001), pp. 23–32.

[16] R.M. Silverstein, T.C. Morill, and G.C. Bassler, *Spectrometric Identification of Organic Compounds*, John Wiley & Sons Inc, New York, 1998.

[17] G. Fetter, A. Botello, V.H. Lara, and P. Bosch, *Detrital Mg(OH)2 and Al(OH)3 in microwaved hydrotalcites*, J. Porous Mater. 8(3) (2001), pp. 227–232.

[18] P. Benito, F.M. Labajos, and V. Rives, *'Microwave-assisted synthesis of layered double hydroxide*, in *Solid State Chemistry*, R.W. Buckley, ed., Nova Science Publisher Inc., New York, 2007, pp. 173–225.

[19] H. Nijs, A. Clearfield, and E.F. Vansant, *The intercalation of phenylphosphonic acid in layered double hydroxides*, Microporous Mesoporous Mater. 23(1–2) (1998), pp. 97–108.

[20] F. Malherbe and J.P. Besse, *Structures and physical properties of films deposited by simultaneous DC sputtering of ZnO and In2O3 or ITO*, J. Solid State Chem. 155(2) (2000), pp. 332–319.

13

Murraya koenigii-mediated synthesis of silver nanoparticles and its activity against three human pathogenic bacteria

S.R. Bonde, D.P. Rathod, A.P. Ingle, R.B. Ade, A.K. Gade and M.K. Rai*

Department of Biotechnology, SGB Amravati University, Amravati 444602, Maharashtra, India

Synthesis of silver nanoparticles (Ag NPs) by the leaf extract of *Murraya koenigii* (Indian curry leaf tree) is reported in this study. The colour of the leaf extract prepared by grinding turned from green to brown after treatment with AgNO$_3$ (1 mM). The UV–visible spectroscopic analysis showed the absorbance peak at about 420 nm, which indicates the synthesis of Ag NPs. Further characterisation by Fourier transform infrared spectroscopy showed the presence of proteins as capping agents, which increase the stability of Ag NPs in the colloids. Scanning electron microscopy demonstrated the presence of spherical Ag NPs in the range of 40–80 nm. The bactericidal activity of the standard antibiotics was significantly increased in the presence of Ag NPs against pathogenic bacteria, *viz.*, *Escherichia coli*-JM-103 (ATCC 39403), *Staphylococcus aureus* (ATCC 25923) and *Pseudomonas aeruginosa* (MTCC 424). Ag NPs in combination with gentamicin showed the maximum activity against *E. coli* (increase in fold area −4.06), followed by *P. aeruginosa* (1.11) and *S. aureus* (0.09), while tetracycline showed maximum activity against *S. aureus* (2.16) followed by *P. aeruginosa* (0.24) and *E. coli* (0.21). The Ag NPs thus obtained demonstrated remarkable antibacterial activity against three human pathogenic bacteria when used in combination with commercially available antibiotics.

Keywords: extracellular; silver nanoparticles; plants; antibiotic; *Murraya koenigii*

1. Introduction

Nanotechnology is a highly multidisciplinary field, drawing from applied physics, materials science, colloid science and interface. The metal nanoparticles have novel magnetic, electronic and optical properties, which vary in their size, shape and composition. Metal nanoparticles (Ag NPs) are more effective, partly because of the high surface volume fraction so that a large proportion of silver atoms are in direct contact with their environment [1]. Due to these properties, metal nanoparticles have many applications in different fields like medicine, electronics, agriculture, etc.

The biosynthetic method employing plant extracts has received attention as being simple, eco-friendly and economically viable compared to the microbial systems like bacteria and fungi because of their pathogenicity, and also the chemical and physical

*Corresponding author. Email: mkrai123@rediffmail.com; pmkrai@hotmail.com

methods used for synthesis of metal nanoparticles [2–4]. Thus, understanding biochemical processes that lead to the formation of nanoscale inorganic material is potentially appealing as an environment-friendly alternative to chemical methods [5].

Gardea-Torresdey et al. [6,7], for the first time, demonstrated synthesis of gold nanoparticles inside live plant. Similarly, alfalfa sprouts were also used for the synthesis of Ag NPs. Biosynthesis of nanoparticles using different plant species like *Aloe vera* [8], *Cinnamomum camphora* [9], *Capsicum annuum* L. [10], *Medicago sativa* and *Brassica juncea* [2], *Brassica chicory* [2], *Azadiracta indica* [11] and *Cymbopogon flexuosus* [12] has been successfully carried out.

Armendariz et al. [13] studied the synthesis of gold nanoparticles of different sizes and shapes using oat and wheat biomass. Similar work on synthesis of gold nanotriangles was performed using tamarind leaf extract [4]. Mude et al. [14], for the first time, used *in vitro*-generated callus extract of *Carica papaya* for the synthesis of Ag NPs. These authors reported formation of spherical nanoparticles, when aqueous silver ions ($AgNO_3$, 1 mM) were treated with callus extract.

In 2009, Kumar and Yadav [15] reported the synthesis of silver and gold nanoparticles using different plants and also found that plant system was more advantageous over other environmentally benign biological processes as it eliminates the elaborate process of maintaining cell cultures. Song and Kim [5] studied the extracellular synthesis of metallic Ag NPs using five plant leaf extracts (Pine, Persimmon, Ginkgo, Magnolia and Platanus). They reported that the stable Ag NPs were formed by treating aqueous solution of $AgNO_3$ with all the above plant leaf extracts as reducing agents of Ag^+ to Ag^0. Magnolia leaf broth was the best reducing agent in terms of synthesis rate and conversion to Ag NPs as compared to other plants. The average particle size they reported was in the range of 15–500 nm. Bar et al. [16] developed an eco-friendly method for rapid synthesis of Ag NPs using aqueous seed extract of *Jatropha curcas*. The Ag NPs formed at different concentrations of $AgNO_3$ were spherical in shape with an average diameter of 15–50 nm.

The emerging multi-drug resistance in microbes is a matter of great concern as these human pathogens are reported to be the leading cause of death worldwide [17,18]. Multiple surveillance studies have demonstrated that resistance among prevalent pathogens is increasing at an alarming rate, leading to greater patient morbidity and mortality. The widespread use of chemicals has resulted in bacterial resistance to antibiotics; in such cases, Ag NPs were found to be effective against multi-drug resistant bacteria [19]. As a result of their small size, nanoparticles have a relatively large surface area in comparison to their volume; it has been suggested that the smaller the silver particles, the more effective the colloid. This larger surface area enables Ag NPs to interact more easily with other substances and therefore increases their antibacterial efficiency.

Ag NPs are supposed to be the new generation of antimicrobials [20] as they are being used in many antimicrobial preparations. Gade et al. [21] and Ingle et al. [19] reported the antibacterial activity of Ag NPs synthesised by fungi. Duran et al. [22] successfully developed Ag NP-impregnated wound dressings and textile fabrics that are useful for burn patients. Ag NPs are also used for surgical masks [23].

Murraya koenigii, an Indian culinary and medicinal herb, is used as a food additive for the purpose of flavour, and sometimes as a preservative for the prevention of growth of pathogenic bacteria [24].

In this study, *M. koenigii* was used for the synthesis of Ag NPs and the activity of synthesised Ag NPs was evaluated against three human pathogenic bacteria in combination with four antibiotics.

2. Materials and methods

2.1. *The test plant*

The young and healthy leaves of *M. koenigii* were collected from the field of Department of Biotechnology, Sant Gadge Baba Amravati University, Amravati, Maharashtra, India.

2.2. *Test bacteria*

Escherichia coli-JM-103 (ATCC 39403), *Staphylococcus aureus* (ATCC 25923) and *Pseudomonas aeruginosa* (MTCC 424) were used to evaluate the activity of Ag NPs in combination with standard antibiotics (gentamicin, ampicillin, tetracycline and streptomycin).

2.3. *Extraction*

The leaves (20 g) were washed twice in tap water and rinsed thrice in distilled water. Then they were surface sterilised by $HgCl_2$ (0.1%) for 1 min, cut into small pieces and ground with 100 mL of sterilised distilled water in an omnimixer. Later, crude extract was filtered through muslin cloth and centrifuged at 10,000 rpm for 15 min to obtain clear leaf extract, which was later used for the synthesis of Ag NPs.

2.4. *Synthesis of Ag NPs*

For the synthesis of Ag NPs, leaf extract was challenged with $AgNO_3$ (1 mM) solution and incubated at room temperature. Control (without treatment with $AgNO_3$, i.e. only leaf extract, 1 mM) was also maintained. After the reduction of aqueous silver ions into Ag NPs, residual silver ions (unreacted silver ions) were removed from the reaction mixture by treatment with sodium chloride (NaCl). As a result, white precipitate of silver chloride (AgCl) was formed after reacting with Ag^+ ion. The precipitate was removed by filtration. The resulting reaction mixture was subjected to different analytical analyses. Triplicates of each treatment were maintained.

2.5. *Detection and characterisation of silver nanoparticles*

2.5.1. *Visual observation*

After the treatment of leaf extract with $AgNO_3$ (1 mM), the colour change of the reaction mixture was visually observed.

2.5.2. *UV–vis spectrophotometric analysis*

The aliquotes of reaction mixture were subjected to the measurement of absorbance by UV–visible spectrophotometer (Perkin Elmer, Lambda-25) at a resolution of 1 nm from 250 to 800 nm for the detection of Ag NPs.

2.5.3. *Fourier transform infrared spectroscopy analysis*

The silver nitrate solutions, after complete reduction and formation of Ag NPs, were centrifuged at 10,000 rpm for 10 min; the pellets obtained were further washed with distilled water twice to remove unbound and free proteins or other compounds present in the solution. Fourier transform-infrared spectroscopy (FT-IR) (Perkin Elmer) used for analysis of Ag NPs by scanning the spectrum in the range 450–4000 cm^{-1} at a resolution of 4 cm^{-1} was carried out.

2.5.4. *Scanning electron microscopy*

Scanning electron microscopy (SEM) was performed to find out the shapes and sizes of the synthesised Ag NPs (JEOL-6380A-version 1.1).

2.5.5. *X-ray diffraction analysis*

X-ray diffraction (XRD) spectra was recorded on a PAN analytical X-PRT PRO, D-8, Advanced Brucker instrument, Netherlands and depicted number of Bragg reflections indexed on the basis of the face-centred cubic (FCC) structure of metallic silver.

2.5.6. *Nanosight LM-20 analysis*

Liquid samples of Ag NPs at the concentration range of 10^7–10^9/mL were introduced into a scattering cell through which a laser beam (approximately 40 mW at $k = 635$ nm) was passed. Particles present within the path of the laser beam were observed *via* a dedicated non-microscope optical instrument (LM-20, NanoSight Pvt. Ltd., UK) having charge-coupled device camera. The motion of the particles in the field of view (approximately $100 \times 100\ \mu m$) was recorded (at 30 fps) and the subsequent video and images were analysed.

2.6. *Assessment of antibacterial activity*

The disc diffusion method was applied to evaluate the combined effect of Ag NPs with four antibiotics against human pathogenic bacteria: *E. coli*-JM-103 (ATCC 39403), *S. aureus* (ATCC 25923) and *P. aeruginosa* (MTCC 424) grown on nutrient agar plates. The overnight grown bacterial culture having 10^5 CFU/mL was used to assess the activity. The test bacterial cultures were inoculated onto solidified agar plates. The different standard antibiotic discs (*viz.* gentamicin, ampicillin, tetracycline and streptomycin) purchased from Hi-Media, Mumbai were used. To evaluate the combined effects, each standard antibiotic disc impregnated with 15 μL solution of Ag NPs was placed on to the agar surface inoculated with test bacteria. The plates were then incubated at 37°C for 24 h.

After incubation, the zones of inhibition were measured and its activities were evaluated by increase in fold area of the activity. The assays were performed in triplicate.

2.7. *Assessment of increase in fold area*

The increase in fold area was assessed by calculating the mean surface area of the inhibition zone of each tested antibiotic. The fold increase area of different antibiotics was calculated using the equation $(y^2 - x^2)/x^2$, where 'x' and 'y' were zone of inhibitions for respective antibiotic (x) and antibiotic + Ag NPs (y), respectively.

3. Results and discussion

The synthesis of Ag NPs by leaf extract of *M. koenigii* was performed in this study. When the leaf extract was treated with $AgNO_3$ (1 mM) and incubated in darkness at room temperature, within 1 h of the reaction, colour changes from green to brown (Figure 1), indicating the formation of Ag NPs. It is an efficient and rapid method, which showed similarity with the results obtained by other researchers who worked with different plant systems [8,10,11]. Change in colour was due to the excitation of surface plasmon vibrations in the metal nanoparticles [1].

Optical absorption spectroscopy has proved to be a very useful technique for the analysis of nanoparticles. In order to verify the synthesis of Ag NPs, the test samples were subjected to UV–vis spectrophotometric analysis. The test samples (leaf extract treated

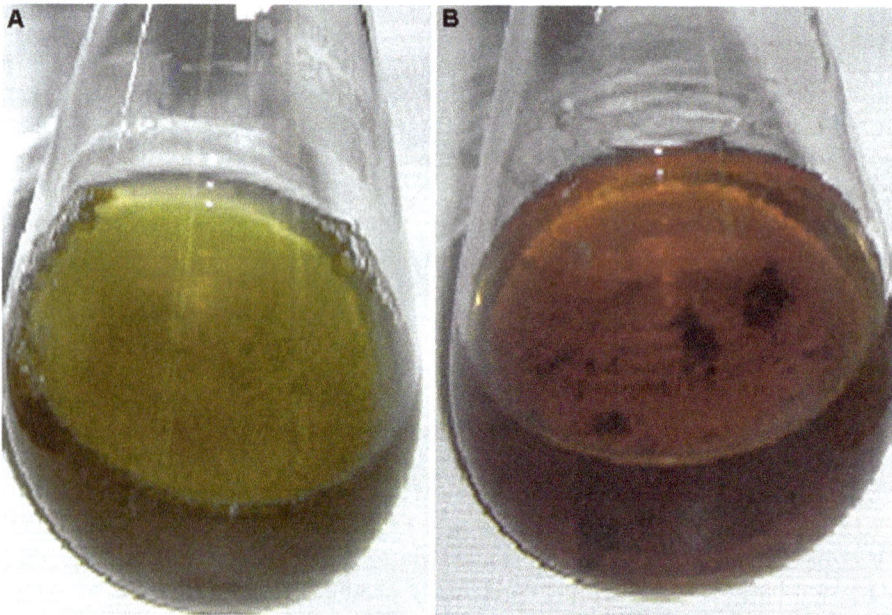

Figure 1. Leaf extract: left (control) before and right (experimental) after treatment with 1 mM of $AgNO_3$ solution.

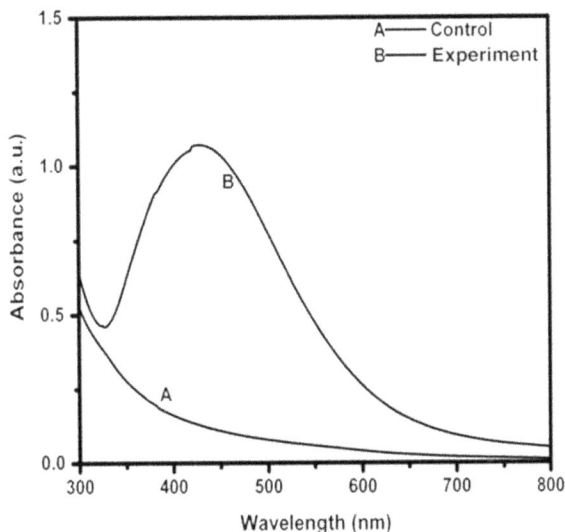

Figure 2. UV–vis spectra of: (A) leaf extract (control) and (B) Ag NPs showing absorbance at about 420 nm.

with 1 mM of AgNO$_3$) were collected in aliquots from the reaction mixture and analysed to record their absorbance by UV–vis spectrophotometer. This analysis showed the sharp absorbance at around 420 nm (Figure 2), which was specific for Ag NPs [8,11]. Huang et al. [9] and Li et al. [10] reported similar results, in which they observed that when sun-dried leaf extracts of C. camphora and C. annuum were challenged with aqueous silver ions, the reaction mixture containing Ag NPs showed the absorption peak at about 420 nm due to the excitation of plasmon resonance vibration.

FT-IR measurements were carried out to identify the possible biomolecules responsible for capping and efficient stabilisation of the metal nanoparticles synthesised by M. koenigii extract. The FT-IR spectrum showed the presence of eight different peaks at 1041, 1285, 1357, 1384, 1450, 1587, 1628 and 1735 cm^{-1} (Figure 3). The FT-IR spectra reveal the presence of different functional groups like C–N, C–O–C, amide linkages and –COO–, these may be between amino acid residues in protein and synthesised Ag NPs, which gave rise to the well-known signatures in the infrared region of the electromagnetic spectrum. The peak at 1041 was associated with stretch vibration of –C–O–C– [9], while other peaks at 1357, 1384, 1450 and 1587 cm^{-1} were aroused for C–N stretching vibrations of aromatic amines [9], residual –NO$_3$ [25], symmetric stretching vibrations of –COO– present in carboxylic acid present in amino acids and aromatic C–C skeletal vibrations, respectively [25], which are normally present and associated with protein biomolecules. The peak at 1735 cm^{-1} was associated with stretch C=C and is assigned to the aldehydes, ketones, carboxylic acids and ester compounds. FT-IR spectroscopic study has confirmed the association between peptides of proteins and biosynthesised Ag NPs, which have the stronger ability to bind silver metal, so that the proteins could most possibly form a coat covering on the Ag NPs (i.e. capping of Ag NPs) and stabilise the nanoparticles in the medium [26].

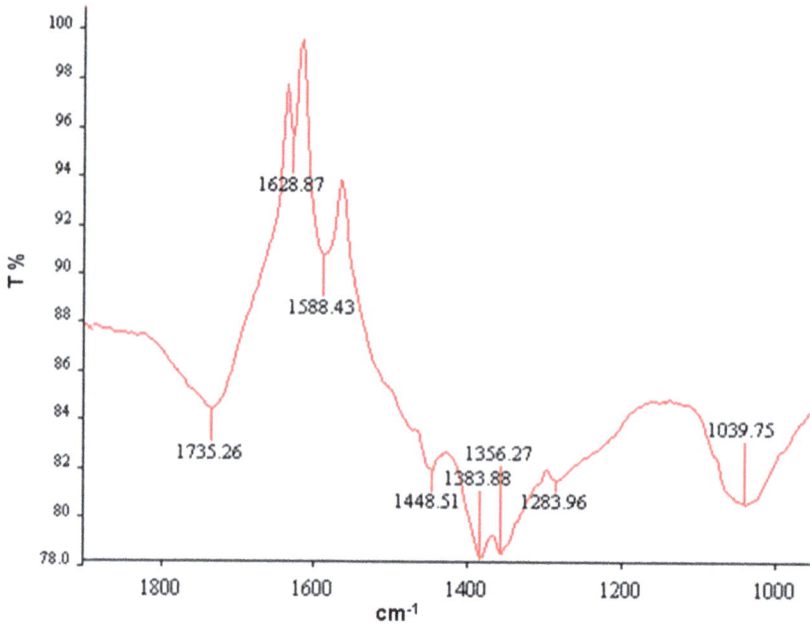

Figure 3. FT-IR spectrum for extract of leaves after treatment with 1 mM silver nitrate solution.

SEM (Figure 4) analysis finally confirmed the synthesis of spherical and polydispersive Ag NPs in the reaction mixture. A particle size distribution histogram determined from the SEM images showed the large variation in the particle size. The particles are in the range of 40–80 nm with average diameter size of 58.44 nm (Figure 5). XRD analysis was carried out to find the structure of Ag NPs. The XRD patterns were observed at positions 37°, 45°, 67° and 79° of 2θ, depicting the presence of (1 1 1), (2 0 0), (2 2 0) and (3 1 1) facets of FCC structure of Ag NPs (Figure 6). The values were in agreement with Joint Committee on powder diffraction (JCPD, standard file no. 04-0783).

Nanoparticle tracking and analysis (NTA) was used to measure the dispersion characteristics, i.e. size and size distribution. In particular, it is the most recently developed system; NTA was assessed in-depth due to its ability to see and size of particles individually on a particle-by-particle basis. NTA allows individual nanoparticles in a suspension to be microscopically visualised and their Brownian motion to be separately but simultaneously analysed, from which the particle size distribution can be obtained on a particle-by-particle basis. Figure 7 shows the particle populations by size and intensity. These results corroborate the results obtained by Montes-Burgos and group [27].

We report synthesis of Ag NPs by *M. koenigii* for the first time. Moreover, for the first time, phytosynthesised Ag NPs were used for the evaluation of their antibacterial efficacy in combination with commercially available antibiotics. From this study, it was observed that the efficacy of Ag NPs against test bacteria (*E. coli*, *S. aureus* and *P. aeruginosa*) increased when assessed in combination with antibiotics. It is evidenced by the data

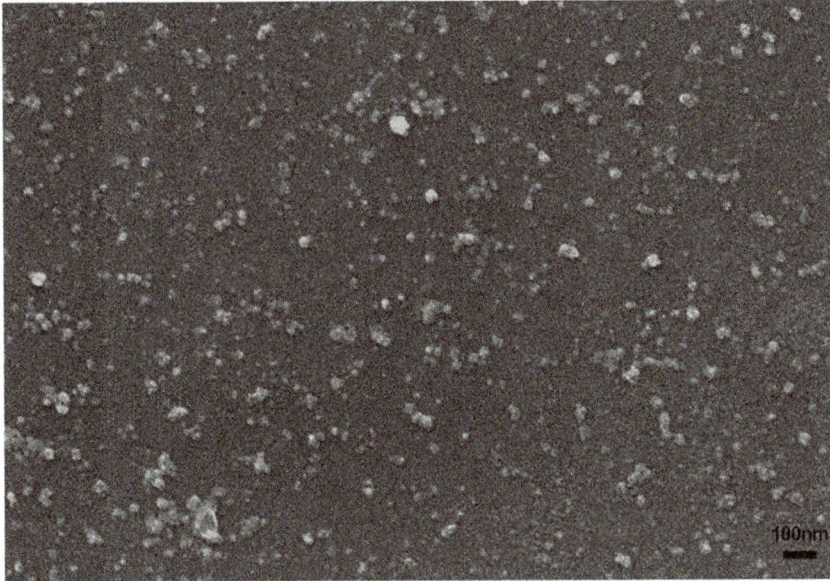

Figure 4. SEM micrograph of Ag NPs (40–80 nm, scale bar 100 nm).

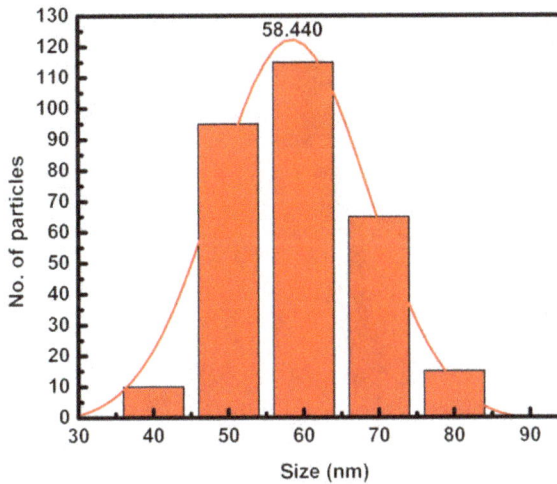

Figure 5. A particle size distribution histogram determined from the SEM images.

provided in Table 1, which showed increase in activity fold area for each antibiotic. The activity of gentamicin (4.06) was significantly enhanced in the presence of Ag NPs against *E. coli* followed by *P. aeruginosa* (1.11). On the other hand, tetracycline showed increase in the activity in the presence of Ag NPs against *S. aureus* (2.16), while other antibiotics did not show significant inhibitory activity against the test bacteria. These findings

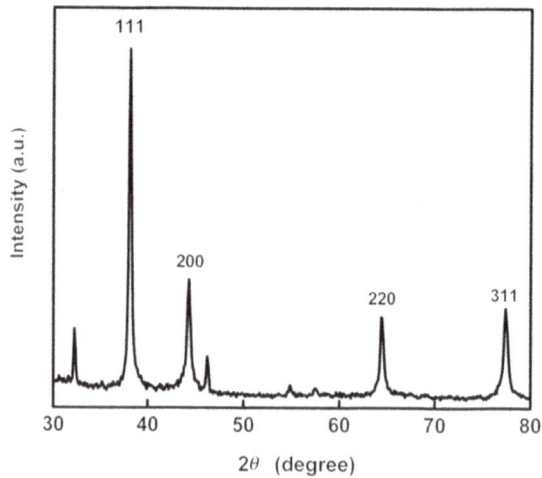

Figure 6. XRD analysis of Ag NPs.

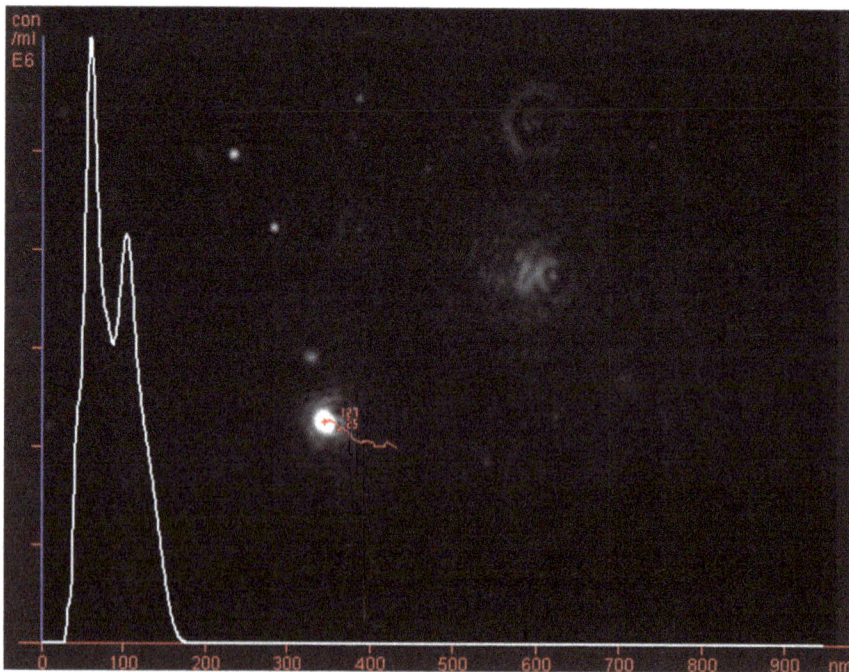

Figure 7. Particle populations by size and intensity analysis of Ag NPs using NanoSight LM-20.

substantiate with the findings reported by Shahverdi and his coworkers [28], who found that *S. aureus* showed the maximum sensitivity to Ag NPs, whereas *E. coli* was less sensitive. On the contrary, Birla et al. [26] reported that the activity of commercially available antibiotics with Ag NPs synthesised by a fungus *Phoma glomerata* was more

Table 1. Comparison of the increase in fold area zone of activity of different antibiotics against *S. aureus*, *E. coli* and *P. aeruginosa* (in the absence and in the presence of Ag NPs at the concentration of 20 μL/disc).

	S. aureus			E. coli			P. aeruginosa		
Antibiotics	Ab (x)	Ab+Ag NPs (y)	Increase in fold area[a]	Ab (x)	Ab+Ag NPs (y)	Increase in fold area	Ab (x)	Ab+Ag NPs (y)	Increase in fold area
Gentamicin	21	22	0.09	8	18	4.06	11	16	1.11
Ampicillin	15	16	0.13	14	18	0.65	12	14	0.36
Tetracycline	18	32	2.16	20	22	0.21	17	19	0.24
Streptomycin	18	19	0.11	12	16	0.77	10	13	0.69

Notes: Ab, antibiotics and Ab + Ag NPs, antibiotics + Ag NPs.
[a]Increase in fold area $= (y^2 - x^2)/x^2$.

against Gram-negative bacteria compared to the Gram-positive bacteria. Therefore, these findings would help us to predict that the effect of Ag NPs depends upon the source of their synthesis. The variations in the bactericidal activity of Ag NPs were also due to the different sizes and shapes of Ag NPs produced by various biological systems. The interactions of Ag NPs with biosystems (microbes and plants) are yet to be understood fully. However, these particles are increasingly being used as antimicrobial agents. It may be speculated that Ag NPs with the same surface areas but with different shapes and sizes may also have different effective surface areas in terms of active facets, though, at present, we are unable to give an estimation of how the surface areas of different nanoparticles affect antimicrobial activity. But the results of Pal et al. [29] provide the basis for the measurement of shape-dependent bacterial activity of Ag NPs. However, it is imperative that the flexibility of nanoparticle preparation methods and their influence on other bacterial strains along with the antibiotics may be explored.

4. Conclusion

It has been demonstrated that the extract of plant *M. koenigii* is capable of producing Ag NPs extracellularly and these Ag NPs are quite stable in solution due to capping likely by the proteins present in the extract. This is an efficient, eco-friendly and simple process. The Ag NPs in combination with commercially available antibiotics showed synergistic activity against multi-drug resistant bacteria, and therefore could be used as an antimicrobial agent after further trials on experimental animals.

References

[1] A. Ahmad, P. Mukherjee, S. Senapati, D. Mandal, M.I. Khan, and R. Kumar, *Extracellular biosynthesis of silver nanoparticles using the fungus Fusarium oxysporum*, Colloids Surf., B. 28 (2003), pp. 313–318.

[2] A.E. Lamb, W.N. Anderson, and R.G. Haverkamp, *The induced accumulation of gold in the plants Brassica juncea, Berkheya coddii and Brassica chicory*, Chem. New Zeal. 9 (2001), pp. 34–36.

[3] A.T. Harris and R. Bali, *On the formation and extent of uptake of silver nanoparticles by live plants*, J. Nanopart. Res. 11051 (2007), pp. 9288–9293.

[4] B. Ankamwar, M. Chaudhary, and M. Sastry, *Gold nanotriangles biologically synthesized using Tamarind leaf extract and potential application in vapour sensing*, Synth. React. Inorg. Met.-Org. Nano-Metal Chem. 35 (2005), pp. 19–26.

[5] J.Y. Song and B.S. Kim, *Rapid biological synthesis of silver nanoparticles using plant leaf extracts*, Bioprocess Biosyst. Eng. 32 (2009), pp. 79–84.

[6] J.L. Gardea-Torresedey, E. Gombez, J.G. Parsons, J. Peralta-Videa, P. Santiago, K.J. Torresday, H.E. Troiani, and J.S. Yacaman, *Formation and growth of Au nanoparticles inside live alfalfa plants*, Nano Lett. 2 (2002), pp. 397–401.

[7] J.L. Gardea-Torresedey, E. Gomez, M. Jose-Yacaman, J.G. Parsons, J.R. Peralta-Videa, and H. Tioani, *Alfalfa sprouts: A natural source for the synthesis of silver nanoparticles*, Langmuir 19 (2003), pp. 1357–1361.

[8] S.P. Chandran, M. Chaudhary, R. Pasricha, A. Ahmad, and M. Sastry, *Synthesis of gold nanotriangles and silver nanotriangles using Aloe vera plant extract*, Biotechnol. Prog. 22 (2006), pp. 577–579.

[9] J. Huang, C. Chen, N. He, J. Hong, Y. Lu, L. Qingbiao, W. Shao, D. Sun, X.H. Wang, Y. Wang, and X. Yiang, *Biosynthesis of silver and gold nanoparticles by novel sun dried Cinnamomum camphora leaf*, Nanotechnology 18 (2007), pp. 105–106.

[10] S. Li, Y. Shen, A. Xie, X. Yu, L. Qiu, L. Zhang, and Q. Zhang, *Green synthesis of silver nanoparticles using Capsicum annuum L. extract*, Green Chem. 9 (2007), pp. 825–858.

[11] S.S. Shankar, A. Ahmad, A. Rai, and M. Sastry, *Rapid synthesis of Au, Ag and bimetallic Au core-Ag shell nanoparticles by using neem (Azadirachta indica) leaf broth*, J. Colloid Interface Sci. 275 (2004), pp. 496–502.

[12] S.S. Shankar, A. Ahmed, B. Akkamwar, M. Sastry, A. Rai, and A. Singh, *Biological synthesis of triangular gold nanoprisms*, Nature 3 (2004), pp. 482–488.

[13] V. Armendariz, J.L. Gardea-Torresedey, J. Gonzalez, I. Herrera, M. Jose-Yacaman, and J.G. Parsons, *Gold nanoparticle formation by oat and wheat biomasses*, Waste Research Technology, Proceedings of 2002 Conference on Application of Waste Resources, Kansas City, MO, 2002, pp. 224–232.

[14] N. Mude, A. Ingle, A. Gade, and M. Rai, *Synthesis of silver nanoparticles by the callus extract of Carica papaya: A first report*, J. Plant Biochem. Biotechnol. 18 (2009), pp. 83–86.

[15] V. Kumar and S.K. Yadav, *Plant-mediated synthesis of silver and gold nanoparticles and their applications*, J. Chem. Technol. Biotechnol. 84 (2009), pp. 151–157.

[16] H. Bar, D.K. Bhui, G.P. Sahoo, P. Sarkar, S. Pyne, and A. Misra, *Green synthesis of silver nanoparticles using seed extract of Jatropha curcas*, Colloids Surf. A: Physicochem. Eng. Asp. 348 (2009), pp. 212–216.

[17] G.D. Wright, *Bacterial resistance to antibiotics: Enzymatic degradation and modifications*, Adv. Drug Delivery Rev. 57 (2005), pp. 1451–1470.

[18] J.E. Bandow, H. Brotz, L.I.O. Leichert, H. Labischinski, and M. Hecker, *Proteomic approach to understanding antibiotic action*, Antimicrob. Agents Chemother. 47 (2003), pp. 948–955.

[19] A. Ingle, A. Gade, S. Pierrat, C. Sonnichsen, and M. Rai, *Mycosynthesis of silver nanoparticles using the fungus Fusarium acuminatum and its activity against some human pathogenic bacteria*, Curr. Nanosci. 4 (2008), pp. 141–144.

[20] M.K. Rai, A.P. Yadav, and A.K. Gade, *Silver nanoparticles as a new generation of antimicrobials*, Biotechnol. Adv. 27(1) (2009), pp. 76–82.

[21] A.K. Gade, P. Bonde, A.P. Ingle, P.D. Marcato, N. Duran, and M.K. Rai, *Exploitation of Aspergillus niger for synthesis of silver nanoparticles*, J. Biobased Mater. Bioenergy 2 (2008), pp. 243–247.

[22] N. Duran, P.D. Marcarto, G.I.H. De Souza, O.L. Alves, and E. Esposito, *Antibacterial effect of silver nanoparticles produced by fungal process on textile fabrics and their effluent treatment*, J. Biomed. Nanotechnol. 3 (2007), pp. 203–208.

[23] Y. Li, P. Leung, Q.W. Song, and E. Newton, *Antimicrobial effects of surgical masks coated with 869 nanoparticles*, J. Hosp. Infect. 62 (2006), pp. 58–63.

[24] A.N. Kesari, R.K. Gupta, and G. Watal, *Hypoglycemic effects of Murraya koenigii on normal and alloxan-diabetic rabbits*, J. Ethnopharmacol. 97(2) (2005), pp. 247–251.

[25] R. Sanghi and P. Verma, *Biomimetic synthesis and characterization of protein capped silver nanoparticles*, Bioresour. Technol. 100 (2009), pp. 501–504.

[26] S.S. Birla, V.V. Tiwari, A.K. Gade, A.P. Ingle, A.P. Yadav, and M.K. Rai, *Fabrication of silver nanoparticles by Phoma glomerata and its combined effect against Escherichia coli, Pseudomonas aeruginosa and Staphylococcus aureus*, Lett. Appl. Microbiol. 48 (2009), pp. 173–179.

[27] I. Montes-Burgos, D. Walczyk, P. Hole, J. Smith, I. Lynch, and K. Dawson, *Characterisation of nanoparticle size and state prior to nanotoxicological studies*, J. Nanopart. Res. 12 (2010), pp. 47–53.

[28] A.R. Shahverdi, A. Fakhimi, H.R. Shahverdi, and S. Minanian, *Synthesis and effect of silver nanoparticles on the antibacterial activity of different antibiotics against S. aureus and E. coli*, Nanomedicine 3 (2007), pp. 168–171.

[29] S. Pal, Y.K. Tak, and J.M. Song, *Does the antibacterial activity of silver nanoparticles depend on the shape of the nanoparticle? A study of the Gram-negative bacterium Escherichia coli*, Appl. Environ. Microbiol. 73 (2007), pp. 1712–1720.

14

Large-scale synthesis of flower-like Te nanocrystals with uniform branches by a surfactant-assisted method

Xingbao Wang[a]* and Weipeng Guan[b]

[a]*Xinhai Senior High School, Lianyungang, Jiangsu 222006, China;*
[b]*Zhejiang Great Southeast Plastic Group Corp., Zhuji, Zhejiang 311809, China*

Flower-like tellurium (Te) nanocrystals with uniform branches were successfully prepared in large quantities by a surfactant-assisted method. The product was characterised with X-ray diffraction (XRD), scanning electron microscopy (SEM), transmission electron microscopy (TEM), selected area electronic diffraction (SAED) and energy dispersive X-ray spectroscopy (EDS). The characterisation results revealed that the branch of flower-like Te was grown along [001] direction. Furthermore, the Raman spectrum of the flower-like Te with uniform branches was studied. The result shows that the flower-like Te exhibits three typical vibration peaks. This study provides a simple method to prepare three-dimensional Te nanostructures in large scale, which broadens their practical applications.

Keywords: tellurium; semiconductors; microstructure; morphology; surfactant-assisted

1. Introduction

In past decades, two- or three-dimensional nanostructures have been the focus of current research on micro-/nanomaterials because properties and applications of the materials are dependent on the spatial arrangement and orientation of the nanocrystals [1,2]. To date, synthesis of complex micro/nanostructures, including fractal patterns, dendrites and other hierarchical structures, has attracted great attention because studies on these hyper-branched structures are useful to fabricate electronic or photonic nanodevices [3]. Recently, many effective methods of synthesising functional materials with complex three-dimensional nanostructures have been developed [4]. However, in most cases, for the fabrication of such superstructures, the employment of appropriate organic ligand or surfactant is generally necessary as the shape modifier [5].

One of the most attractive materials is tellurium (Te) because it exhibits many fascinating properties, for example, semiconductivity, thermoelectricity, nonlinear optical responses, unique photoconductivity, high piezoelectricity and catalytic activity toward some reactions [6–10]. In the past few decades, Te with different morphologies such as

*Corresponding author. Email: xingbaowang@126.com

nanotubes, nanorods, nanowires, nanobelts, feather-like and flower-like nanostructures, etc. have been synthesised [11–18]. However, the exploration of 3D complex Te structures based on 1D nanostructures is still an attractive field, and it is still challenging to control the morphologies of 3D Te nanostructures over a wide range. To our best knowledge, flower-like Te with uniform branches has not been prepared in large scale. Herein, 3D flower-like Te nanocrystals with uniform branches have been synthesised on a large scale by a surfactant-assisted biphasic solvothermal approach. Our synthesis employs diethyldithiocarbamato tellurium (IV) (TDEC) dissolved in chloroform as Te source, with 2, 2′-dithiodibenzoic acid (DTBA) and cetyltrimethyl ammonium bromide (CTAB) dissolved in NaOH solution layered above the aqueous solution as reducing agent and surfactant, respectively.

2. Experimental section

2.1. *Synthesis of flower-like Te with uniform branches*

Flower-like Te nanocrystals with uniform branches were prepared using a chemical process similar to that described by Wang et al. [16]. The difference lies in the use of surfactant CTAB. A cationic surfactant (CTAB) was chosen because it was expected to have strong interaction with the anionic surfaces of Te crystals [17]. Diethyldithiocarbamato tellurium (IV) (TDEC) used in this experiment was of commercial grade and was recrystallised twice from high-purity chloroform prior to use. DTBA, chloroform ($CHCl_3$), sodium hydroxide (NaOH) and cetyltrimethyl ammonium bromide (CTAB) of analytical grade were used in this study without further purification. In a typical procedure, 0.13 mmol of TDEC, 27.7 mL of $CHCl_3$, 0.65 mmol of DTBA, 4.3 mL of 0.93 mol L^{-1} NaOH and 0.4 g of CTAB were added in sequence up to 62% of the total volume without stirring. The autoclave was sealed and maintained at 130°C for 3 h, and then cooled to room temperature naturally. The obtained black solids were collected from the interface of two immiscible solvents by centrifuging the reaction mixture. The products were then washed with distilled water and absolute ethanol several times and dried in vacuum at room temperature for several hours before further characterisation.

2.2. *Characterisation*

The morphology was examined with a FEI Nova Nanosem 200 scanning electron microscope (SEM), a transmission electron microscope (TEM) performed on a Hitachi H-800 TEM at an accelerating voltage of 75 kV, and a high-resolution transmission electron microscope (HRTEM) (JEOL-2010) operated at an acceleration voltage of 200 kV. The Chemical composition was confirmed by recording energy dispersive X-ray analysis (EDS), while the structural information was obtained using a Bruker D8 Advance X-ray diffractometer (XRD) employing a Cu Kα radiation source. The Raman spectrum was recorded on a Renishaw Model 1000 Raman spectrometer, with 514.5 nm radiation from a 20 mW air-cooled argon ion laser being used as the exciting source.

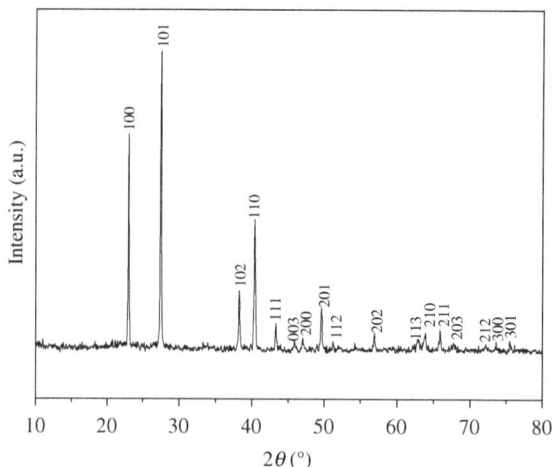

Figure 1. XRD pattern of the as-prepared sample for the reaction time of 3 h.

3. Results and discussion

Figure 1 shows that the product is well crystallised elemental Te. All the diffraction peaks can be indexed as hexagonal Te phase (JCPDS 36-1452), and no other peaks can be found. It can be concluded that high purity of hexagonal Te was obtained by the present synthetic method.

Figure 2(a) shows a panoramic SEM image of the as-synthesised products. As shown in Figure 2(a), the sample is composed of 3D nanomaterials with abundant yield. Further observation based on the high-magnification image (Figure 2(b)) reveals that the flower-like nanostructures are composed of uniform nanorods with lengths of $4 \sim 7 \,\mu m$ and diameters of about 200 nm. It should be noted that the structure of nanorod is different from the previous report (i.e. needle-like nanorods) [16]. It is well known that the presence of CTAB can influence the nanocrystal growth through chemically selective adsorption onto certain crystal faces of target materials [17]. In the absence of surfactant, needle-like t-Te nanorods grow radically from the core because of the inherent anisotropic growth of Te [16]. The surfactant was absolutely necessary for the generation of flower-like Te with uniform branches. A representative TEM image of a small flower-like Te is shown in Figure 2(c). The SAED pattern of single nanorod (inset of Figure 2(d)) reveals several diffraction points, which can be indexed to hexagonal Te. Figure 2(d) shows the HRTEM image recorded on the nanorod, which could be verified, the preferred growth direction and the single crystalline nature of the nanorods. Observed interplanar spacing is about 0.60 nm, which corresponds to the separation between (001) lattice planes of hexagonal Te. This means that the nanorods might grow along a preferred direction of [001].

EDS analysis was used to analyse the composition of the nanostuctures (Figure 3). Strong Te peaks undoubtedly confirmed that the product is Te. Cu and C peaks come from the carbon-coated copper grid, which is used as sample supporter during measurement process.

The Raman scattering spectrum performed on the synthesised flower-like Te nanocrystals is depicted in Figure 4. The sharp Raman lines for the sample shows the

Figure 2. (a) Low-magnification SEM image of the as-prepared flower-like Te nanocrystals with uniform branches. (b) High-magnification SEM image of flower-like Te nanocrystals. (c) TEM image of flower-like Te nanocrystals. (d) HRTEM image of a single nanorod. The inset shows corresponding SAED pattern obtained from the single nanorod.

presence of hexagonal Te with a characteristic band at 96.5, 124.8 and 144.9 cm^{-1}, respectively, which are close to the results reported previously on the Te nanotubes [18]. No other bands are detected by the Raman spectra. It is worth noting that the changes of Raman peaks positions and line-shapes have not been observed, which is different from the needle-like Te nanorods crystallinity [16]. This might be attributed to the diameter of Te nanorods are uniform in the present case.

4. Conclusions

In summary, flower-like Te nanocrystals with uniform branches in large scale were prepared through a surfactant-assisted method using CTAB as the surfactant. According to SEM and TEM observations, the as-obtained Te crystals are flower-shaped. Based on the determination of XRD, HRTEM and EDS analysis, it was confirmed that the obtained sample was mainly Te in hexagonal phase with a preferential orientation along

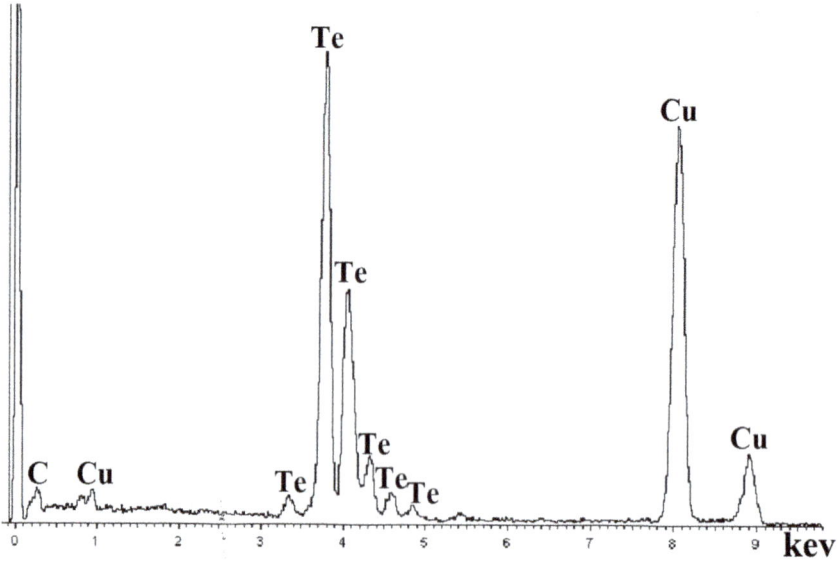

Figure 3. EDS analysis of the flower-like Te nanocrystals.

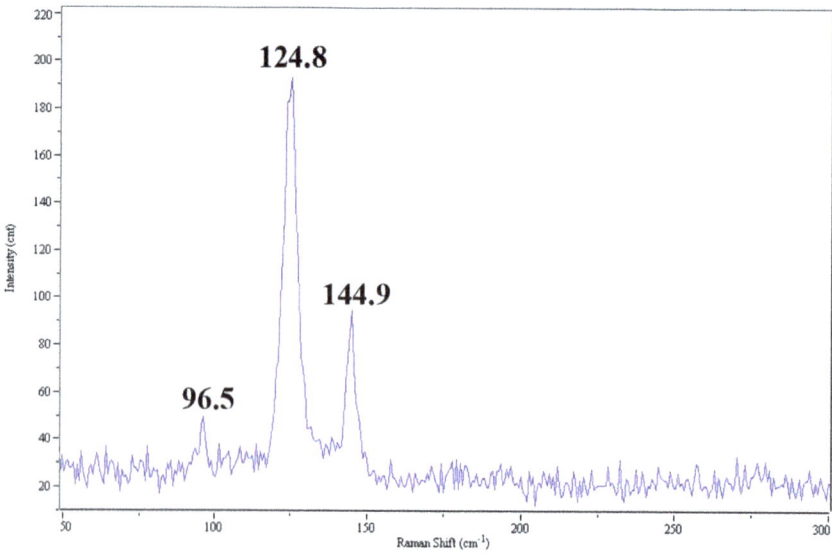

Figure 4. Raman scattering spectrum of flower-like Te nanocrystals prepared by biphasic solvothermal method at 130°C for 3 h.

[001] direction. In addition, Raman spectroscopy performed on one branch also shows that the Te nanorod displayed three characteristic peaks. The obtained Te crystals with special morphology can be potentially useful for the fabrication of nanodevices with novel property.

Acknowledgements

This work was supported through Graduate Foundation of Wenzhou University (No. 3160603601010724).

References

[1] R. Maoz, E. Frydman, S.R. Cohen, and J. Sagiv, *Constructive nanolithography: Site-defined silver self-assembly on nanoelectrochemically patterned monolayer templates*, Adv. Mater. 12 (2000), pp. 424–429.

[2] F. Gao, Q.Y. Lu, and D.Y. Zhao, *Controllable assembly of ordered semiconductor Ag_2S nanostructures*, Nano Lett. 3 (2003), pp. 85–88.

[3] R. Qiu, X.L. Zhang, R. Qiao, Y. Li, Y.I. Kim, and Y.S. Kang, *CuNi dendritic material: Synthesis, mechanism discussion, and application as glucose sensor*, Chem. Mater. 19 (2007), pp. 4174–4180.

[4] C. Burda, X. Chen, R. Narayanan, and M.A. El-Sayed, *Chemistry and properties of nanocrystals of different shapes*, Chem. Rev. 105 (2005), pp. 1025–1102.

[5] W.P. Lim, H.Y. Low, and W.S. Chin, *From winter snowflakes to spring blossoms: Manipulating the growth of copper sulfide dendrites*, Cryst. Growth Des. 7 (2007), pp. 2429–2435.

[6] P. Tangney and S. Fahy, *Density-functional theory approach to ultrafast laser excitation of semiconductors: Application to the A1 phonon in tellurium*, Phys. Rev. B 65 (2002), pp. 054302–054314.

[7] I. Shih and C.H. Champness, *Czochralski growth of tellurium single crystals*, J. Cryst. Growth 44 (1978), pp. 492–498.

[8] J. Beauvais, R.A. Lessard, P. Galarneau, and E.J. Knystautas, *Self-developing holographic recording in Li-implanted Te thin films*, Appl. Phys. Lett. 57 (1990), pp. 1354–1356.

[9] A.W. Zhao, C.H. Ye, G.W. Meng, L.D. Zhang, and P.M. Ajayan, *Tellurium nanowire arrays synthesized by electrochemical and electrophoretic deposition*, J. Mater. Res. 18 (2003), pp. 2318–2322.

[10] J. Lu, Y. Xie, F. Xu, and L. Zhu, *Study of the dissolution behavior of selenium and tellurium in different solvents-a novel route to Se, Te tubular bulk single crystals*, J. Mater. Chem. 12 (2002), pp. 2755–2761.

[11] B. Zhang, W.Y. Hou, X.C. Ye, S.Q. Fu, and Y. Xie, *1D tellurium nanostructures: Photothermally assisted morphology-controlled synthesis and applications in preparing functional nanoscale materials*, Adv. Funct. Mater. 17 (2007), pp. 486–492.

[12] Q. Wang, G. Li, Y. Liu, S. Xu, K. Wang, and J. Chen, *Fabrication and growth mechanism of selenium and tellurium nanobelts through a vacuum vapor deposition route*, J. Phys. Chem. C 111 (2007), pp. 12926–12932.

[13] P. Mohanty, T. Kang, B. Kim, and J. Park, *Synthesis of single crystalline tellurium nanotubes with triangular and hexagonal cross sections*, J. Phys. Chem. B 110 (2006), pp. 791–795.

[14] Z. Lin, Z. Yang, and H. Chang, *Preparation of fluorescent tellurium nanowires at room temperature*, Cryst. Growth Des. 8 (2008), pp. 351–357.

[15] U.K. Gautama and C.N.R. Rao, *Controlled synthesis of crystalline tellurium nanorods, nanowires, nanobelts and related structures by a self-seeding solution process*, J. Mater. Chem. 14 (2004), pp. 2530–2535.

[16] S. Wang, W. Guan, X. Chen, D. Ma, L. Wang, S. Huang, and J. Wang, *Synthesis, characterization and optical properties of flower-like tellurium*, Cryst. Eng. Comm. 12 (2010), pp. 166–171.

[17] J. Li, J. Zhang, and Y. Qian, *Surfactant-assisted synthesis of bundle-like nanostructures with well-aligned Te nanorods*, Solid State Sci. 10 (2008), pp. 1549–1555.

[18] J.M. Song, Y.Z. Lin, Y.J. Zhan, Y.C. Tian, G. Liu, and S.H. Yu, *Superlong high-quality tellurium nanotubes: Synthesis, characterization, and optical property*, Cryst. Growth Des. 8 (2008), pp. 1902–1908.

15

Growth control of TiO$_2$ nanotubes in different physical environments

Patricia M. Perillo* and Daniel F. Rodriguez

Comisión Nacional de Energía Atómica, CAC, Grupo MEMS, Av. Gral. Paz 1499 (1650) Bs. As., Argentina

TiO$_2$ nanotubes were fabricated by anodic oxidation of pure titanium substrate in glycerol containing fluoride. Anodisation of the TiO$_2$ foils was carried out in three different conditions: stirring the electrolyte, in a stationary state and air bubbling. For the three conditions mentioned above the formation and dimensions of the resulting nanotubes are detailed. Stirring the electrolyte has proved to be advantageous in comparison with other conditions like air bubbling and stationary state.

Keywords: TiO$_2$; nanotubes; anodic oxidation

1. Introduction

TiO$_2$ nanotubes have received significant technological interest due to the wide range of applications and the feasibility to produce highly ordered TiO$_2$ in recent years, for example their use in gas sensors [1–3] and photovoltaic cells [4].

Many approaches have been developed for fabricating TiO$_2$ nanotubes such as templating synthesis [5], hydrothermal treatment [6], sol–gel methods [7,8] and anodic oxidisation [9,10]. Among them, the anodic oxidisation is a relatively simple technique that can be easily automated to fabricate high-oriented uniform nanotube arrays [11–14].

The titania nanotubes with a high crystallinity and high surface area are expected to be promising materials with noble photocatalytical [15–19], electrical and optical properties [20,21]. Their wettability and biocompatibility further makes them important for biomedicine applications [22–24].

Recently some articles reported on the formation of self-organised TiO$_2$ nanotubes arrays with an ultrahigh aspect ratio [25,26] in a viscous organic electrolyte such as ethylene glycol or glycerol by anodisation of Ti in fluorine containing electrolyte [27,28].

2. Experimental techniques

Sheets of commercially pure grade titanium (99.8%) have been used as electrode/substrate for the nanotube growth [9,10]. The samples have dimensions of 15 mm × 10 mm × 0.5 mm thick. Prior to the experiments, the titanium foils were polished using 320, 400 and 600 mesh sandpaper, then the foils were etched some seconds in HF 1:20 diluted solution and cleaned in HCl diluted solution, afterwards rinsed with deionised water and dried in a nitrogen stream. The process was conducted in a Teflon electrochemical cell with a platinum foil (1 cm^2) as cathode and titanium foil as anode at a constant potential with a dc power supply. The growth of the nanotube arrays

*Corresponding author. Email: perillo@cnea.gov.ar

has been obtained using a glycerol solution with 0.25% ammonium fluoride [27,28]. An epoxi lacquer was used to mask the samples, leaving an area of 1 cm^2 in contact with the solution. The electrolyte temperature was maintained at room temperature. Potentiostatic anodisation of the TiO_2 foils was carried out in three different conditions: (a) with stirring using a magnetic stirring bar (20 mm long) at 100 rpm (referred as SAT), (b) no stirring in a stationary electrolyte and (c) with synthetic air bubbling into the solution.

Anodising was performed by applying a sweep rate of 13.8 mVs^{-1} from 0 to 50 V and then held at a constant potential for 3 h using a 6032A Hewlett Packard system power supply. A Keithley 2000 multimeter was used to measure the current. The distance between the two electrodes was kept at 2 cm in all the experiments.

After the electrochemical treatment, the samples were rinsed with deionised water and dried with a stream of nitrogen. Then the nanotube films were annealed in air at 550°C for 2 h.

The samples were characterised by scanning electron microscopy (SEM), X-ray diffraction (XRD) and Raman spectroscopy. SEM (FEI model QUANTA 200, 25 HV) was employed for the morphological charactherisation of the TiO_2 nanotubes samples.

XRD patterns were obtained with a Philips diffractometer model PW3710 using Co-Kα radiation of 0.1789 nm in wavelength at 40 kV, 30 mA. Raman measurements were carried out in order to examine the crystalline form of TiO_2 nanotube arrays at room temperature using a Horiba Jovin Yvon model Labram HR Raman system. The Raman spectra were recorded in a backscattering configuration using an exciting wavelength of 633 nm from HeNe laser complemented by Ar$^+$.

3. Results and discussion

In samples without stirring (stationary state), it is observed that the surface of the film with nanotubes is not uniform. The size of the nanotubes are different: near the surface of the liquid, the nanotubes are larger (in this case the size are similar to those prepared with stirring), but farther from the surface of the liquid, the size of the nanotubes are lower. Figure 1(a) shows the film near the surface of the liquid, where the size of the nanotubes was of 180 nm, Figure 1(b) corresponds to the middle of the sample, the size of the nanotubes was a diameter approximated of 130 nm and in Figure 1(c) far away of the surface of the liquid, the size of the nanotubes was 45 nm. Near the surface of the liquid there is greater amount of oxygen due to the contact with the air, therefore nanotubes are compact. Only porous rather than tubular structures could be observed at the position far from this interface without stirring.

In experiments prepared with stirring, we can see that the nanotubes have the same size in all the surface of the foil exposed. The size of the nanotubes was a diameter of 150 nm and the wall thickness has a dimension of 30 nm (Figure 2). Self ordered arrays of titanium oxide nanotubes were obtained by stirring the solution. Similar results were obtained by other authors [29,30]. It is highly possible that the effect of stirring is to improve the dissolve or transfer the precipitation from the electrodes into electrolyte. This might be consistent with what Figure 6 demonstrates that the resistance is reduced as stirring employed in the anodising process, which is due to the reduction of precipitation amount inside electrodes with stirring [31].

In experiments with air bubbles, the surface is uniform and the size of the nanotubes is similar but are more compact than stirring. The average size of the nanotubes is 120 nm. A sponge-like porous structure is observed in certain parts of the surface (Figure 3). The wall thickness of nanotubes is essentially the same regardless of stirring and bubbling. In the case of nanotubes fabricated with bubbling, it is difficult to estimate the thickness of the wall, because the structure is compact. Stirring of the electrolyte can have a remarkable effect on the anodic oxidation and tubular features compared with those produced from stationary electrolyte.

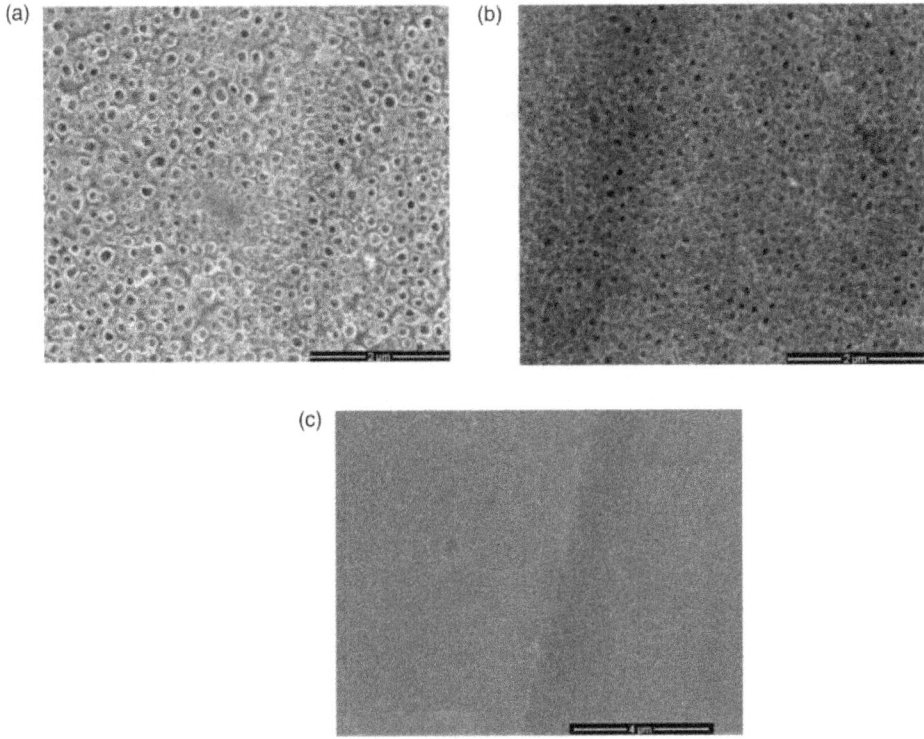

Figure 1.   SEM top images of anodised TiO_2 grown in an stationary state (a) near the surface of the liquid, (b) middle of the sample and (c) far away of the surface of the liquid.

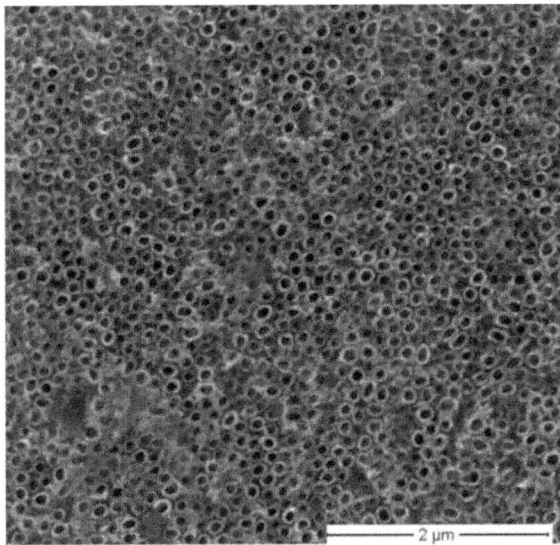

Figure 2.   Nanotubes prepared with stirring.

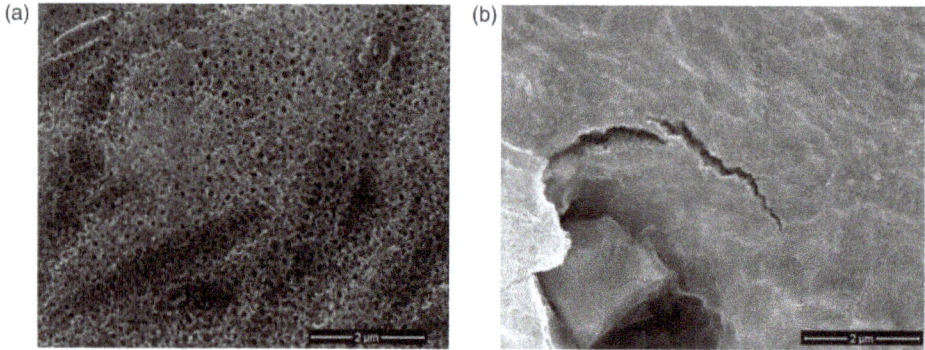

Figure 3.   Nanotubes prepared with bubbling (a) compact structure and (b) spongelike porous structure.

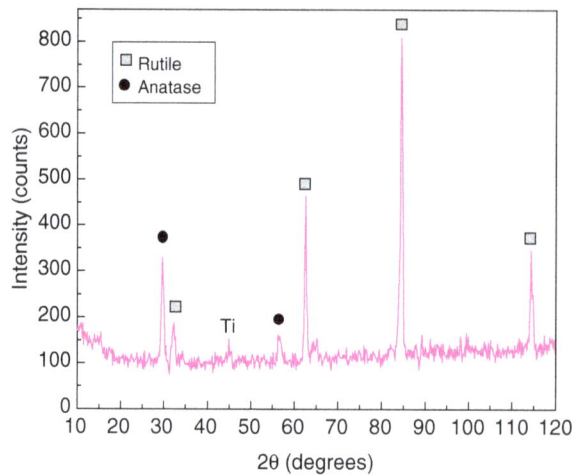

Figure 4.   XRD patterns of the TiO$_2$ anodised.

The corresponding XRD patterns of the Ti anodised after heat treatment are shown in Figure 4. The structure evidently consists of a mixed phase of anatase and rutile. This mixture of phases depends strongly on heat treatment and these results are consistent with the literature [32,33].

Raman spectra of specimens anodised is shown in Figure 5. The Raman peaks exhibited at 143, 196, 197, 393–395, 517–521 and 636–638 cm^{-1} demonstrate the formation of anatase TiO$_2$ after annealing at 550°C. The rutile phase is not present in the spectrum because it is present in small amounts with respect to the anatase.

Figure 6 shows the current density versus time for the anodising process in three different conditions: (a) stirring, (b) no stirring and (c) bubbling. The current density during anodising varies in the range of 0.03–0.5 mA cm^{-2}. The oscillations of current density during the process is due to passivation and repassivation [34]. The current density is higher with stirring than without it. The current density decreases at the beginning of the plateau (constant voltage), that is the resistance increases due to the growth of nanotubes. This variation is larger in the case of bubbling possibly because the amount of dissolved oxygen is higher in the solution, and this explains why obtained nanotubes are more compact [35].

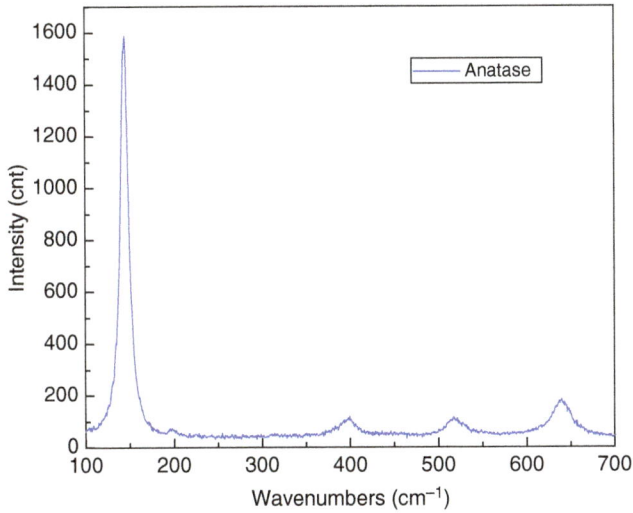

Figure 5. Raman spectra of annealed TiO$_2$ nanotube arrays.

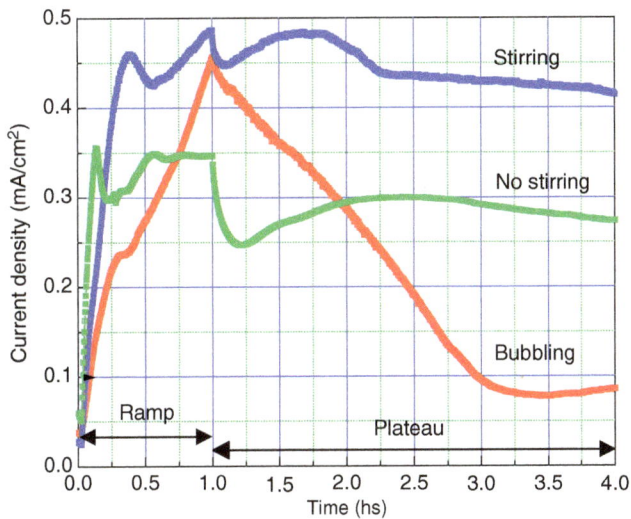

Figure 6. Comparison of the current density in function of the time.

4. Conclusions

It can be concluded that: anodic oxidation should be made with stirring, so that the film of nanotubes are of uniform size. To summarise, stirring of the glycerol-based electrolyte has proved to be advantageous in retaining the tubular structure and providing smooth tubes in comparison with other oxidation conditions like (a) unstirred or (b) bubbling with synthetic air. In this work, agitation of the electrolyte improved the currents and the quality of the tubes.

Acknowledgements

The authors would like to acknowledge Adriana Dominguez and Pablo Reynoso Peitsch for technical help.

References

[1] M. Paulose, O K. Varghese, G.K. Mor, C A. Grimes, and K.G. Ong, *Backside illuminated dye-sensitized solar cells based on titania nanotube array electrode*, Nanotechnology 17 (2006), pp. 398–402.

[2] S.L. Dongrong Li, J.W. Xiaogan Li, and S.A. Akbar, *A selective room temperature formaldehyde gas sensor using TiO₂ nanotube*, Sensor Actuat B 156 (2011), pp. 505–509.

[3] E. Sxennik, Z. Colak, N. Kilinc, and Z. Ziya Oüzturk, *Synthesis of highly-ordered TiO₂ nanotubes for a hydrogen sensor*, Int. J. Hydr. Energy 35 (2010), pp. 4420–4427.

[4] G.K. Mor, K. Shankar, M. Paulose, O.K. Varghese, and C A. Grimes, *Use of highly-ordered TiO₂ nanotube arrays in dye-sensitized solar cells*, Nano Lett. 6 (2006), pp. 215–218.

[5] S. Lee, Ch. Jeon, and Y. Park, *Fabrication of TiO₂ tubules by template synthesis and hydrolysis with water vapor*, Chem. Mater. 16(22) (2004), pp. 4292–4295.

[6] D. Wang, F. Zhou, Y. Liu, and W. Liu, *Synthesis and characterization of anatase TiO₂ nanotubes with uniform diameter from titanium powder*, Mater. Lett. 62 (2008), pp. 1819–1822.

[7] T. Kasuga, M. Hiramatsu, A. Hoson, T. Sekino, and K. Niihara, *Formation of titanium oxide nanotube*, Langmuir 14 (1998), pp. 3160–3163.

[8] J.H. Jung, H. Kobayashi, K J.C. van Bommel, S. Shinkai, and T. Shimizu, *Creation of novel helical ribbon and double-layered nanotube TiO₂ structures using an organogel template*, Chem. Mater. 14 (2002), pp. 1445–1447.

[9] D. Gong, C.A. Grimes, O.K. Varghese, W. Hu, R.S. Singh, Z. Chen, and E.C. Dickey, *Titanium oxide nanotube arrays prepared by anodic oxidation*, J. Mater. Res. 16 (2001), pp. 3331–3334.

[10] R. Ge, W. Fu, H. Yang, Y. Zhang, W. Zhao, Z. Liu, C. Wang, H. Zhu, Q. Yu, and G. Zou, *Fabrication and characterization of highly-ordered titania nanotubes via electrochemical anodization*, Mater. Lett. 62 (2008), pp. 2688–2691.

[11] F. Mura, A. Masci, M. Pasquali, and A. Pozio, *Effect of a galvanostatic treatment on the preparation of highly ordered TiO₂ nanotubes*, Electrochim. Acta 54 (2009), pp. 3794–3798.

[12] J.M. Macak, H. Tsuchiya, and P. Schmuki, *High-aspect-ratio TiO₂ nanotubes by anodization of titanium*, Angew. Chem. Int. Ed. 44 (2005), pp. 2100–2102.

[13] S.K. Mohapatra, K.S. Raja, M. Misra, V.K. Mahajan, and M. Ahmadian, *Synthesis of self-organized mixed oxide nanotubes by sonoelectrochemical anodization of Ti–8Mn alloy*, Electrochim. Acta 53 (2007), pp. 590–597.

[14] H. Park and H.-G. Kim, *Characterizations of highly ordered TiO₂ nanotube arrays obtained by anodic oxidation*, Trans. Electr. Electron. Mater. 11 (3) (2010), pp. 112–115.

[15] O K. Varghese, M. Paulose, T J. LaTempa, and C A. Grimes, *High-rate solar photocatalytic conversion of CO₂ and water vapor to hydrocarbon fuels*, Nano Lett. 9(2) (2009), pp. 731–737.

[16] M. Qamar, C.R. Yoon, H J. Oh, N.H. Lee, K. Park, D H. Kim, K.S. Lee, W J. Lee, and S J. Kim, *Preparation and photocatalytic activity of nanotubes obtained from titanium dioxide*, Catal. Today 131 (2008), pp. 3–14.

[17] S. Sreekantan, R. Hazan, and Z. Lockman, *Photoactivity of anatase/rutile TiO₂ nanotubes formed by anodization method*, Thin Solid Films 518 (2009), pp. 16–21.

[18] A. Fujishima, H. Irie, and K. Hashimoto, *TiO₂ photocatalysis: A historical overview and future prospects*, Jpn. J. Appl. Phys. 44 (2005), pp. 8269–8285.

[19] Y. Zhang, J. Chen, and X. Li, *Preparation and photocatalytic performance of anatase/rutile mixed-phase TiO₂ nanotubes*, Catal. Lett. 139 (2010), pp. 129–133.

[20] F.M. Hossain, A.V. Evteev, I V. Belova, J. Nowotny, and G.E. Murch, *Electronic and optical properties of anatase TiO₂ nanotubes*, Comput. Mater. Sci. 48(4) (2010), pp. 854–858.

[21] F. Mura, A. Masci, M. Pasquali, and A. Pozio, *Stable TiO₂ nanotube arrays with high UV photoconversion efficiency*, Electrochim. Acta 55 (2010), pp. 2246–2251.

[22] L. Peng, A.D. Mendelsohn, T.J. Latempa, S. Yoriya, C.A. Grimes, and T.A. Desai, *Long-term small molecule and protein elution from TiO₂ nanotubes*, Nano Lett. 6(5) (2009), pp. 1932–1936.

[23] L. Peng, M.L. Eltgroth, T.J. LaTempa, C A. Grimes, and T A. Desai, *The effect of TiO₂ nanotubes on endothelial function and smooth muscle proliferation*, Biomaterials 30 (2009), pp. 1268–1272.

[24] S. Oh, Ch. Daraio, L.-H. Chen, T.R. Pisanic, R.R. Fiñones, and S. Jin, *Significantly accelerated osteoblast cell growth on aligned TiO₂ nanotubes*, J. Biomed. Mater. Res. Part A 78A (2006), pp. 97–103.

[25] X.-Y. Fan, Y.-H. Zhang, P. Xiao, F. Hu, and H. Zhang, *Preparation of high-orderly TiO₂ nanotubes in organic solution and characterization of C-doped TiO₂*, Chin. J. Chem. Phys. 20(6) (2007), pp. 753–758.

[26] X. Wang, S. Zhang, and L. Sun, *A Two-step anodization to grow high-aspect-ratio TiO₂ nanotubes*, Thin Solid Films 519 (2011), pp. 4694–4698.

[27] A. Valota, D J. LeClere, T. Hashimoto, P. Skeldon, G.E. Thompson, S. Berger, J. Kunze, and P. Schmuki, *The efficiency of nanotube formation on titanium anodized under voltage and current control in fluoride/glycerol electrolyte*, Nanotechnology 19 (2008), 7pp.

[28] A. Jaroenworaluck, D. Regonini, C. R. Bowen, and R. Stevens, *A microscopy study of the effect of heat treatment on the structure and properties of anodised TiO₂ nanotubes*, Appl. Surf. Sci. 256 (2010), pp. 2672–2679.

[29] R. Narayanan, T.-Y. Kwon, and K.-H. Kim, *Anodic TiO₂ from stirred Na₂SO₄/NaF electrolytes: Effect of applied voltage and stirring*, Mater. Lett. 63 (2009), pp. 2003–2006.

[30] R. Narayanan, J.-Y. Ha, T.-Y. Kwon, and K.-H. Kim, *Structure and properties of self-organized TiO₂ nanotubes from stirred baths*, Metall. Mater. Trans. B 39(3) (2008), pp. 493–499.

[31] G. Patermarakis and K. Moussoutzanis, *Mathematical models for the anodization conditions and structural features of porous anodic Al_2O_3 films on aluminum*, J. Electrochem. Soc. 142 (1995), pp. 737–743.

[32] J.M. Macak, S. Aldabergerova, A. Ghicov, and P. Schmuki, *Smooth anodic TiO_2 nanotubes: Annealing and structure*, Phys. Status Solidi (a) 203 (2006), pp. 67–77.

[33] D. Regonini, A. Jaroenworaluck, R. Stevens, and C.R. Bowen, *Effect of heat treatment on the properties and structure of TiO_2 nanotubes: Phase composition and chemical composition*, Surf. Interf. Anal. 42 (2010), pp. 139–144.

[34] Y.-X. Tang, J. Tao, Y.-Y. Zhang, T. Wu, H.-J. Tao, and Y.-R. Zhu, *Preparation of TiO_2 nanotube on glass by anodization of Ti films at room temperature*, Trans. Nonferrous Met. Soc. China 19 (2009), pp. 192–198.

[35] H.E. Prakasam, K. Shankar, M. Paulose, O.K. Varghese, and C.A. Grimes, *A new benchmark for TiO_2 nanotube array growth by anodization*, J. Phys. Chem. C111 (2007), pp. 7235–7241.

16

A facile and general route to size-controlled synthesis of metal (Ni, Co and Mn) oxide nanoparticle and their optical behavior

S. Chakrabarty and K. Chatterjee*

Department of Physics and TechnoPhysics, Vidyasagar University, Midnapore 721102, West Bengal, India

In this article, we report a general and facile route to prepare transitional metal (Ni, Co and Mn) oxide nanoparticles via soft chemical route followed by the controlled heat treatment. The method has been successfully involved to produce uniform and well-crystalline phase of the different metal oxide nanoparticles. The product materials were characterised by the X-ray powder diffraction and transmission electron microscopy to investigate the structural and morphological details. Optical absorption study reveals that the NiO and Mn_2O_3 nanoparticles show single and sharp absorption peak at 327 and 270 nm, respectively, while Co_3O_4 nanoparticles show two broad peaks centred at 524 and 770 nm. From the photoluminescence spectroscopy study, it is observed that NiO and Mn_2O_3 nanoparticles show strong emissivity at 324 and 340 nm, respectively, and Co_3O_4 nanoparticles show weak emission peak at 586 nm. The size variation of the nanoparticle has been presented taking NiO as the typical member of the transitional metal oxide family. The effect of particle size on the emission intensity of NiO nanoparticle has also been discussed.

Keywords: nanostructures; oxides; chemical synthesis; structural characterisation; optical properties

1. Introduction

Much attention is being paid in developing nanomaterials based on transitional metal oxides due to their structural flexibility combined with novel chemical and physical properties [1,2]. These properties, coming out from the electronic structure and surface to volume ratio in the nanometre regime, are strongly related to their size [3,4]. This gives the transitional metal oxide nanostructured materials great advantages over their bulk counterpart in many applications such as gas sensors, catalysts, magnetic materials, electrochromic devices and high-temperature solar selective absorbers [5–9]. Large number of potential applications of NiO, Co_3O_4 and Mn_2O_3 lead many material scientists to investigate the introduction of metal oxides into novel nano-materials with valuable properties for future applications. Realisation of these potential applications depends on the availability of size-controlled, uniform, mono-dispersed and wellcrystalline nanomaterials. Considerable efforts have been paid to synthesise individual metal oxide like NiO [10–12], Mn_2O_3 [13–15] and Co_3O_4 [16–18]. But, the work to integrate the synthesis route of different transitional metal oxide nanoparticle in a general approach, as per our knowledge, is very less attempted and those attempts have their own limitations. For example, Peng and his group introduced the method based on the pyrolysis of metal fatty acid salts to produce different metal oxide nanoparticles with size variation flexibility [19], and very recently Stein and his team reported the template-directed

---

*Corresponding author. Email: kuntal@mail.vidyasagar.ac.in

synthesis and organisation of metal oxide nanoparticles employing top-down approach [20]. But in these cases, the complexity of organo-metallic compound, the phosphate removal steps and the role of the residual ligands/surfactants in the measurement of the physical properties of those elements will always raise questions. Gedanken introduced sonochemical synthesis of transitional metal oxide nanoparticles from respective metal acetates [21]. But the process of sonochemistry arising from acoustic cavitation phenomenon is inherently complex in nature and the process needs a high-end instrument to be employed. Scientists have also followed the nonaqueous sol–gel routes to synthesise metal oxide nanoparticles to achieve uniform particle morphologies and crystallite sizes in the range of just a few nanometres utilising the manifold role of the organic components in the reaction mixture [22–24]. But, parallel to the formation of the inorganic nanoparticles, the initial organic species (i.e. solvent and organic constituent of the precursor) undergo transformation reactions that are often troublesome to be understood properly. Identification and quantification of these organic by-products and the chemical reaction mechanisms leading to the nanoparticles demand in-depth understanding of the underneath chemistry. Thomas and his group have reported a generalised hydrothermal method to prepare metal oxide, but that is also of hollow type having dimension of few micrometres [25] which lacks the novelty of nanodimension. In contrast, some easy to achieve and cost effective general method has to be achieved which will not only be very easy to understand but also provides uniform, and wellcrystalline metal oxide nanoparticle with controllable size. Another key point is that very limited work has been done to present the comparative study of the optical behaviour of transitional metal oxide nanoparticles synthesised in an identical route. The knowledge of optical and magnetic properties as a whole will help to understand the the feasibility of designing multifunctional materials out of these metal oxide nanoparticles. Though several other aspects of the transitional metal oxide nanoparticles have attracted much attention, the possibilities in their optical responses are somehow very less attempted.

Here we report an easy to achieve, cost effective, surfactant-free and generalised synthesis route of NiO, Mn_2O_3 and Co_3O_4 nanoparticles and comparative study of optical properties coming out from those metal oxide nanoparticles. In our discussion, we have demonstrated the synthesis of well-crystalline, well-dispersed, uniform nanostructured metal oxide (NiO, Mn_2O_3 and Co_3O_4) nanoparticles by a facile soft chemical route followed by controlled heat treatment. The crystalline phases of all as-prepared samples were identified by X-ray diffraction (XRD). The structure and morphological properties were investigated by transmission electron microscopy (TEM) and high-resolution TEM (HRTEM). The absorbance spectra were obtained from UV–Vis spectrophotometer. In optical absorption spectra, NiO and Mn_2O_3 nanoparticles exhibit strong absorption at 327 (3.78 eV) and 270 nm (5.54 eV), respectively, and Co_3O_4 nanoparticle shows two broad absorption peaks at 524 (2.357 eV) and 770 nm (1.6 eV). Photoluminescence spectroscopy reveals the strong emission capability of NiO and Mn_2O_3 at UV region but feeble emissivity for Co_3O_4 at visible region. Here we also report how to tailor the size of these transitional metal oxide nanoparticles taking NiO as a typical one. The effect of size on the emission property of NiO nanoparticles has also been discussed.

2. Experimental

$NiCl_2$, $MnSO_4$ and $CoCl_2$ were taken as the precursor of Ni, Mn and Co, respectively. In a typical synthesis process, metal precursor and $NaHCO_3$ were taken in equal (1:1) molar ratio and dissolved in distilled water to make two separate solutions. Metal precursor solution was stirred by magnetic stirrer for 15 min. Then the $NaHCO_3$ solution was added to the metal precursor solution dropwise with constant stirring. During the whole process, the system was kept in ice bath.

After 15 min, the products were collected by centrifugation followed by proper rinsing with distilled water. The products were dried at 100°C. The dried sample was heattreated at 600°C for 2 h in a programmable furnace to produce well-crystalline, pure and uniform metal oxide (NiO, Mn_2O_3 and Co_3O_4) nanoparticles. This annealing temperature is optimised to get well-crystalline product materials for different metal oxide samples. The temperature is optimised on the basis of the TGA analysis taken from the reported literature on NiO [26], Mn_2O_3 [27] and Co_3O_4 [28]. It is evident that 600°C temperature is in full support to provide complete oxidative decomposition of all the corresponding metal hydroxide phase in a general form. For the size variation of the particle, the metal precursor and the $NaHCO_3$ were taken in different molar ratio (1:2 and 1:5) at the very first step of the chemical reaction. The rest of the procedure is same as stated above.

All the powdered samples were characterised both by Rigaku Mini-Flex X-ray diffractometer using Cu-Kα radiation ($\lambda = 1.54178$ Å) source as well as by JEM 2010 TEM at an accelerating voltage of 200 keV. Selected area electron diffraction (SAED) was employed to examine the crystal lattice structure of the nanoparticles. Optical absorption spectra of the powdered samples were recorded in UV–Vis 1700 Shimadzu spectrophotometer. The powdered samples were dispersed in ethyl alcohol and mounted in the sample chamber, while pure ethyl alcohol was taken in the reference beam position. For photoluminescence measurement, the samples were also dispersed in ethyl alcohol and the measurements were carried out in F-7000 Hitachi PL spectrophotometer at room temperature. In the study of particle size effect on emissivity, equal amount of NiO sample was taken in 20 mL of ethyl alcohol in each case to execute photoluminescence measurement.

3. Result and discussions

Figure 1 shows the details of XRD patterns of the as-prepared NiO, Mn_2O_3 and Co_3O_4 samples in (a), (b) and (c), respectively. All the reflection peaks of the crystalline phases of the oxide samples have been indexed from the ICDD file. From the figure it is evident that the product materials are well crystalline and as no peaks remain unmatched, it can be said that all the metal oxides are in pure form without having any impurity. All the three metal oxide phases are identified as cubic system having the cell parameter $a = 4.18$ Å for NiO (bunsenite), $a = 9.4091$ Å for Mn_2O_3 (byxbyite-c) and $a = 8.056$ Å for Co_3O_4 (cobalt oxide). The ICDD card numbers are 71-4751, 41-1442 and 65-3103, respectively. The grain size of the respective metal oxide particles were calculated using the Debye–Scherrer formula:

$$D = \frac{0.9 \times \lambda}{\beta \times \cos(\theta)} \qquad (1)$$

where D: grain size, λ: wavelength of X-rays, β: full width of half maximum of diffraction peak and θ: angle corresponding to the peak. From the calculations it is found that the average grain size of NiO is 20 nm; Mn_2O_3 is 90 nm and Co_3O_4 is 45 nm.

The photo gallery of Figure 2 shows the typical TEM study of the synthesised metal oxide nanoparticles. TEM image, SAED pattern and HRTEM images of all the product materials have been shown in this figure as a whole. Figure 2(a), (d) and (g) show the distribution of different nanoparticles and from these images it is seen that the average size of the nanoparticle of NiO is $\sim$20 nm and Co_3O_4 is $\sim$40 nm, whereas the average size of the Mn_2O_3 nanoparticle is little bigger, $\sim$100 nm. The results are in well agreement for NiO and Mn_2O_3 with the calculated size of the respective nanoparticles from the XRD result using the Debye–Scherrer formula.

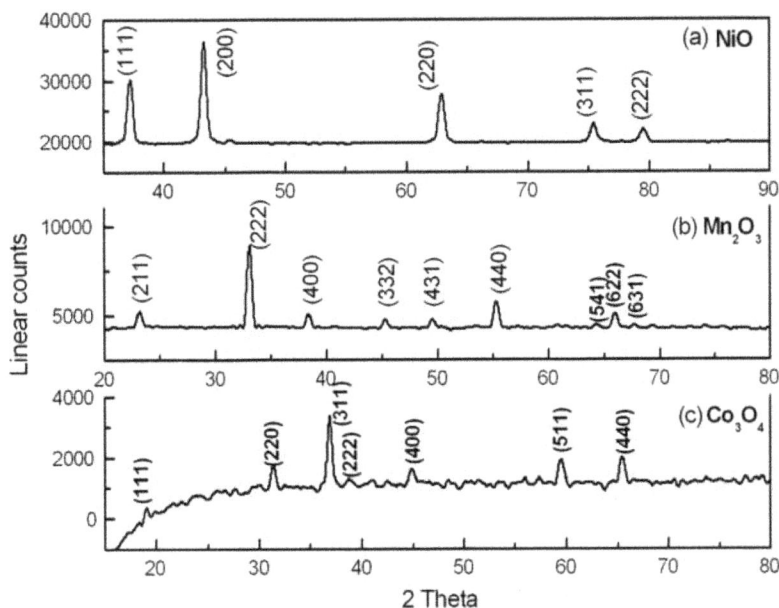

Figure 1.   XRD pattern of as synthesised metal oxide nanoparticles: (a) NiO; (b) Mn_2O_3 and (c) Co_3O_4.

The insets in Figure 2(d) and (g) show two distinct Mn_2O_3 nanoparticles and a single Co_3O_4 nanoparticle to make the sense clear about the morphology of the prepared particles. Here, it is interesting to note that the synthesised particles are in general irregular-shaped nanoparticles having sharp crystal edge which demands further study to understand the effect of reaction parameters on their growth mechanism. These kind of nonspherical sharp-edged nanoparticles have potential possibilities in the field of morphology-dependent physical properties being tuned into different shapes by changing the reaction parameter. SAED patterns, taken from a single nanoparticle in each case, shown in Figure 2(b), (e) and (h), evidently present the well-crystalline nature. The corresponding lattice planes have been shown in the respective figures. Figure 2(e), (f) and (i) represent the typical HRTEM images where the lattice fringes are shown for the three different metal oxide nanoparticles. The lattice planes calculated from the respective lattice spacing are shown within the figure. From the HRTEM result it is clear that the nanoparticles are structurally uniform, perfect and well crystalline, and have only one growth direction over their entire dimension. The crystalline uniformity, which is even better than the product obtained from the nonaqueous sol–gel techniques [23], proves the potential of this simplistic synthesis procedure.

To investigate the comparative optical properties of these much familiar magnetic metal oxide nanoparticles, we have carried out the absorption and luminescence study. Figure 3 shows the optical absorption spectra of NiO, Mn_2O_3 and Co_3O_4 nanoparticles. From Figure 3(a) and (b) it is seen that the prepared uniform, well-crystalline NiO and Mn_2O_3 nanoparticles exhibit sharp increase in the absorption spectra in the UV region (peak at $\sim$327 and 270 nm, respectively) and the materials are almost transparent for the visible range. This absorption is attributed to band gap absorption for both the cases. It is noteworthy that the absorption band edge of Mn_2O_3 nanoparticle gives about 341 nm (3.64 eV) which is even blue shifted from very recent findings of Mn_2O_3 nanoparticles prepared by sol–gel technique [29]. In the case of Co_3O_4 nanoparticles, it is seen that two broad peaks are originated at around 524 and 770 nm. These two band gap

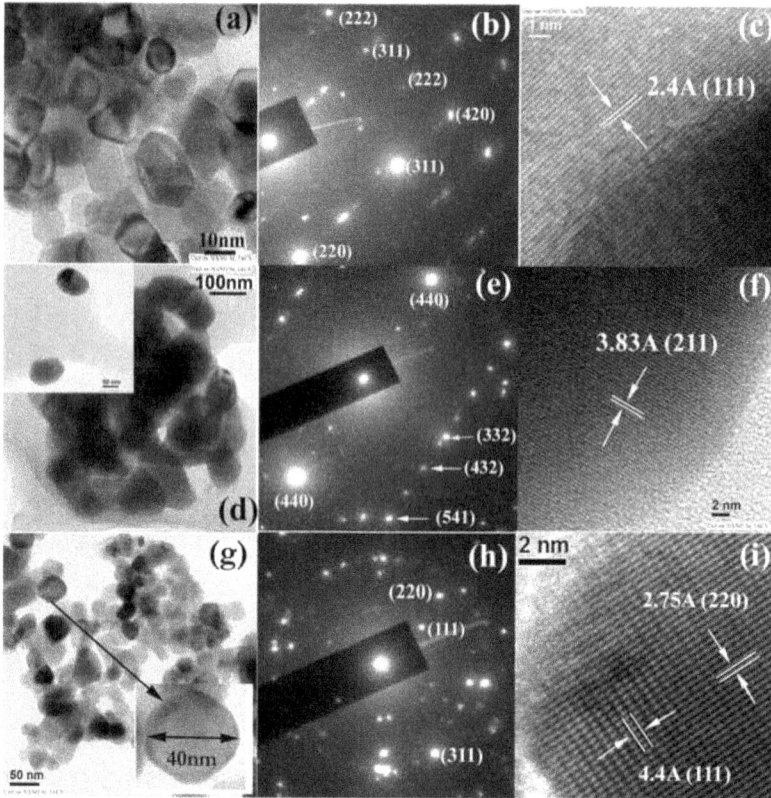

Figure 2.   Typical TEM image of the as-prepared NiO (a), Mn_2O_3 (d) and Co_3O_4 (g) nanoparticles. (d) Inset shows two distinct Mn_2O_3 nanoparticles. (b), (e) and (h) are the SAED pattern of NiO, Mn_2O_3 and Co_3O_4, respectively, taken from the nanoparticle of the respective oxide. Typical HRTEM image from a single nanoparticle of NiO (c), Mn_2O_3 (f) and Co_3O_4 (i).

absorption peaks closely support the previously recorded band energies of Co_3O_4 nanoparticles by Zhang et al. [30]. The multiple band gaps for the Co_3O_4 nanoparticles indicate the possibility of $O^{2-} - Co^{2+}$ and $O^{2-} - Co^{3+}$ charge-transfer processes in Co_3O_4 nanoparticles, as observed by Xu and Zeng [31] for Co_3O_4 nanostructure. For semiconductor material the quantum confinement comes into picture if the size of the material becomes comparable to the Bhor radius of exciton and the absorption edge is shifted to a higher frequency. Here we also get the absorption peak or the absorption edge of all the three direct band gap semiconductor materials in the wavelength range distinctly different from their bulk counterpart. In our findings, the blue-shifted absorption peak positions of NiO, Mn_2O_3 and Co_3O_4 confirm that the optical band gap energies become larger as the crystallite size decreases.

The emission nature of the prepared metal oxide nanoparticles has been examined by photoluminescence study. Figure 4 shows the emission spectra taken from three different samples. Here, to have the clear idea of excitation dependencies of emission, two emission spectra taken at two different excitation energies have been drawn for each sample. Figure 4(a) and (b) show that NiO and Mn_2O_3 nanoparticles exhibit strong emission characteristic at 324 and 340 nm, respectively. Co_3O_4 nanoparticle does not exhibit strong emission nature except a weak peak at 586 nm. As it is evident from the figure that the emission peaks are not being shifted with the

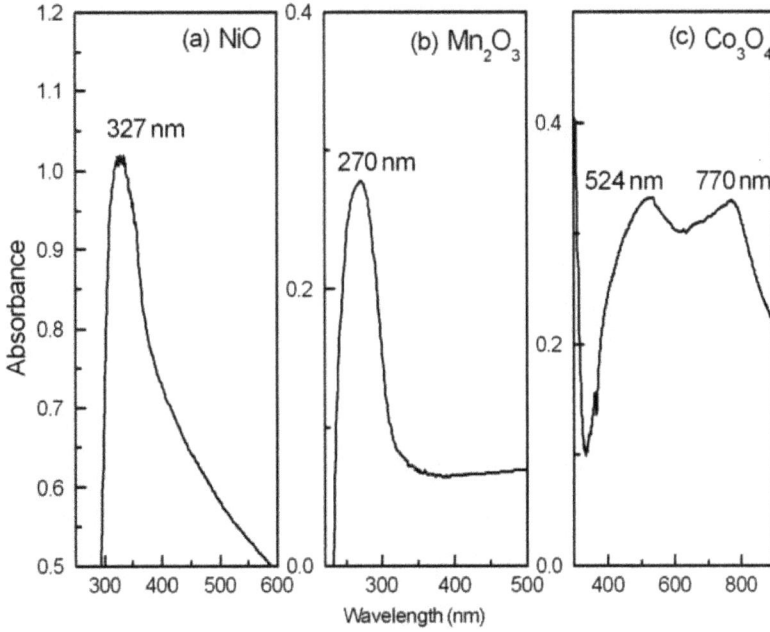

Figure 3.   Optical absorption spectra for three different as-prepared metal oxide nanoparticles: (a) NiO; (b) Mn_2O_3; and (c) Co_3O_4 taken at room temperature.

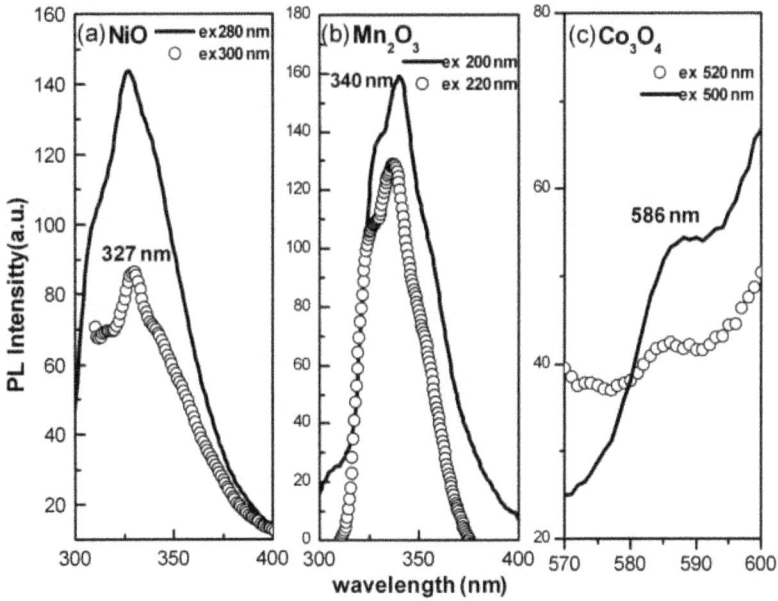

Figure 4.   Room temperature photoluminescence spectra for three different metal oxide nanoparticles taken at two different excitation energies: (a) NiO; (b) Mn_2O_3 and (c) Co_3O_4.

Molar ratio	Size from Debye - Scherrer formula	TEM images for different molar ratio	Emission spectra for NiO prepared at different molar ratio
(1:1)	17.8 nm		
(1:2)	24 nm		
(1:5)	30 nm		

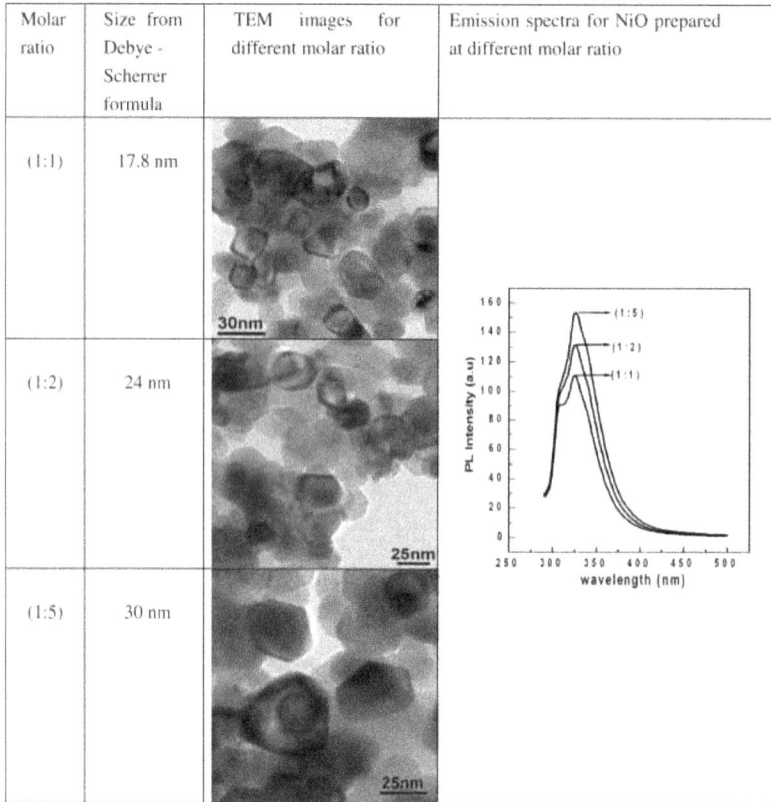

Figure 5. NiO nanoparticles with three different grain size and the respective emission spectra.

change of excitation energy, the emission are real and the origin of this emission from the 3d transitional metal oxide nanoparticles is attributed to the electronic transition from the d shell of the respective metal. Handsome number of literatures about the transition of the $3d^8$ electrons in NiO [32–34], which supports our assumption, are available. In the case of Mn_2O_3, the strong UV emission peak at 340 nm is marked differently from the emission value obtained by Gnanam and Rajendran [29]. It is commonly accepted that the UV emission should be attributed to the radiative annihilation of excitons. But it needs further study to conclude about the exact nature of these emission behaviours in the present case. It is true that highquality crystals emit UV light. The property of strong UV emission should be attributed to the high purity and perfect crystallinity of our prepared NiO and Mn_2O_3 nanoparticles. The emission of the prepared NiO and Mn_2O_3 nanoparticles in the UV region may play an effective role in the perspective of nanophotonics technology.

Next, we have extended our study to show the size variation of the metal oxide nanoparticles and their effect in emission property. For this purpose, NiO was chosen as a typical member of the transitional metal oxide family. This size variation morphology can be controlled by changing the molar ratio of the initial reactants for other metal oxides as well, but they have not been included in this report.

Figure 5 shows the size variation of the NiO nanoparticle by changing the molar ratio of the starting reactants. It is evident from the figure that the particle size increases as the amount of $NaHCO_3$ increases in the reaction. The particle size measured from the TEM is in good agreement with the calculated value of the grain size using Equation (1) from their respective XRD pattern. Figure 5 shows the emission spectra of NiO nanoparticles with different particle size. It is seen that the emission peak intensity decreases with decreasing grain size. The same kind of response for the Mn_2O_3 nanocrystals was observed earlier [35]. This kind of variation of the emission intensity is probably attributed to the fact that as the size of the NiO decreases, the surface structure increases. Consequently, the oxygen dangling bonds increases resulting stronger reactive force. Therefore, the number of intermediate energy states increases effectively. It is quite reasonable to think that the relaxation at some intermediate energy levels (such as the surface, the crystal boundary, the oxygen hole energy levels, etc.), which has higher possibility as the dimension of the nanoparticle decreases, may be responsible for the fact that emission intensity decreases with decreasing grain size in NiO nanoparticles.

This simple and general synthesis approach can be extended to the formation of other metal (Fe, Cr and Cu) oxide nanocrystals as well. Moreover, the size of the nanoparticle can be varied by changing the molar ratio of the precursor materials and the NaHCO3, which is successfully established by taking one typical example of NiO. In comparison with the synthesis of metal oxides, in the presence of surfactants, our two-step approach is considerably simpler. The initial reaction mixture just consists of two components, the metal oxide precursor and a common inorganic reactant. The small number of reactants in this synthesis technique simplifies the characterisation of the final product. The annealing temperature is typically around 600°C, which is not so high and the whole procedure does not require any high-precision instrument. But the main advantage of this surfactant-free synthesis method clearly lies in the improvement of product purity. Whereas surface-adsorbed surfactants influence the toxicity of nanoparticles and lower the accessibility of the nanoparticle surface in catalytic and sensing applications, these problems are not an issue in nanopowders obtained by the presented route.

4. Conclusions

In conclusion, a very useful, surfactant-free, generalised two-step method has been established to synthesise 3d transitional metal oxide nanoparticle. On the basis of this approach, structurally uniform, well-crystalline metal (Ni, Mn and Co) oxide nanoparticles have been grown successfully. XRD and TEM/HRTEM studies reveal that the product materials are pure, wellcrystalline sharp-edged nanoparticles having only one growth direction. The comparative optical properties of these magnetic metal oxide nanoparticles have been investigated. The optical absorption study shows sharp absorption for NiO and Mn_2O_3 nanoparticles in the UV range and two broad absorption peaks for Co_3O_4 in the visible region. The emission spectra exhibit the strong emission peak for NiO and Mn_2O_3 nanoparticles at 324 and 340 nm, respectively, while Co_3O_4 reveals feeble emission at 586 nm. The way to tailor the size of the nanoparticles has been reported taking NiO as the typical one. The effect of particle size on the emission property has been demonstrated with theoretical justification.

Acknowledgements

The authors gratefully acknowledge the financial support from FastTrack, Department of Science and Technology, India and SAP, University Grant Commission, India. The authors are also thankful to the Unit of NanoScience, IACS, Kolkata for providing the TEM facility and to the USIC, Vidyasagar University, for providing the photoluminescence spectrophotometer.

References

[1] G.J. Moore, R. Portal, A. Le Gal La Salle, and D. Guyomard, *Synthesis of nanocrystalline layered manganese oxides by the electrochemical reduction of AMnO4 (A = K, Li)*, J. Power Sources 97 (2001), pp. 393–397.

[2] M.M. Thackeray, S.H. Kang, C.S. Johnson, J.T. Vaughey, R. Benedek, and S.A. Hackney, *Li2MnO3-stabilized LiMO2 (M = Mn, Ni, Co) electrodes for high energy lithium-ion batteries*, J. Mater. Chem. 17 (2007), pp. 3112–3125.

[3] M.L. Steigerwald and L.E. Brus, *Semiconductor crystallites: A class of large molecules*, Acc. Chem. Res. 23 (1990), pp. 183–188.

[4] A.P. Alivisatos, *Perspectives on the physical chemistry of semiconductor nanocrystals*, J. Phys. Chem. 100 (1996), pp. 13226–13239.

[5] M. Ando, T. Kobayashi, S. Iijima, and M. Haruta, *Optical recognition of CO and H2 by use of gas-sensitive Au-Co3O4 composite Olms*, J. Mater. Chem. 7 (1997), pp. 1779–1783.

[6] I. Porqueras and E. Bertran, *Electrochromic behaviour of nickel oxide thin films deposited by thermal evaporation*, Thin Solid Films 398–399 (2001), pp. 41–44.

[7] Y. Ichiyanagi, N. Wakabayashi, J. Yamazaki, S. Yamada, Y. Kimishima, E. Komatsu, and H. Tajima, *Magnetic properties of NiO nanoparticles*, Physica B 329–333 (2003), pp. 862–863.

[8] Y. Zhang, Y. Chen, T. Wang, J. Zhou, and Y. Zhao, *Synthesis and magnetic properties of nanoporous Co3O4 nanoflowers*, Microporous Mesoporous Mater. 114 (2008), pp. 257–261.

[9] H. Bi, S. Li, Y. Zhang, and Y. Du, *Ferromagnetic-like behavior of ultrafine NiO nanocrystallites*, J. Magn. Magn. Mater. 277 (2004), pp. 363–367.

[10] X. Wang, J. Song, L. Gao, J. Jin, H. Zhengand, and Z. Zhang, *Optical and electrochemical properties of nanosized NiO via thermal decomposition of nickel oxalate nanofibres*, Nanotechnology 16 (2005), pp. 37–39.

[11] T. Nathan, A. Aziz, A.F. Noor, and S.R.S. Prabaharan, *Nanostructured NiO for electrochemical capacitors: Synthesis and electrochemical properties*, J. Solid State Electrochem. 12 (2008), pp. 1003–1009.

[12] B. Sasi, K.G. Gopchandran, P.K. Manoj, P. Koshy, P.P. Rao, and V.K. Vaidyan, *Preparation of transparent and semiconducting NiO films*, Vacuum 68 (2002), pp. 149–154.

[13] W.L. He, Y. Zhang, X. Zhang, H. Wang, and H. Yan, *Low temperature preparation of nanocrystalline Mn2O3 via ethanol-thermal reduction of MnO2*, J. Cryst. Growth 252 (2003), pp. 285–288.

[14] Z. Yang, W. Zhang, Q. Wang, X. Song, and Y. Qian, *Synthesis of porous and hollow microspheres of nanocrystalline Mn2O3*, Chem. Phys. Lett. 418 (2006), pp. 46–49.

[15] S. Ashoka, P. Chithaiah, C.N. Tharamani, and G.T. Chandrappa, *Synthesis and characterisation of microstructural a- Mn2O3 materials*, J. Exp. Nanosci. 5 (2010), pp. 285–293.

[16] D. Zou, C. Xu, H. Luo, L. Wang, and T. Ying, *Synthesis of Co3O4 nanoparticles via an ionic liquid-assisted methodology at room temperature*, Mater. Lett. 62 (2008), pp. 1976–1978.

[17] A.A. Athawale, V. Singh, B.R. Mehta, and K. Navinkiran, *Solvent mediated morphological control of aniline stabilized cobalt oxide nanoparticles*, J Alloys Compd. 492 (2010), pp. 331–338.

[18] N.N. Binitha, P.V. Suraja, Z. Yaakob, M.R. Resmi, and P.P. Silija, *Simple synthesis of Co3O4 nanoflakes using a low-temperature sol–gel method suitable for photodegradation of dyes*, J. Sol–Gel Sci. Technol. 53 (2010), pp. 466–469.

[19] N.R. Jana, Y. Chen, and X. Peng, *Size- and shape-controlled magnetic (Cr, Mn, Fe, Co, Ni) oxide nanocrystals via a simple and general approach*, Chem. Mater. 16 (2004), pp. 3931–3935.

[20] F. Li, Y. Qian, and A. Stein, *Template-directed synthesis and organization of shaped oxide/phosphate nanoparticles*, Chem. Mater. 22 (2010), pp. 3226–3235.

[21] R. Vijaya Kumar, Y. Diamant, and A. Gedanken, *Sonochemical synthesis and characterization of nanometer-size transition metal oxides from metal acetates*, Chem. Mater. 12 (2000), pp. 2301–2305.

[22] M. Niederberger and G. Garnweitner, *Organic reaction pathways in the nonaqueous synthesis of metal oxide nanoparticles*, Chem. Eur. J. 12 (2006), pp. 7282–7302.

[23] M. Niederberger, *Nonaqueous sol–gel routes to metal oxide nanoparticles*, Acc. Chem. Res. 40 (2007), pp. 793–800.

[24] Y.W. Jun, J.S. Choiand, and J. Cheon, *Shape control of semi-conductor and metal oxide nanocrystals through nonhydrolytic colloidal routes*, Angew. Chem. Int. Ed. 45 (2006), pp. 3414–3439.

[25] M.M. Titirici, M. Antonietti, and A. Thomas, *A generalized synthesis of metal oxide hollow spheres using a hydrothermal approach*, Chem. Mater. 18 (2006), pp. 3808–3812.

[26] D. Song, C. Song, Z. Hu, and X. Fu, *Fabrication of hollow spheres and thin films of nickel oxide with hierarchical structures*, J. Phys. Chem. 109 (2005), pp. 1125–1129.

[27] Y-F Han, L. Chen, K. Ramesh, Z. Zhong, F. Chen, J. Chin, and H. Mook, *Coral like nanostructured alpha-Mn2O3 nanocrystals for catalytic combustion of methane Part I: Preparation and characterization*, Catal. Today 131 (2008), pp. 35–41.

[28] X. Wang, X. Chen, L. Gao, H. Zhang, and Y. Qian, *One dimensional arrays of Co3O4 nanoparticles: Synthesis, characterization and optical and electrical properties*, J. Phys. Chem. B 108 (2004), pp. 16401–16404.

[29] S. Gnanam and V. Rajendran, *Synthesis of CeO2 or α − Mn2O3 nanoparticles via sol–gel process and their optical properties*, J. Sol–Gel Sci. Technol. 58 (2011), pp. 62–69.

[30] X. Wang, X. Chen, L. Gao, H. Zheng, Z. Zhang, and Y. Qian, *One-dimensional arrays of Co3O4 nanoparticles: Synthesis, characterization, and optical and electrochemical properties*, J. Phys. Chem. B 108 (2004), pp. 16401–16404.

[31] R. Xu and H.C. Zeng, *Self-generation of tiered surfactant superstructures for one-pot synthesis of Co₃O₄ nanocubes and their close and non-close-packed organizations*, Langmuir 20 (2004), pp. 9780–9790.

[32] D. Adler and J. Feinleib, *Electrical and optical properties of narrow-band materials*, Phys. Rev. B 2 (1970), pp. 3112–3134.

[33] A. Gorschluter and H. Merz, *Localized d–d excitations in NiO (100) and CoO (100)*, Phys. Rev. B 49 (1994), pp. 17293–17302.

[34] B. Fromme, M. Moller, Th. Anschutz, C. Bethke, and E. Kisker, *Electron-exchange processes in the excitations of NiO (100) surface d states*, Phys. Rev. Lett. 77 (1996), pp. 1548–1551.

[35] Z. Chen, S. Zhang, S. Tan, J. Wang, S. Jin, and J. Hou, *Room-temperature green and ultraviolet emission from different grain-sized Mn sub (2)O sub (3) nanocrystals*, J. Mat. Sci. Lett. 21 (2002), pp. 411–413.

17

Carbon nanotube-filled conductive adhesives for electronic applications

Irfan Ahmad Mir and D. Kumar*

Department of Applied Chemistry & Polymer Technology, Delhi Technological University (Delhi College of Engineering), Shahbad Daulatpur, Bawana Road, Delhi-110042, India

Carbon nanotubes (CNTs) are being used as filler in epoxy matrix to produce isotropically conductive adhesives (ICAs). Different loadings of multiwalled carbon nanotubes were used to produce composites of different concentrations. These composites have been studied for their thermal behaviour, conductivity and impact properties. Percolation threshold is very low and conductivity value of $10^{-2}\,\mathrm{S\,cm^{-1}}$ was obtained at a filler concentration of only 0.3%. Higher percentage of the epoxy shear strength is retained. Differential scanning calorimetry results do not show any major influence on typical curing peaks of epoxy. Scanning electron microscopy confirmed that CNTs are easily dispersed in the epoxy matrix and there is a strong phase interaction. These composites with very low filler loadings showed a great prospect of being used as ICAs.

Keywords: carbon nanotubes; electrically conductive adhesives; composites

1. Introduction

Electronic interconnections have been dominated by tin/lead soldering since decades. But in response to environmental legislations, it is being replaced by other metallic alloys or electrically conductive adhesives [1]. Among electrically conductive adhesives, isotropically conductive adhesives (ICAs) with metallic fillers have gained a lot of commercial significance. ICAs, however, require high filler loading to achieve desired conductivity and hence the mechanical properties of the matrix get degraded. Also, due to the presence of metallic fillers, such adhesives show decrease in conductivity after exposure to moisture because of corrosion at the interface of substrate and component. Replacing metallic fillers with carbon nanotubes (CNTs) can help in overcoming these problems. We have already developed ICAs where metallic filler was replaced with organic conducting polymers like polyaniline [2] and polypyrrole [3]. Due to small size and better compatibility of CNTs, maximum impact strength of matrix can be retained. Also, loss of conductivity due to corrosion can be prevented. One of the most widely used matrices for ICAs are epoxies. So, in this article, novel ICAs have been formulated by adding CNTs as filler to an epoxy/anhydride system.

Epoxy/CNT composites have been prepared by a number of groups [4–6]. In fact, the earliest CNT composite developed was based on epoxy matrix [7]. In that study, purified nanotubes were embedded in epoxy to study the cross-sectional images of nanotubes. Since then, more such composites have been reported mostly using *in situ* polymerisation. Martin et al. [8] dispersed multiwalled carbon nanotubes (MWCNTs) in an epoxy system based on bisphenol A and studied

---

*Corresponding author. Email: drdkumar@yahoo.co.uk, dkumar@dce.ac.in

the influence of AC and DC electric fields on alignment of conductive networks. Gojny et al. [9] tried to use calendaring for dispersing CNTs in a viscous epoxy matrix. A depression in glass transition temperature of conducting composites obtained by blending CNTs with epoxy resins has been reported by Barrau et al. [10] near the percolation threshold. Based on the studies carried out by Zhou et al. [11] on CNT/epoxy composites, a linear damage model has been combined with the Weibull distribution function to establish a constitutive equation for neat and nano-phased carbon/epoxy. Similarly, Li and Lumpp [12] used MWCNT of various aspect ratios to develop conductive adhesives for aerospace applications. Composites with very high conductivity have been developed through resin transfer moulding technique recently [13].

Varying the concentration of CNT in epoxy matrix, different composites were prepared and studied for conductivity, impact properties, thermal and morphological characteristics. ICAs with better stability and impact properties were developed and reported in this article.

2. Experimental

2.1. *Materials*

The matrix polymer used was an epoxy, named Epon-862, which is based on diglycidyl ether of bisphenol-F, manufactured by Hexion Speciality Chemicals, Inc. Houston, Texas, USA, and procured from Miller Stephenson Chemical Company, Inc., Danbury, Connecticut, USA. The epoxy equivalent weight of this resin is approximately 170 g/equivalent. Anhydride hardener hexahydrophthalic anhydride (HHPA) and the catalyst 2-ethyl-4-methylimidazole (2E4MZ) were supplied by Sigma-Aldrich Chemicals Pvt. Ltd, Bangalore, India. MWCNTs used were procured from Cheap Tubes Inc., Brattleboro, VT, USA. They had a diameter of 40–70 nm and a length of 100 mm.

2.2. *Preparation of CNT/epoxy composites*

The pre-cure resin composites were formulated as per the following procedure.

(1) Curing agent (HHPA) which is solid at room temperature was heated slightly to melt it. Calculated amounts of CNTs were added to equal amounts of HHPA so that the final concentrations of composite will be 0.1%, 0.2%, 0.3%, 0.4% and 0.5%. The CNT concentration is described as weight percent. Each sample was sonicated in an ultrasonicator (model USB-2.25 from Accumax India, Delhi) for 30 min at 40°C to obtain dispersion.

(2) After the dispersion, the sample was allowed to cool down to room temperature and appropriate quantity of epoxy resin was added so that the epoxy/hardener was in a ratio of 1: 0.85. A certain amount of 2E4MZ, i.e. 0.1 parts per hundred parts of epoxy resin was added and mixed thoroughly.

(3) The mixture was stirred by a glass rod with heat, if necessary, until a homogenous mixture was formed. The mixture was left undisturbed for some time to remove air bubbles before further use.

3. Characterisation

3.1. *Cure study*

A differential scanning calorimeter (DSC) from TA Instruments, New castle, Delaware, USA (model Q20), was used to study curing of samples. Dynamic scans were done on samples of about 10 mg, at a heating rate of 5°C/min from room temperature to 250°C. Freshly mixed samples were placed in an aluminium hermetic DSC pan and heated under a nitrogen purge. After the dynamic scan, samples were cooled to room temperature and scanned again at the same rate. Glass transition temperature (T_g) of the samples was derived from the curve of reversible heat flow versus temperature.

3.2. *Thermogravimetric analysis*

TGA (thermogravimetric analysis) thermograms of the composites were recorded using TGA instrument of TA instruments, New Castle, Delaware, USA (model Q50), under nitrogen environment up to 600°C at a heating rate of 10°C/min.

3.3. *Conductivity measurement*

Conductivity was measured by four-probe technique on cured films of the ICAs. The detailed procedure for laying films of the ICAs has been reported earlier [3]. Conductivity was measured by means of the standard in line four-probe method using semiconductor characterisation system of Keithley Instruments, Inc., Cleveland, Ohio, USA (model 4200). Both surfaces of the films were scratched using flint paper before measurements to ensure proper contact.

3.4. *Impact performance*

Drop tests were conducted based on the standard established by National Centre of Manufacturing Sciences (NCMS), USA as per procedure reported earlier [14]. In this test, a mounted chip carrier and circuit board assembly are dropped onto hard surfaces from a height of 1.5 m (60 in) and it is necessary for a conductive adhesive to pass six drops for application as a solder replacement.

3.5. *Environmental ageing study*

Effect of environmental ageing on the cured samples was studied by conditioning at 85°C/ ∼ 100% RH until 500 h and measuring the weight gain after various time periods. Five samples of each adhesive were laid on a standard square glass cover slip and cured for 1 h at 150°C in a preheated oven. The cured samples were weighed on a Mettler instruments, Greifensee, Switzerland balance (AE-240) and placed on a plastic mesh above the water level in a temperature controlled water bath. Selected samples were periodically removed and weighed at ageing times. The water bath was maintained at 85°C and tightly closed except when samples were removed for testing so that the relative humidity in the water bath chamber was nearly 100%. Unfortunately, no attempt was made to determine whether moisture equilibrium was achieved [15].

3.6. *Lap shear strength test*

Lap shear determines the strength of adhesives for bonding materials. The test method is primarily comparative. The test is applicable for determining adhesive strengths, surface preparation parameters and adhesive environmental durability. Lap shear test was performed as per ASTM D3163 specifications. Two specimens with polyimide material on one side and copper surface on the other and dimensions 1×4 in are bonded together with adhesive so that the overlap area is 1×1 in. The overlap area was etched by flint paper prior to bonding with the adhesive to be tested. The adhesive was applied between etched panels and clamped in place. The thickness was maintained using end strands of a lead wire of diameter 0.1 mm. After curing, the cooled specimens were pulled apart by an Electronic Universal Testing Machine (UTM) of

Instron, UK (model 3369) at a pull rate of 0.05 in/min and peak stress was determined. Two groups of specimens were prepared. In each group, five specimen were prepared for every sample of ICA. One group was tested after cure, second was tested after conditioning for 200 h at 85°C/ ~ 100% RH.

3.7. *Scanning electron microscopy*

Scanning electron micrographs (SEM) were obtained with a ZEISS EVO series SEM (model EV050) of Carl Zeiss SMT Ltd, Cambridge, UK, at an acceleration voltage of 10 kV. All samples were plasma coated with a thin layer of gold to provide electrical conduction and reduce surface charging.

4. Result and discussion

4.1. *Conductivity measurement*

Electrical conductivity of a material corresponds to the ease at which the charge flows in it. For a composite material, it depends on the characteristics of the filler component and its subsequent arrangement inside the matrix. Epoxy matrix is highly insulating with conductivity of the order of 10^{-14} S cm^{-1}. When the conductive filler is incorporated, the insulating matrix is traversed by the conductive filler which forms clusters. When these clusters align themselves throughout the matrix, they form conductive channels due to contact between adjacent particles. Composites containing conductive filler in insulating polymers become electrically conductive when the filler content exceeds a critical value, known as a percolation threshold. The percolation threshold is characterised by a sharp jump in the conductivity by many orders of magnitude, which is attributed to the formation of a three-dimensional conductive network of the fillers within the matrix. Figure 1 shows the electrical conductivity of the epoxy/CNT composites as a function of CNT concentration. There is a sharp increase in conductivity around 0.1% concentration which is due to high conductivity of CNT. The percolation threshold of these composites taken at 10^{-6} S cm^{-1} is around a very low filler concentration of 0.2%. This very low percolation threshold is due to the high aspect ratio and nano-dimensions of CNTs. These findings were consistent with the

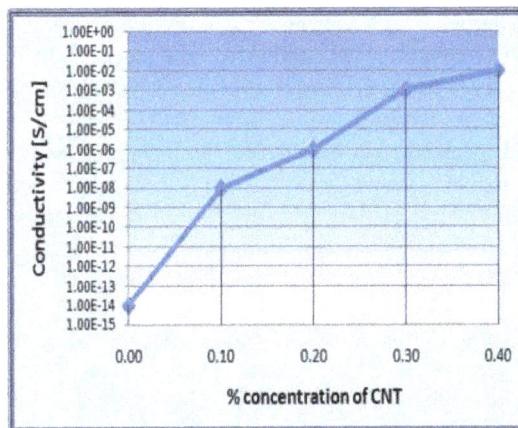

Figure 1.   Electrical conductivity at different CNT concentrations.

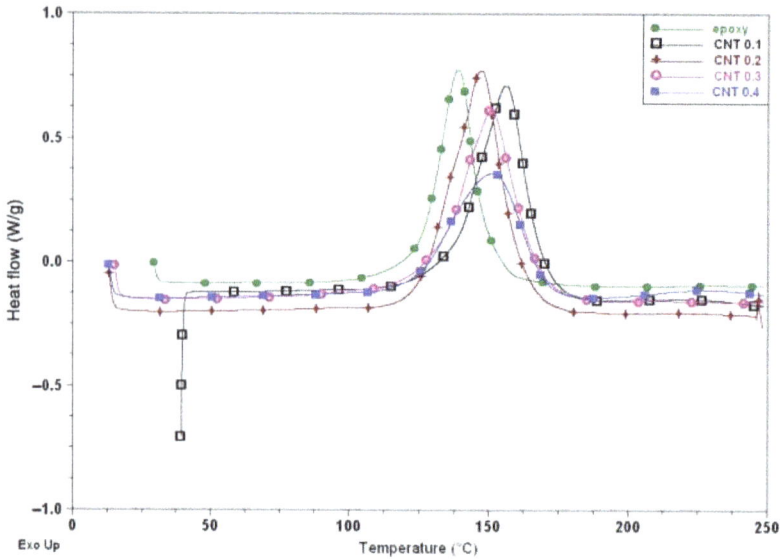

Figure 2.   DSC thermograms of neat epoxy and CNT/epoxy composites.

previous reports of such CNT/epoxy composites [8,16,17]. After this value, conductivity showed a gradual increase till it almost became saturated at 10^{-2} S cm^{-1}. These results are highly significant compared to our previous studies with conducting polymer fillers, where high filler loadings were required to produce the same effect. The smooth propagation of conductivity indicates that the dispersion of CNTs inside the matrix is uniform and proper conductive channels have been formed. No serious effort was taken to align CNTs which might have increased the conductivity values.

4.2.   Curing behaviour

DSC is a tool for observing the curing profile of epoxy system, both neat and when blended. DSC thermograms of neat epoxy and CNT/epoxy composites are shown in Figure 2. The exothermic peak of epoxy/anhydride system appears at 138°C, depicting the complete cross- linking reaction of the epoxy system. A single curing peak is observed in all the samples, which depicts that the basic curing profile of epoxy remains unaffected by the presence of CNTs. With the increase in CNT content, the peak slightly shifts towards higher temperature. This is due to an increase in uncured part in the matrix. However, as compared to our previous observations with conducting polymer fillers, peaks are sharp. The peak corresponding to the sample with 0.2% CNT concentration shows higher ΔH characteristics. It is worthwhile to note that percolation threshold of the composites is around this concentration. The reason may lie in depression in T_g which will be discussed in next section. The sample with 0.4% CNT concentration was viscous during mixing, which indicates that the cured part, i.e. epoxy had decreased substantially. That is why it shows a broadening of cure peak and lesser ΔH values. All the results clearly show that due to smaller size and high aspect ratio of CNTs, they provide least hindrance to cross-linking or formation of network in epoxy/anhydride systems. This holds great value for such composites as the mechanical properties of the matrix may be retained to maximum level.

   Glass transition behaviour for the composites were determined by cooling the samples after DSC to room temperature and then heating the cured samples up to 250°C at the same rate. The T_g value

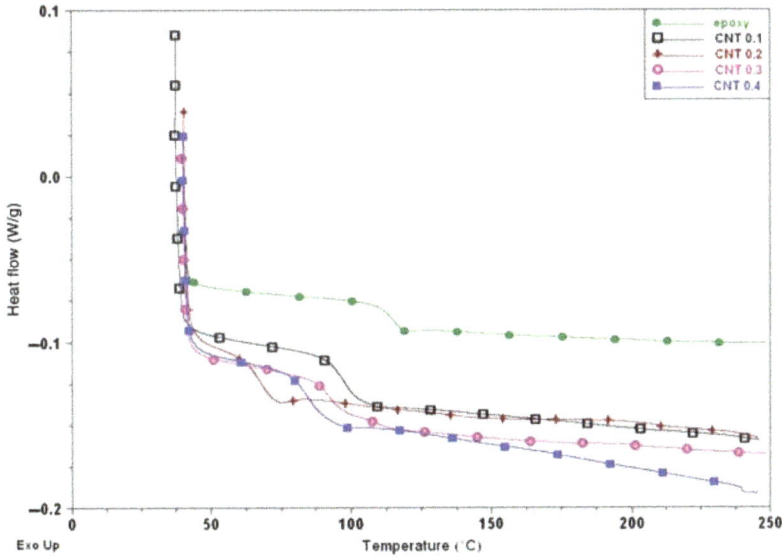

Figure 3. Variation of glass transition temperature versus CNT concentration.

of neat epoxy system was found to be around 115°C. The CNT content dependence of the T_g is presented in Figure 3. It can be observed that T_g values are affected in almost all samples, however, sample with 0.2% CNT concentration shows an unexpected behaviour. There is a sharp depression of T_g corresponding to this concentration. This behaviour was confirmed with repeated scans and seems to be independent of epoxy curing. The concentration is around percolation threshold and such behaviour has also been observed by Barrau et al. [10]. It is interpreted that conduction percolation is associated with a particular configuration of the conducting particles in the matrix. At the percolation threshold particles align themselves in definite infinite patterns, while before that they exist in finite size clusters. In the percolation range, the free volume accessible to the molecular motion of epoxy chain segments is maximal.

4.3. *Thermogravimetric analysis*

The weight loss versus temperature curve in Figure 4 shows that the overall degradation characteristics of the epoxy system are retained. There is a characteristic two-step transition. Neat epoxy shows a small weight loss before 150°C which may be due to loss of volatiles and then undergoes a complete degradation around 600°C. It is observed that the concentrations from 0.1% to 0.3% show little variation in weight loss characteristics, while beyond that weight loss around 150°C increases. The increase in weight loss may be due to loss of volatiles or some embedded moisture. Hence, it is concluded from the results obtained that CNT incorporation does not have a significant effect on the thermal degradation of epoxy systems.

4.4. *Environmental ageing*

Resistance to moisture present in the service environment is an important property of ICAs. Moisture may decrease their adhesive property or cause corrosion to metallic component joints. CNT/epoxy composite samples were subjected to 80°C/ ∼ 100% RH ageing and the effect

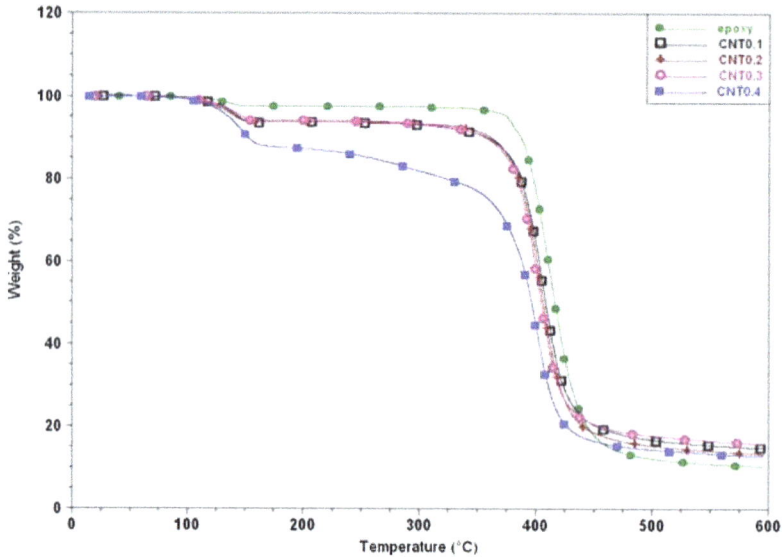

Figure 4.   TGA thermograms of CNT/epoxy composites.

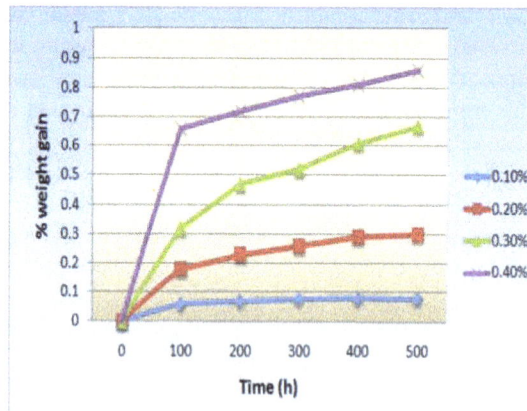

Figure 5.   Moisture absorption behaviour with increasing CNT content.

was studied as weight gain over a period of time. The results obtained are plotted against time and presented in Figure 5. It is observed that the samples are almost unaffected by the moisture.

There is a slight increase in weight till first 100 h and thereafter it is almost stable. The moisture absorbed may be bound to filler particles not completely adhered to epoxy network. It is evident from the results obtained that there is not even 1% increase of weight in the samples. Thus, these composites are highly stable under environment ageing. CNT due to their small size and lower loading levels are totally engulfed inside the matrix and hence do not bind water molecules. Although, as the CNT concentration increased to 0.4%, moisture absorption increased. But considering the overall effect, it is pretty clear that these samples can withstand extreme conditions of heat and moisture.

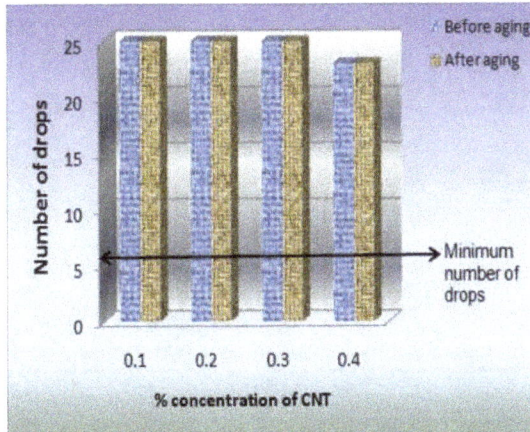

Figure 6.   Drop test performance of ICAs.

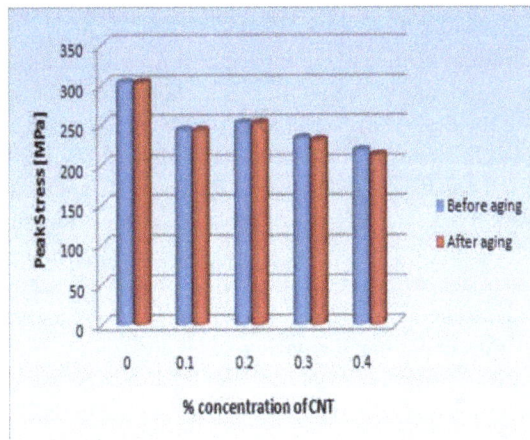

Figure 7.   Lap shear strength of ICAs.

4.5. *Drop test*

As per NCMS guidelines, an adhesive on a test assembly must pass six drops from a height of 60 in. The samples were studied for drop performance by recording the number of drops required to detach the chip from drop assembly and the results are shown in Figure 6. Due to good impact properties, the drop performance of all the samples was extraordinarily good. There was no visible effect even after 20 drops. The contacts formed had a good smooth surface and no cracks could be found even under a simple microscope. However, assemblies joined by sample with 0.4% CNT show little weathering at the edges after 20 drops. These observations are a direct consequence of retention of maximum impact properties of strong epoxy matrix as evidenced by lap shear studies. The small size and uniform dimensions of CNTs had a very little effect on network formation of the epoxy matrix. When the drop assemblies were subjected to 80°C/ $\sim$ 100% RH ageing, drop test performance observed shows no significant change.

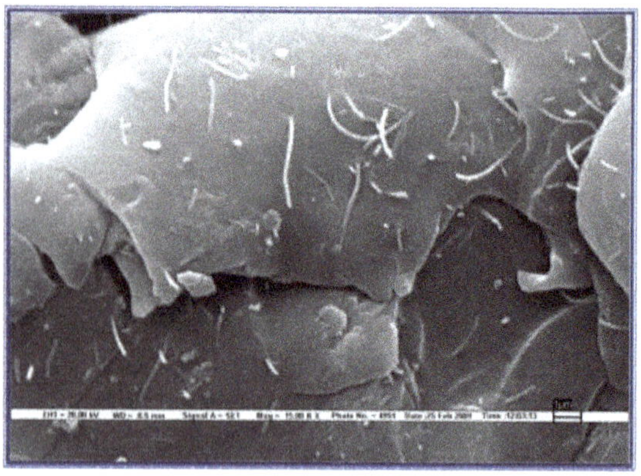

Figure 8.   SEM micrograph of CNT/epoxy surface.

4.6. *Lap shear strength*

Lap shear strength is another critical parameter which defines the binding strength of ICAs. As shown in Figure 7, lap shear of adhesives decreases with increase in CNT concentration. This is the outcome of proliferation of filler phase into matrix network and decrease in transmission of mechanical energy from the matrix to the filler. But, the decrease is very small and the shear strength is far greater than the desired impact strength for ICAs. The detached test coupons show that the fracture occurred at the interface of adhesive and metal surface of the test coupons rather than in the adhesive itself. It means that the observed values of lap shear are lesser than the actual impact strength of the adhesives. The etched copper surface does not undergo effective wetting hence the adhesion at that surface is poor. No cracking was observed in the adhesive material and toughness of the ICA was intact. After ageing at 85°C/ ∼ 100% RH until 200 h, samples show a small decrease in lap shear strength, thus depicting that these adhesives do not show a significant damping of adhesive strength under harsh environmental conditions. Although, epoxy matrix is not affected by moisture, filler particles loosely adhered to matrix network may absorb water and detach causing voids and crack propagation under impact. The effect of ageing on lap shear also shows a slight increasing trend with increase in CNT concentration. Hence, proper dispersion of filler particles and high aspect ratio of CNTs have significantly improved the impact performance of ICAs.

4.7. *Scanning electron microscopy*

To understand the dispersion of CNTs within epoxy matrix, the surface morphology was studied using SEM. Figure 8 shows the peel fractured surface of the ICA with 0.4% CNT. The surface is rough and CNTs are randomly distributed in the matrix. It is evident that the CNTs are well adhered to the surface, indicating that the interface was strong. Strong filler/matrix interface is another reason for good impact performance of these composites. No distinct agglomeration is evident, so the dispersion of CNTs in the curing agent HHPA is an efficient method of preventing re-agglomeration. All these observations clearly establish that CNTs are easily distributed within epoxy matrix and there is no distinct phase separation. This is the reason for good impact properties because the stress is easily propagated through filler moieties. Also, because of the smooth distribution, conductivity is established at lower filler concentrations due to the formation of conductive channels which act as paths for charge propagation.

5. Conclusions

MWCNTs were used as filler in epoxy/anhydride systems and studied for their application as ICAs. This study reveals that small size and uniformity of dimensions play an important role in defining the suitability of a material for being used as filler in ICAs. CNTs have high impact strength and good aspect ratio; hence, a higher percentage of the epoxy shear strength is retained. Conductivity is established at a very low filler loading and conductivity value as high as $10^{-2}\,\text{S}\,\text{cm}^{-1}$ was obtained at a filler concentration of only 0.3%. The ICAs so formed not only show good adhesion but also show least effect of ageing. The surface characteristics show that there is a strong interface between matrix and filler phases, which defines the worth of an efficient composite system. Hence, it is concluded that incorporation of CNTs as filler inside the epoxy matrix produces ICAs with high electrical conductivity, great impact strength at very low filler loadings. The overall properties of these samples establish their use as ICAs for electronic interconnections. However, the properties may further be improved by devising methods for incorporating higher concentrations of CNTs through more efficient dispersion agents. Also, introducing functionalisation may improve adhesion and thus increase the impact performance further.

Acknowledgements

The authors express their sincere gratitude to the vice chancellor, Delhi Technological University, Delhi, for his kind encouragement and support. One of the authors, Irfan Ahmad Mir, is thankful to the Council of Scientific and Industrial Research, Government of India, for providing financial support as senior research fellowship.

References

[1] I. Mir and D. Kumar, *Recent advances in isotropic conductive adhesives for electronics packaging applications*, Int. J. Adhes. Adhes. 28 (2008), pp. 362–371.
[2] I.A. Mir and D. Kumar, *Development of polyaniline/epoxy composite as a prospective solder replacement material*, Int. J. Polym. Mater. 59 (2010), pp. 994–1007.
[3] I.A. Mir and D. Kumar, *Development of polypyrrole/epoxy composites as isotropically conductive adhesives*, J. Adhes. 86 (2010), pp. 447–462.
[4] A. Allaoui, S. Bai, H.M. Cheng, and J.B. Bai, *Mechanical and electrical properties of a MWNT/epoxy composite*, Compos. Sci. Technol. 62 (2002), pp. 1993–1998.
[5] X. Xu, M.M. Thwe, C. Shearwood, and K. Liao, *Mechanical properties and interfacial characteristics of carbon-nanotube-reinforced epoxy thin films*, Appl. Phys. Lett. 81 (2002), pp. 2833–2835.
[6] D. Puglia, L. Valentini, and J.M. Kenny, *Analysis of the cure reaction of carbon nanotubes/epoxy resin composites through thermal analysis and Raman spectroscopy*, J. Appl. Polym. Sci. 88 (2003), pp. 452–458.
[7] P.M. Ajayan, O. Stephan, C. Colliex, and D. Trauth, *Aligned carbon nanotube arrays formed by cutting a polymer resin-nanotube composite*, Science 265 (1994), pp. 1212–1214.
[8] C.A. Martin, J. Sandler, A.H. Windle, M. Schwarz, W. Bauhofer, and K. Schulte, *Electric field-induced aligned multi-wall carbon nanotube networks in epoxy composites*, Polymer 46 (2005), pp. 877–886.
[9] F.H. Gojny, M. Wichmann, U. Kopke, B. Fiedler, and K. Schulte, *Carbon nanotube-reinforced epoxy-composites: Enhanced stiffness and fracture toughness at low nanotube content*, Compos. Sci. Technol. 64 (2004), pp. 2363–2371.
[10] S. Barrau, P. Demont, C. Maraval, A. Bernes, and C. Lacabanne, *Glass transition temperature depression at the percolation threshold in carbon nanotube-epoxy resin and polypyrrole-epoxy resin composites*, Macromol. Rapid Commun. 26 (2005), pp. 390–394.
[11] Y. Zhou, F. Pervin, L. Lewis, and S. Jeelani, *Fabrication and characterization of carbon/epoxy composites mixed with multi-walled carbon nanotubes*, Mater. Sci. Eng. A 475 (2008), pp. 157–165.
[12] J. Li and J.K. Lumpp, *Carbon nanotube filled conductive adhesives for aerospace applications*, Proceedings of the Aerospace Conference, IEEE, Big Sky, MT, USA, 2007, pp. 1–6.
[13] Q.F. Cheng, J.P. Wang, J.J. Wen, C.H. Liu, K.L. Jiang, and Q.Q. Li, *Carbon nanotube/epoxy composites fabricated by resin transfer molding*, Carbon 48 (2010), pp. 260–266.
[14] Y. Rao, D. Lu, and C.P. Wong, *A study of impact performance of conductive adhesives*, Int. J. Adhes. Adhes. 24 (2004), pp. 449–453.

[15] S. Xu, DA. Dillard, and J.G. Dillard, *Environmental aging effects on the durability of electrically conductive adhesive joints*, Int. J. Adhes. Adhes. 23 (2003), pp. 235–250.

[16] J.K.W. Sandler, J.E. Kirk, I.A. Kinloch, M.S.P. Shaffer, and A.H. Windle, *Ultra-low electrical percolation threshold in carbon-nanotube-epoxy composites*, Polymer 44 (2003), pp. 5893–5899.

[17] A. Moisala, Q. Li, I.A. Kinloch, and A.H. Windle, *Thermal and electrical conductivity of single- and multi-walled carbon nanotube-epoxy composites*, Compos. Sci. Technol. 66 (2006), pp. 1285-1288.

18

Determination of absolute quantum yields of luminescing nanomaterials over a broad spectral range: from the integrating sphere theory to the correct methodology

Jan Valenta*

Department of Chemical Physics & Optics, Faculty of Mathematics & Physics, Charles University, Ke Karlovu 3, Prague 2 CZ-121 16, Czech Republic

Determination of fluorescence or photo- and electro-luminescence quantum yield (QY) is of increasing importance for standardization of novel light-emitting materials like semiconductor quantum dots. The most straightforward spectroscopic QY measurement technique is based on the exploitation of an integrating sphere (IS). We propose a compact set-up and procedure for reliable measurements of QY over a broad excitation spectral range which uses excitation by light-emitting diodes. Starting from the general IS theory we show that (i) a variable excitation pattern has no fatal influence on the QY determination and (ii) a simple two-configuration experiment is an adequate basis of the correct experimental procedure. Then we give guidelines for the QY experiment and calculations as well as a few practical examples and error analysis. Our comprehensive description of absolute luminescence QY techniques based on the IS theory has a general applicability not restricted to the apparatus described in this paper.

Keywords: luminescence; quantum yield; efficiency; radiometry; nanocrystals

1. Introduction

The quantum yield (QY) of luminescence (or fluorescence/phosphorescence) is defined as the ratio of number of emitted photons and number of absorbed photons during the excitation of a sample under investigation. Alternative term "luminescence quantum efficiency" should be — according to the recommendation by International Union of Pure and Applied Chemistry (IUPAC) [1] — reserved to the fraction of molecules in a particular excited state that emit luminescence.

In any case, the knowledge of luminescence QY is very useful for advanced characterization and application of luminescing materials (light-converting phosphors, fluorescence labels, etc.). For example, the knowledge of both the luminescence decay time τ_L and QY η enables to separate the radiative τ_r and non-radiative τ_{nr} lifetimes of an excited state using the simple relation

$$\eta = \frac{\tau_L}{\tau_r} = \frac{1/\tau_r}{1/\tau_r + 1/\tau_{nr}}. \tag{1}$$

This is a unique way to obtain these important parameters from relatively simple optical measurements.[2] (Note, that in some cases, when properties of luminescing objects in an ensemble have broad distribution, Equation (1) cannot be applied as measurements of η and τ_L, could be dominated by different subsets of the ensemble.)

*Email: jan.valenta@mff.cuni.cz

We will restrict our description to measurements of luminescence excited by light, i.e. photoluminescence (PL). In general, there are two approaches to determine PL QY [3,4]:

(1) *Comparative methods* – based on comparison of PL measurements of an investigated sample and a luminescence standard under identical conditions. This technique is the most often used one (sometimes called the Parker–Rees method).

(2) *Absolute methods* – directly measure a portion of absorbed photons emitted as luminescence (optical methods) or portion of photons lost by non-radiative recombinations (calorimetric methods: photoacoustic, thermal-lensing, etc.).

The approach discussed in this paper is an absolute optical technique based on exploitation of an integrating sphere (IS). Different ISs are applied in various radiometry set-ups with the aim to minimize problems related to angular dependence of reflection, emission and scattering. Several commercial instruments with an IS dedicated to PL QY measurements have been introduced recently, e.g. Quantaurus-QY by Hamamatsu or Quanta-Phi module for FluoroMax® by Horiba Scientific.[5] These devices are reliable but quite expensive and their users often struggle with lack of detailed information on the QY determination procedure and the technique background. Fortunately, a series of valuable methodological papers by Resch-Genger et al. appeared recently.[6,7] However, these papers are oriented mostly to determination of fluorescence QY in solutions of organic dyes and do not cover fully the theoretical background of radiometry using IS.

The aim of the present paper is to provide a consistent theoretical background of the PL QY methodology for testing solid samples. This description should respond to many practical questions, like: How the position and size of a sample and the excitation spot size within an IS influence the methodology or what is the precision of determined PL QY values? We believe that our work can help readers to shorten their way to establish a correct method for determining PL QY of their (nano)materials.

Second aim of this paper is to extend the PL QY measurement to a broad spectral range. Most of the PL QY set-ups are limited to application of just one or a few excitation wavelengths using lasers or lamps attached to a monochromator. This is supposed to be sufficient because the so-called *Kasha–Vavilov* (KV) *rule* is often well fulfilled. The KV rule (which was formulated for organic fluorophores) says that both the luminescence spectral shape and its QY do not depend on the applied excitation wavelength.[8] However, some advanced materials (e.g. semiconductor quantum dots showing efficient carrier multiplication [9]) violate the KV rule and investigation of PL QY over extended excitation range is necessary.

When measuring QY over an extremely broad excitation range, the wide tuneability and perfect stability of an excitation source become fundamental. The only continuously emitting excitation sources which are widely available and provide enough radiation from the UV-C to the near infrared (NIR) region are xenon and halogen lamps (however, the Xe lamps have low emission in the red–NIR region while the halogen lamps have low UV emission). In the case of a pulsed excitation, optical parametric oscillators pumped by pulsed lasers and also some fibre lasers can be applied. Recently, new broad-range light sources based on the laser-induced plasma were introduced and might be relevant for broader application in PL QY measurements in future.[10]

The development of light-emitting diodes (LEDs) during the last 50 years has led to a tremendous increase in their efficiency and decrease in the price per output power which enable the current "lighting revolution".[11] Especially the advent of GaN-based LEDs (in 1993) and application of hetero-junctions enormously increased the spectral range of commercially available high-brightness diodes from IR to the UV-C (down to 240 nm) spectral range. The spectroscopical use of LEDs is rapidly developing taking advantage from the high stability, small size, simple driving circuits, long lifetime, easy modulation, etc.

Based on our long-term experience, we designed a relatively simple and cheap luminescence spectrometer with IS and excitation by LEDs described in the next paragraph. Starting from the general IS theory, we find the correct procedure to determine PL QY. A practical application is demonstrated using model materials, namely semiconductor quantum dots. Finally, the error analysis gives a general approach to estimate uncertainty of obtained PL QY values.

2. Experimental equipment

The principal scheme of our PL QY set-up is shown in Figure 1. The central component is an IS. Excitation sources – LEDs – are inserted in an adapter on one of the IS ports and driven by a stabilized power supply. A sample is mounted on an IS port – most often the one in front of the excitation port. Output signal is collected by a fibre bundle placed in the direction perpendicular to the excitation axis and it is shielded by baffles against the direct visibility of both the LED excitation source and the sample.[1] The end of the fibre bundle (advantageously with the stripe-like shape, see the inset image on the left side of Figure 1) is coupled to the input slit of a spectrometer. This fibre-coupling system is convenient as the IS can be easily moved or rotated (if needed, e.g. for calibration measurements) without affecting the coupling efficiency and consequently the set-up response.

The spectroscopic detection part should be designed according to the required sensitivity, spectral range and resolution. Almost any standard spectrometer can be exploited, even the compact low-cost fixed-range spectrometers. For the present work we applied the 15-cm or 30-cm focal length imaging spectrographs and the liquid-nitrogen-cooled CCD camera for detection. Coupling of the fibre output into the spectrometer is done with a double lens coupler specially adapted for the numerical apertures of both the fibre waveguide and the spectrometer. For details on luminescence spectroscopy techniques see, e.g. the textbook by Pelant and Valenta.[12] In the following sections we are going to describe only the two key parts of our set-up: LEDs and IS.

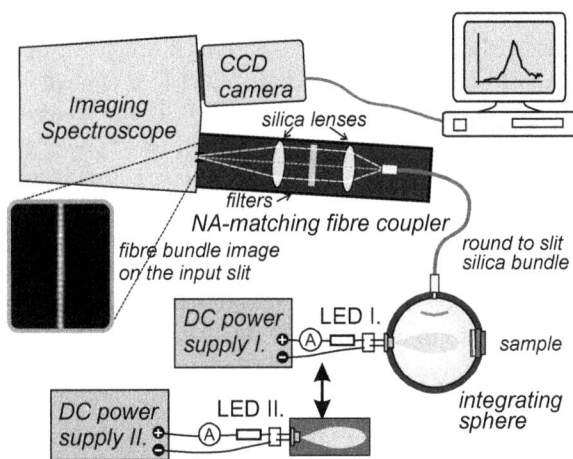

Figure 1. The principal scheme of the apparatus for measuring absolute PL QY using an IS and LEDs as excitation sources. The second power supply for LED is applied in order to enable emission stabilization of an LED before its application in experiments.

2.1. *Excitation sources – LEDs*

From the point of view of PL QY measurements, the following parameters of LEDs are the most important: an electro-luminescence (EL) *emission band peak* and *width*, *radiation power* and its *temporal stability*, and *emission pattern*. Spectral parameters (the position and width of an emission band) depend strongly on the internal semiconductor composition and band engineering (homo- or hetero-junctions, quantum wells, etc.), the full-width at half-maximum being usually between 20 and 40 nm.[13] Therefore, the LEDs can be often used to excite PL in QY measurement directly, i.e. without passing any monochromator or filters, if the spectral resolution of 20–40 nm is acceptable. This significantly simplifies an experimental set-up and decreases its price. If needed, an LED can be combined with an appropriate band-pass filter to narrow the excitation band-width and/or limit the long-wavelength tail of EL.

When selecting appropriate LEDs using catalogue data (or data sheets) one has to keep in mind that the actual parameters of real devices (especially the peak wavelength and output power) can deviate significantly from the nominal values. Therefore, all diodes must be thoroughly tested (and selected) prior to their application in PL QY measurements.

An LED emission radiation pattern (an angle distribution of radiant flux) depends mainly on an LED package. In case of packages with a flat plane-parallel window (e.g. deep-UV diodes UV-TOP® produced by Sensor Electronics Technology, Inc.) the viewing angle is very wide, about 120°, and can be eventually reduced by a small lens or by a mirror concentrator. Most LED packages have the shape of a hemispherical lens which reduces the apex angle of an emission cone down to 10° or even below. Often, there is a significant variation of the area and shape of IS surface which is directly excited by different LEDs in the set covering a broad spectral range. As we want to keep the apparatus simple, without any complicated optical system at the excitation side, the PL QY experimental procedure must be designed insensitive to these variations of an LED emission pattern (see Section 3).

Note that the broad emission angle of most LEDs excludes a straightforward application of the PL QY measurement procedure proposed by de Mello et al. [14] which requires a narrow excitation beam (laser) with its profile smaller than an investigated sample. This method is based on a comparison of the direct excitation with the indirect excitation by diffused light (with a sample shifted out of an excitation beam).

Finally, we have to discuss the light-emission stability of LEDs. This characteristic is of special importance as the PL QY technique is based on comparisons of several measurements which cannot be done simultaneously (at least not in the described simple set-up). We strongly recommend testing of stability of LEDs under the same conditions (e.g. applied voltage and current, the LED holder – influencing thermal exchange between an LED and its environment) as expected to be used during PL QY experiments. Obviously, a sophisticated power supply with a feedback could ensure more stable LED output (but such an option is not included in the simple set-up described here).

One has to take into account that the EL signal always contains shot noise which is described by the Poissonian distribution – it means that the standard deviation is a square root of the average signal counts N. Consequently, there would be no observable effect of EL temporal instability on the QY precision if the change of integrated EL signal during PL QY experiment was smaller than the shot noise. (In general, the acceptable LED emission intensity drift is about 1% in 1 h and the fast fluctuations should be below the shot noise.)

For many LEDs we found that the necessary stabilization takes about 10–15 min. Then the necessity to stabilize each diode after its mounting on the IS and switching on the power supply could cause a huge increase in experiment duration. Therefore, we adopted a simple solution: two independent power supplies are used in parallel, which allows stabilizing of one LED while the other one is used in experiments (see Figure 1).

2.2. *Integrating sphere*

An IS is a spherical cavity with highly reflective (diffusive) surfaces which enable a spatial integration of incoming light flux. It is widely applied in radiometry, photometry and other optical experiments.[15]

Let us consider a *perfect IS* with the following parameters: the inner surface area S_{is} with reflectance ρ_{is} is smaller than the complete sphere surface because part of it is occupied by ports and sample areas with reduced (or zero) reflectance ρ. Size of each one from n ports is characterized by the *fill factor f* – ratio of the port area to the area of IS $f_k = S_k/S_{is}$. An ideal IS distributes the *input radiant flux* Φ_{in} evenly over the whole inner surface and the *output flux* Φ_{out} through a port with area S_{out} is equal to [16]

$$\Phi_{out} = \Phi_{in} \cdot \frac{f_{out} \cdot \rho_0}{1 - \rho_{is}(1 - \sum_{k=1}^{n} f_k) - \sum_{k=1}^{n} f_k \cdot \rho_k}, \tag{2}$$

where ρ_0 is reflectance of the first surface encountered by incoming radiation. The fraction on the right side of Equation (2) (which is equal to the ratio of output and input fluxes Φ_{out}/Φ_{in}) is called the *IS efficiency* χ_{is}. Then χ_{is}/f_{out} is the *sphere multiplier M*

$$M = \frac{\rho_0}{1 - \rho_{is}(1 - \sum_{k=1}^{n} f_k) - \sum_{k=1}^{n} f_k \cdot \rho_k}, \tag{3}$$

which characterizes how much the IS surface radiance increases compared with a single reflection. For most of ISs reflectance ρ_{is} and the full port fill fraction f are in the range of 0.94–0.99 and 0.02–0.05, respectively, and then the multiplier is between 10 and 50.[17]

In Figure 2(a), we present the reflectance spectrum of the Spectraflect material (based on barium sulphate) which forms a coating of our IS with diameter of 10 cm (purchased from SphereOptics GmbH, Germany). The theoretical sphere multiplier for such IS with different configurations of ports is calculated using Equation (3) and plotted in Figure 2(c).

In the described set-up, an optical fibre bundle is applied to guide a portion of the signal from the IS to the spectral detection system (Figures 1 and 3). Only signal reflected from the baffle in front of the bundle and coming within the fibre acceptance cone (characterized by its numerical aperture $\mathbb{N}$) is coupled to the fibre bundle (Figure 3). Then the output flux is given by

$$\Phi_{out} = L_{is} \cdot S_f \cdot \pi \cdot \mathbb{N}_f^2(1 - \rho_f), \tag{4}$$

where S_f and ρ_f are the area and reflectance of the fibre bundle. L_{is} is the radiance of the IS surface equal to

$$L_{is} = \frac{\Phi_{in}}{\pi \, S_{is}} \cdot M. \tag{5}$$

In our experimental configuration (Figure 3) we have only two ports (tp) – the input (in) and the fibre output (f) ports with fill factors and reflectance of f_{in}, f_f and ρ_{in}, ρ_f, respectively. Then the sphere multiplier, M, takes the following form:

$$M^{tp} = \frac{\rho_0}{1 - \rho_{is}(1 - f_{in} - f_f) - f_{in} \cdot \rho_{in} - f_f \cdot \rho_f}. \tag{6}$$

Note the same description is valid also for the imaging output coupling system which projects an output aperture (hole) onto an input slit of a spectrometer. In this case, the reflectance of the output port is close to zero $\rho_f \sim 0$.

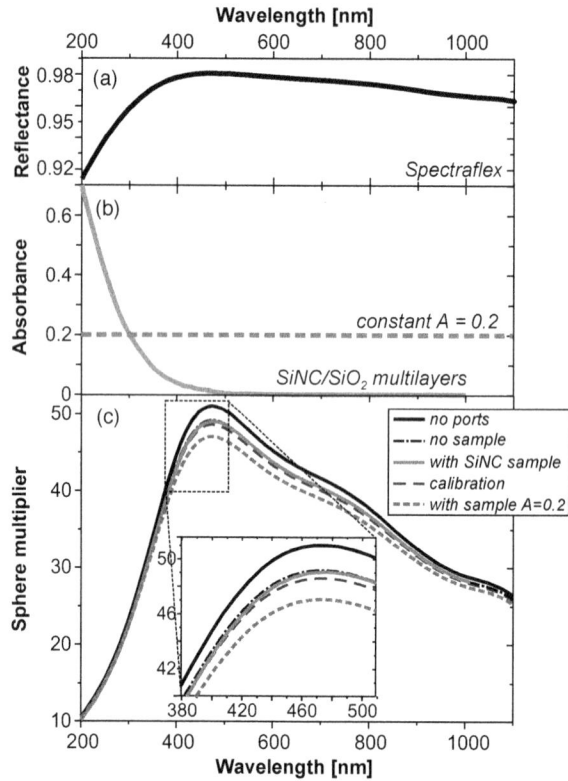

Figure 2.   (a) Diffuse reflectance of the Spectraflect coating. (b) Absorbance of the Si/SiO_2 nanocrystalline multilayer sample (solid line) and the hypothetical sample with flat absorbance of 0.2 (dashed line). (c) IS multipliers calculated using Equation (3) or (6) for the IS with 10-cm diameter covered by the Spectraflect coating and having: no ports (uppermost solid line), two ports (a LED input and a fibre output) but no sample (dotted-dashed black line), two ports with a sample (12×12 mm) of Si/SiO_2 nanocrystalline multilayers (thin line) or a sample of the same size but $A = 0.2$ (lower dashed line) and the calibration configuration (an open input port, a fibre output).

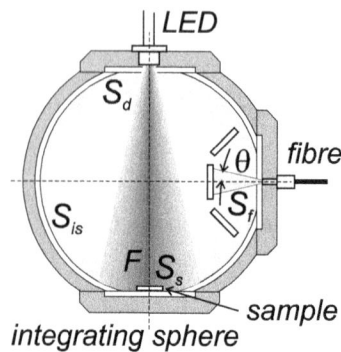

Figure 3.   The basic parameters of the integrated sphere experiment with output signal coupled into a fibre bundle (for meaning of symbols, see Section 2.2).

2.3. *Correction for the spectral response of an experimental apparatus*

The calibration of the spectral response of an experimental system is necessary for any luminescence experiment involving a wide spectral range.[10] In this paper we describe experiments dealing with extremely broad spectral range from 300 to 1000 nm, which includes even the edge regions where the sensitivity is rapidly decreasing. Therefore, the precise calibration of the spectral response is of special importance.

Two calibrated radiance standards were applied: the tungsten halogen filament lamp (Oriel Model No. 63355, calibrated between 250 and 2400 nm) and the deuterium lamp (Oriel Model No. 63945, calibrated between 200 and 400 nm). The lamps were placed at the distance of 50 cm from the entrance port of an IS (Figure 4(b)) because the absolute calibration data are given for this distance. Consequently, we can perform the absolute calibration of a set-up in radiometric units (even if it is not necessary for the PL QY determination – for that knowledge of the correction function spectral shape in arbitrary units is enough as will be described below). By dividing the measured spectral signal (expressed in count/s/pixel) by the calibrated spectral irradiance of the standard ($\Xi_{e\lambda}$ in W/nm/m^2) multiplied by the area of an entrance port (in m^2) we obtain the *sensitivity spectrum of the apparatus* $C(\lambda)$ in (count/s/pixel)/(W/nm). This sensitivity curve is then used to convert experimental spectra (count/s/pixel) into spectral radiance (W/nm). By integrating the area of spectral bands, total radiant power is obtained.

The calibration procedure must be done separately for all configurations of the apparatus which will be used during PL QY experiments. For example, different slit width, diffraction gratings

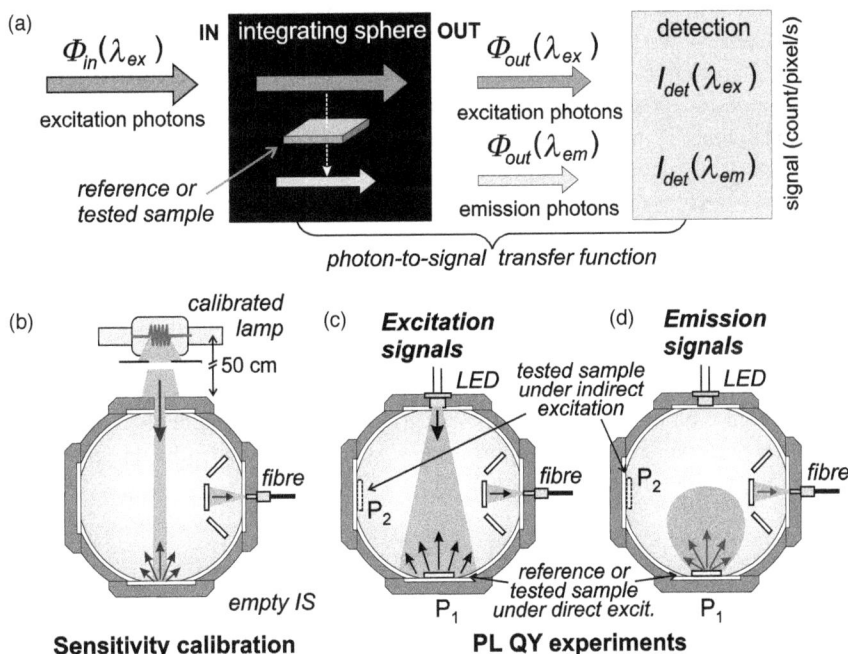

Figure 4. (a) Schematic representation of the PL QY measurement as a sort of "black-box" experiment, where part of the input signal (photon flux) is transformed to the longer-wavelength signal (PL) and then converted to detected signal counts. The lower panel illustrates the three experimental configurations to be used for the PL QY determination: (b) sensitivity calibration, (c) excitation/absorption and (d) emission (PL) signal acquisition. Note that only one (or none) sample is present inside IS during one experimental step – multiple samples in (c) and (d) just illustrate possible positions of a sample in or outside the direct excitation spot (P_1 and P_2, respectively).

(mounted on a turret) or even detectors may be used in order to optimize sensitivity and/or the detected spectral range.

We have to note that for a correct calibration of boundary spectral regions, where the response of an apparatus rapidly drops down, special care must be given to effects like stray light of a spectrometer.[18] Useful recommendations for characterization of a PL apparatus were recently published by Resch-Genger and DeRose.[19]

3. Experimental configuration and procedure

In the following paragraph we describe the PL QY determination for a solid sample in the form of a flat plate (or, possibly, a small cuvette with a liquid sample). We suppose to have two samples: (1) *reference sample* (RS) represented by a substrate or matrix without a studied material and (2) *tested sample* (TS) formed by the same substrate (matrix) but containing a studied active material (dispersed in a matrix or deposited on a substrate).

The experiment consists of several pairs of measurements – one measurement covers the excitation range, i.e. the spectral range of an applied LED (Figure 4(c)), and the second one covers the whole PL emission band of a TS (Figure 4(d)). This pair of measurements is usually performed for three experimental configurations: the RS sample is placed in the incoming flux, the TS placed in the incoming flux and the TS located out of the direct illumination by an LED. We will see below that the last configuration is redundant.

Figure 4(a) represents the PL QY measurements as a sort of a "black-box" experiment. When we insert a sample into an IS (black-box) part of the input, photon flux is converted into longer-wavelength (PL emission[2]) photons. Our task is to determine efficiency of this conversion (QY). An experimental apparatus converts photon fluxes into digital signal (counts/detector pixel/s) with certain photon-to-signal transfer function which we call the *spectral sensitivity* $C(\lambda)$.

The detected signal I^{det} is directly proportional to the IS output flux $I^{\text{det}} = K\Phi^{\text{out}}$, where K is a constant (spectrally dependent) characterizing the detection system sensitivity (without IS). Using Equations (4)–(6) we obtain

$$I^{\text{det}} = K \frac{S_{\text{f}} N_{\text{f}}^2 (1 - \rho_{\text{f}})}{S_{\text{is}}} \cdot M^{\text{tp}} \cdot \Phi_{\text{in}} = K \cdot k \cdot M^{\text{tp}} \cdot \Phi_{\text{in}}. \tag{7}$$

The left fraction in Equation (7) was labelled k, which is the spectrally dependent constant for a given IS with a fibre coupler.

During the calibration measurement (Figure 4(b)), the first encountered surface is an IS wall, $\rho_0 = \rho_{\text{is}}$ and the reflectance of an input port is zero $\rho_{\text{in}} = 0$. Consequently, the sphere multiplier for the calibration measurement is (Figure 2(c))

$$M^{\text{cal}} = \frac{\rho_{\text{is}}}{1 - \rho_{\text{is}}(1 - f_{\text{in}} - f_{\text{f}}) - f_{\text{f}} \cdot \rho_{\text{f}}}. \tag{8}$$

The input flux Φ_{in} in Equation (7) has the form of spectral radiant flux $\Phi_{e\lambda}$ obtained by multiplying the spectral irradiance of a calibrated source $\Xi_{e\lambda}$ by the area of an input port S_{in}. By dividing $I^{\text{det}}(\lambda)$ with $\Phi_{e\lambda}(\lambda)$ we obtain the sensitivity spectrum $C(\lambda)$ (i.e. $C = KkM^{\text{cal}}$), as was mentioned in the previous paragraph.

A sample introduced in the IS plays two roles:

(a) *Semitransparent port* – The sample with area S_{s} and absorbance A_{s} (fraction of incident light which is absorbed during single passage through a sample, i.e. the probability to absorb a passing photon) placed anywhere on an IS wall influences the sphere multiplier M approximately as a port with reflectance $(1 - A_s)^3$.

(b) *Input filter* – When the sample is placed on a IS wall illuminated directly by an LED, it reduces the first reflectance ρ_0 from ρ_{is} to $(1 - A_s)\rho_{is}$, which has the same effect on the IS surface radiance L_{is} as the (double-passage) filtering of incoming photon flux.

In order to provide a general description (in respect to a relation between the sample and the direct excitation spot sizes and shapes), we suppose that the IS surface area directly illuminated by an LED S_{il} is larger than the sample area S_s. Let the fraction of input flux Φ_{in} passing through a sample be $F = (0, 1)$, where $F = 1$ means that the whole flux passes through a sample. Now, we are going to calculate the absorbed power and the detected signal for three experimental configurations (Figure 4(b–d)): (a) the RS under the direct excitation, (b) the TS under the direct excitation and (c) the TS under indirect excitation by diffused light.

In the first case (supposing that RS has no absorption in the investigated spectral range ($A_{RS} = 0$) and the possible specular reflections from sample facets have a negligible influence on the distribution of light inside a sphere) we obtain for the directly absorbed power Φ_{RS}^{abs}, the IS surface radiance L_{RS}^{is} (directly proportional to the detected signal, see Equations (5) and (7)) and the multiplier M_{RS}

$$\Phi_{RS}^{abs} = F \cdot A_{RS} \cdot \Phi^{LED} = 0, \tag{9a}$$

$$L_{RS}^{is} = \frac{(1 - F)\Phi^{LED} + F(1 - A_{RS})\Phi^{LED}}{\pi \cdot S_{is}} M_{RS} = \frac{\Phi^{LED}}{\pi \cdot S_{is}} M_{RS}, \tag{9b}$$

$$M_{RS} = \frac{\rho_{is}}{1 - \rho_{is}(1 - f_{in} - f_f) - f_{in} \cdot \rho_{LED} - f_f \cdot \rho_f}. \tag{9c}$$

For the TS excited directly by the input flux we obtain

$$\Phi_{TS}^{abs} = F \cdot A_{TS} \cdot \Phi^{LED}, \tag{10a}$$

$$L_{TS}^{is} = \frac{(1 - F)\Phi^{LED} + F(1 - A_{TS})\Phi^{LED}}{\pi \cdot S_{is}} M_{TS} \doteq \frac{(1 - F \cdot A_{TS})\Phi^{LED}}{\pi \cdot S_{is}} M_{TS}, \tag{10b}$$

$$M_{TS} = \frac{\rho_{is}}{1 - \rho_{is}(1 - f_{in} - f_f) - f_{in} \cdot \rho_{LED} - f_f \cdot \rho_f + f_{TS} \cdot A_{TS} \cdot \rho_{is}}. \tag{10c}$$

We see that the relative difference between signals measured in the first and second configurations $(I_{RS}^{ex} - I_{TS}^{ex})/I_{RS}^{ex}$ (which is the same as $(L_{RS}^{is} - L_{TS}^{is})/L_{RS}^{is}$) would be equal to the power of absorbed light FA_{TS}, if the multipliers M_{RS} and M_{TS} were equal (but this condition is not fulfilled as we will see below).

Finally, in the third configuration the RS is placed on an IS surface but out of the incoming LED radiation and receives only diffused radiant flux given by the radiance of an IS surface L_{TSind}^{is} multiplied by the solid angle π and the sample surface S_{TS}.

$$\Phi_{TSind}^{abs} = \frac{S_{TS}}{S_{is}} M_{TS} \cdot A_{TS} \cdot \Phi^{LED} = f_{TS} \cdot M_{TS} \cdot A_{TS} \cdot \Phi^{LED}, \tag{11a}$$

$$L_{TSind}^{is} = \frac{\Phi^{LED}}{\pi \cdot S_{is}} M_{TS}, \tag{11b}$$

$$M_{TS} = \frac{\rho_{is}}{1 - \rho_{is}(1 - f_{in} - f_f) - f_{in} \cdot \rho_{LED} - f_f \cdot \rho_f + f_{TS} \cdot A_{TS} \cdot \rho_{is}}. \tag{11c}$$

It is instructive to estimate the proportion of indirectly and directly absorbed power by comparing Equation (11a) with Equation (10a)

$$\frac{\Phi_{\text{TSind}}^{\text{abs}}}{\Phi_{\text{TS}}^{\text{abs}}} = f_{\text{TS}} \cdot \frac{M_{\text{TS}}}{F}. \tag{12}$$

Let us take the following realistic parameters: the IS with an inner diameter of 100 mm and baffles with area of about 3000 mm² ($S_{\text{is}} = 34,400 \text{ mm}^2$), sample 12×12 mm ($S_{\text{TS}} = 144 \text{ mm}^2$), multiplier $M_{\text{TS}} = 40$ and $F = 0.5$, then the indirect/direct absorption ratio is 0.335. It means that indirectly absorbed power is about one-third of the directly absorbed power.

Evidently, the indirect excitation takes place also in the second experimental configuration with the TS under direct excitation. It is also described by Equations (11a)–(11c) but the flux Φ^{LED} must be reduced to $\Phi^{\text{LED}}(1 - FA_{\text{TS}})$ due to the "filtering" of incoming radiation by the direct absorption in TS.

At first sight, it is not clear whether the indirect absorption has an adequate influence on the detected signal. Therefore, we have to show that the total (direct + indirect) absorbance (see Equations (10a) and (11a))

$$\frac{\Phi_{\text{TS}}^{\text{abs}}}{\Phi^{\text{LED}}} = F \cdot A_{\text{TS}} + f_{\text{TS}} \cdot M_{\text{TS}} \cdot A_{\text{TS}} \cdot (1 - F \cdot A_{\text{TS}}) \tag{13}$$

is equal to the relative change of the detected signal $(I_{\text{RS}}^{\text{ex}} - I_{\text{TS}}^{\text{ex}})/I_{\text{RS}}^{\text{ex}}$, which is the same as $(L_{\text{RS}}^{\text{is}} - L_{\text{TS}}^{\text{is}})/L_{\text{RS}}^{\text{is}}$. Using Equations (9b), (10b) and a simple relation between M_{RS} and M_{TS}, $(1/M_{\text{RS}}) = (1/M_{\text{TS}}) - f_{\text{TS}}A_{\text{TS}}$ (derived from Equations (9c) and (10c)), we obtain

$$\frac{L_{\text{RS}}^{\text{is}} - L_{\text{TS}}^{\text{is}}}{L_{\text{RS}}^{\text{is}}} = 1 - (1 - F \cdot A_{\text{TS}})\frac{M_{\text{TS}}}{M_{\text{RS}}} = F \cdot A_{\text{TS}} + f_{\text{TS}} \cdot M_{\text{TS}} \cdot A_{\text{TS}}(1 - F \cdot A_{\text{TS}}), \tag{14}$$

which is equal to Equation (13). By this we proved that *both the direct and the indirect absorption in a TS have adequate influence on measured signals*; the first one through "filtering" of the incoming radiant flux and the second one through changes of the IS multiplier.

Let us briefly discuss the role of the factor F. In Figure 5, we plot the direct and indirect absorbance as function of F for the case of highly absorbing (Figure 5(a)) and weakly absorbing sample (Figure 5(b)). The sample fill factor f_{TS} and the sphere multiplier M_{TS} were taken as 0.0046 and 40, respectively. Obviously, the total absorbance always decreases with reducing F (even if the indirect absorption increases for highly absorbing sample), which can significantly impair precision of the QY determination for samples with low absorbance. For such samples we recommend having the F fraction high (it means the excitation beam illuminating roughly the whole sample surface).

From the above discussion we conclude that the correct absorbed power can be obtained from the first two experiments, i.e. direct excitation of RS (or empty IS) and of TS. The difference of these excitation signals must be corrected by the sensitivity C (in order to avoid confusions, we shall use notation C^{ex} and C^{em} for sensitivity in the excitation and emission bands, respectively, even if these are just different points of the whole sensitivity spectrum $C(\lambda)$) to give the correct spectral radiance

$$\frac{I_{\text{RS}}^{\text{ex}} - I_{\text{TS}}^{\text{ex}}}{C^{\text{ex}}} = \frac{K \cdot k \cdot \Phi^{\text{LED}}(M_{\text{RS}} - (1 - F \cdot A_{\text{TS}}) \cdot M_{\text{TS}})}{K \cdot k \cdot M^{\text{cal}}}. \tag{15}$$

In the case of emission measurements, the PL radiation from a sample plays the role of an input flux, i.e. we have to replace Φ^{LED} by either $\Phi_{\text{TS}}^{\text{PL}}$ or $\Phi_{\text{RS}}^{\text{PL}}$ and derive the relation for emitted spectral

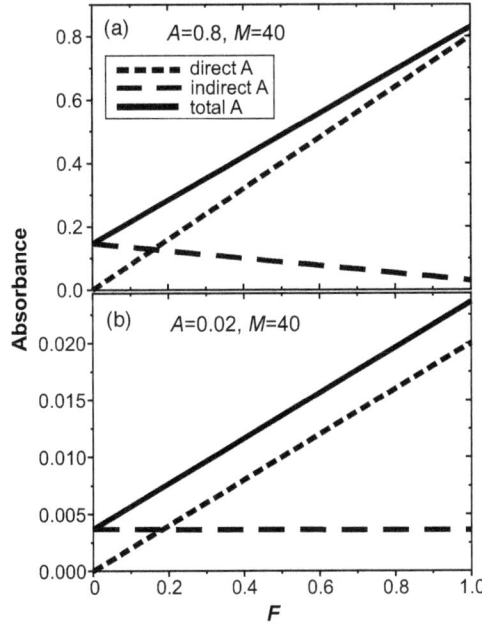

Figure 5.　Relation (Equation (13)) between the direct (short-dashed line), indirect (dashed line) and total (solid line) absorbance as function of the direct flux fill factor F for the case of (a) highly absorbing ($A = 0.8$) and (b) weakly absorbing ($A = 0.02$) sample. The sample area fill factor f_{TS} and the sphere multiplier M_{TS} are 0.0046 and 40, respectively.

radiance (power) analogical to Equation (15). This is a straightforward task if the (re)absorption of PL within the RS and TS samples is supposed to be negligible

$$\frac{I_{TS}^{em} - I_{RS}^{em}}{C^{em}} = \frac{K \cdot k \cdot (\Phi_{TS}^{PL} \cdot M_{TS} - \Phi_{RS}^{PL} \cdot M_{RS})}{K \cdot k \cdot M^{cal}} = \eta_{power} \cdot \frac{(I_{RS}^{ex} - I_{TS}^{ex})}{C^{ex}}. \tag{16}$$

In principle, there should be no PL signal from the RS (before we supposed that its absorption is negligible) but there can be detectable Raman scattering (mostly from the sample matrix, especially for liquid samples). Most often also absorbance by a TS in the emission region is negligible – in this case $M_{RS} = M_{TS}$. However, if the sample shows significant overlap of absorption and luminescence spectra, the reabsorption of PL takes place and can affect the precision of a QY determination. A method for analysing the effects of self-absorption was proposed by Ahn et al.[20]

The emitted power given by Equation (16) is, obviously, equal to the absorbed power (given by Equation (15)) multiplied by the PL power efficiency, which can be determined by dividing Equation (16) by Equation (15). The power efficiency is then converted to QY by dividing of both Equations (15) and (16) with the photon energy.

We have to remind that most of the quantities in Equations (15) and (16) are spectrally dependent and the emission and excitation measurements concern different spectral regions. Therefore, the overall PL QY must include integration over the whole excitation and emission spectral regions

$$\eta = \frac{\int_{em.band} \dfrac{I_{TS}^{em}(\lambda_{em}) - I_{RS}^{em}(\lambda_{em})}{C(\lambda_{em}) \cdot T_{EmF}(\lambda_{em}) \cdot (h \cdot c/\lambda_{em})} \cdot d\lambda_{em}}{\int_{ex.band} \dfrac{I_{RS}^{ex}(\lambda_{ex}) - I_{TS}^{ex}(\lambda_{ex})}{C(\lambda_{ex}) \cdot T_{ExF}(\lambda_{ex}) \cdot (h \cdot c/\lambda_{ex})} \cdot d\lambda_{ex}}. \tag{17}$$

In this equation for QY η, the nominator represents number of PL photons emitted by the TS, while denominator stands for the number of absorbed photons. The functions T_{ExF} and T_{EmF} represent transmittance spectra of filters used during acquisition of excitation and emission signals (e.g. a long-pass filter cutting off excitation photons) and hc/λ is the energy of photon.

The calculation of integrals in Equation (17) can be considerably simplified (and easily automated) when all spectral data are represented at the same spectral points (detector pixels). Then the integral is replaced by a sum of the calculated values in all spectral points multiplied by the span between spectral points $\Delta\lambda$, i.e.

$$\eta = \frac{\sum_{em.band} \dfrac{I_{TS}^{em}(\lambda_{em}) - I_{RS}^{em}(\lambda_{em})}{C(\lambda_{em}) \cdot T_{EmF}(\lambda_{em}) \cdot (h \cdot c/\lambda_{em})} \cdot \Delta\lambda_{em}}{\sum_{ex.band} \dfrac{I_{RS}^{ex}(\lambda_{ex}) - I_{TS}^{ex}(\lambda_{ex})}{C(\lambda_{ex}) \cdot T_{ExF}(\lambda_{ex}) \cdot (h \cdot c/\lambda_{ex})} \cdot \Delta\lambda_{ex}}. \tag{18}$$

This equation represents the essence of our PL QY method, as it includes all necessary input signals, and indicates the most straightforward approach to PL QY experiments.

We used the following *standard experiment procedure* for obtaining PL QY:

(a) Determine the excitation and the emission range for investigated samples.
(b) Consider whether a RS is needed or an empty IS can be used.
(c) Decide the spectral measurement ranges and the use of filters.
(d) Stabilize the first excitation LED and start the stabilization of the second LED.
(e) Perform measurement in both excitation and emission region with the RS.
(f) Then the same for one or more investigated samples (if the duration of experiment is long then repeat from time to time the reference measurement – possibly after each TS – and briefly check that the LED signal is stable, if not stop the measurement for this LED and start again after sufficient LED stabilization).
(g) At the end measure the RS again.
(h) Change the stabilized LED and repeat steps e–g for all LEDs from the desired spectral region.
(i) Data treatment for each LED should start by comparing all reference measurements. If they indicate LED intensity variations higher than the desired precision, disregard these experiments done with the unstable LED.
(j) The corresponding data sets are subtracted, divided by the sensitivity spectrum $C(\lambda)$, by transmittance $T(\lambda)$ of used filters and by photon energy hc/λ. Then all data points within the excitation and the emission range are multiplied by the width of a spectral step $\Delta\lambda$ and summed up (Equation (18)). The final QY is a ratio of the emission/excitation integral signals.

We have to note that the above given procedure is the same for both absolutely and relatively calibrated apparatus. The difference is that the spectral sensitivity $C(\lambda)$ is given only in arbitrary units in the case of relative calibration. Obviously, Equations (18) and (19) contain ratio of $C(\lambda_{ex})/C(\lambda_{em})$, i.e. $C(\lambda)$ can be given in any units as only the spectral shape of $C(\lambda)$ is important.

4. Experimental examples

The above described technique is an absolute method to determine PL QY which does not require comparison with fluorescence standards. However, it can be useful to test the prepared and calibrated set-up by measuring some fluorescence standards if available.[21]

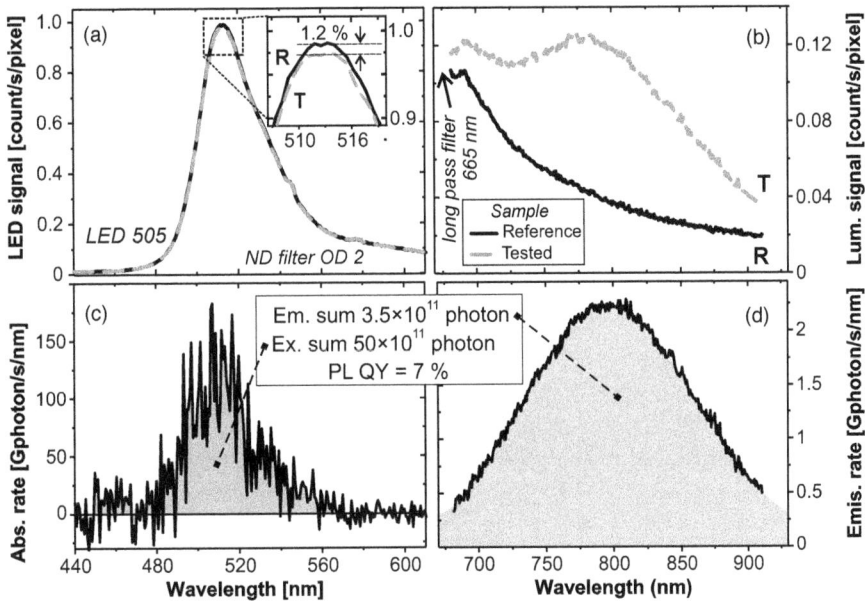

Figure 6. PL QY experiments on the sample of SiNC/SiO$_2$ multilayers on a quartz substrate (its absorbance is given in Figure 2(b)). The upper panels show experimental data for (a) the excitation region (using LED 505 with actual peak around 513 nm) and (b) the emission region. The black solid and grey dashed lines represent the signal from reference and TS, respectively. The inset in the panel (a) gives expanded view on the excitation peak area. The lower panels (c) and (d) show the corrected differences of spectra from panels (a) and (b), respectively. The ratio of integrated emission and excitation signals gives QY of about 7%.

Here, we demonstrate (Figure 6) the method and ability of our set-up using a solid-state material having the form of a substrate with active material deposited on it. More specifically, we used multilayers of silicon nanocrystals in the SiO$_2$ matrix deposited on a quartz substrate (12 × 12 mm).[22] This specific sample can reveal the capability of our set-up because it has rather low absorption (increasing from red towards the UV spectral region, see Figure 2(b)) and large (Stokes) shift between absorption (UV-blue) and emission (red and NIR) regions (Figure 6). The excitation and emission bands are detected separately using one or two detection windows (if we want to fully characterize the very broad emission band) and adequate filters in the detection path. These filters are used either to stop excitation photons which otherwise appear at multiples of excitation wavelength in a grating spectrometer (this is performed by the adequate long-pass filters during an emission window acquisition) or attenuate the detected (non-absorbed) excitation signal (with neutral density filters during an excitation window acquisition) in order to avoid saturation of the detector range. Low absorbance of this type of thin-film samples is demonstrated in Figure 6(a) as very small difference of excitation signals between reference and TS. Despite such low absorption the determination of PL QY is still possible if the excitation source (LED) was sufficiently stable and the direct excitation fill factor F was as high as possible. On the other hand, measurement of the emission spectral band is reliable as there is only weak background signal from the RS and the value of QY is relatively high, about 7% (Figure 6(b)).

5.   Error analysis

In optical spectroscopy we deal with photon signals, i.e. detection events are discrete and therefore described by the Poisson probability function (see, e.g. [23]). Then the standard deviation of a measurement with the mean count value of N is equal to $\sqrt{N}$.

There are four photon signals in Equation (18) ($I_{TS}^{em}, I_{RS}^{em}, I_{TS}^{ex}, I_{RS}^{ex}$) along with photon energy (hc/λ) and transmittance of filters $T(\lambda)$ (which are considered to be precise, i.e. without uncertainty). The last component is the sensitivity spectrum $C(\lambda)$ which has also limited precision (see discussion below).

In order to obtain a fast estimation of the PL QY uncertainty, we shall modify Equation (18) into the following form:

$$
\eta \cong \frac{T_{ExF}(\bar{\lambda}_{ex})\bar{\lambda}_{em} \cdot \Delta\bar{\lambda}_{em}}{T_{EmF}(\bar{\lambda}_{em})\bar{\lambda}_{ex} \cdot \Delta\bar{\lambda}_{ex}} \cdot \frac{C(\bar{\lambda}_{ex})}{C(\bar{\lambda}_{em})} \cdot \frac{\sum_{em.band}(I_{TS}^{em}(\lambda_{em}) - I_{RS}^{em}(\lambda_{em}))}{\sum_{ex.band}(I_{RS}^{ex}(\lambda_{ex}) - I_{TS}^{ex}(\lambda_{ex}))}
$$

$$
\cong const \cdot SR(\lambda_{ex}, \lambda_{em}) \cdot \frac{N_{TS}^{em} - N_{RS}^{em}}{N_{RS}^{ex} - N_{TS}^{ex}},
$$

(19)

where all components, except the light signals, were extracted outside the sum and replaced by their mean value (over the range of emission or excitation band). We assume that this is very good approximation for narrow bands and acceptable estimate for broader bands. The first fraction in Equation (19) is then considered to be precise constant without uncertainty.

The second fraction in Equation (19) (labelled SR – sensitivity ratio), i.e. the mean sensitivity in the excitation band divided by its value for emission band, is also constant but known only with limited precision. This uncertainty does not have a character of a random error but an unknown systematic error. Note that any error in scaling of the sensitivity spectrum is not important as we have here a ratio of two points from this spectrum (in other words, we do not need the absolute calibration of an apparatus response, but the relative calibration – i.e. correcting for the spectral shape deformation – is enough). For PL QY measurements involving a limited spectral range (absorption and emission wavelengths being close together) within the region of optimal response of an apparatus, the uncertainty of the SR could be neglected. But in case of the very broad investigated spectral range, which includes regions of decreasing sensitivity, the uncertainty of sensitivity must be considered (see Figure 7(b)). However, the reliable determination of the uncertainty of a sensitivity calibration is a complicated task which goes beyond the scope of this paper (and will be published separately).

In the last term of Equation (19), we replaced the sum of measured intensity differences by differences of summed intensity. The intensity signals summed over the whole excitation and emission bands (N) are also described by the Poisson statistics, therefore the relative error of the nominator and denominator can be written as

$$
\frac{\alpha_{N_{ex}}}{N_{ex}} = \frac{\sqrt{N_{TS}^{ex} + N_{RS}^{ex}}}{(N_{RS}^{ex} - N_{TS}^{ex})}, \quad \frac{\alpha_{N_{em}}}{N_{em}} = \frac{\sqrt{N_{TS}^{em} + N_{RS}^{em}}}{(N_{TS}^{em} - N_{RS}^{em})}.
$$

(20)

Finally, an estimate of the overall relative uncertainty of PL QY is obtained from the rules for propagation of errors in multivariable functions as

$$
\frac{\alpha_\eta}{\eta} = \sqrt{\left(\frac{\alpha_{N_{em}}}{N_{em}}\right)^2 + \left(\frac{\alpha_{N_{ex}}}{N_{ex}}\right)^2 + \left(\frac{\alpha_{SR}}{SR}\right)^2},
$$

(21)

where the result from Equation (20) must be inserted.

The above described error analysis is applied to a set of PL QY measurements (7) on a pair of samples of SiNC/SiO$_2$ multilayers on the quartz substrate (similar to the sample represented in Figure 6); one of these samples is additionally passivated by hydrogen annealing which increases its PL QY. The absorption spectrum of both samples is almost identical and has the form of a smooth edge starting from the blue-green region and steadily increasing towards shorter wavelengths (similar to the spectrum shown in Figure 2(b)). The estimated relative errors of the three

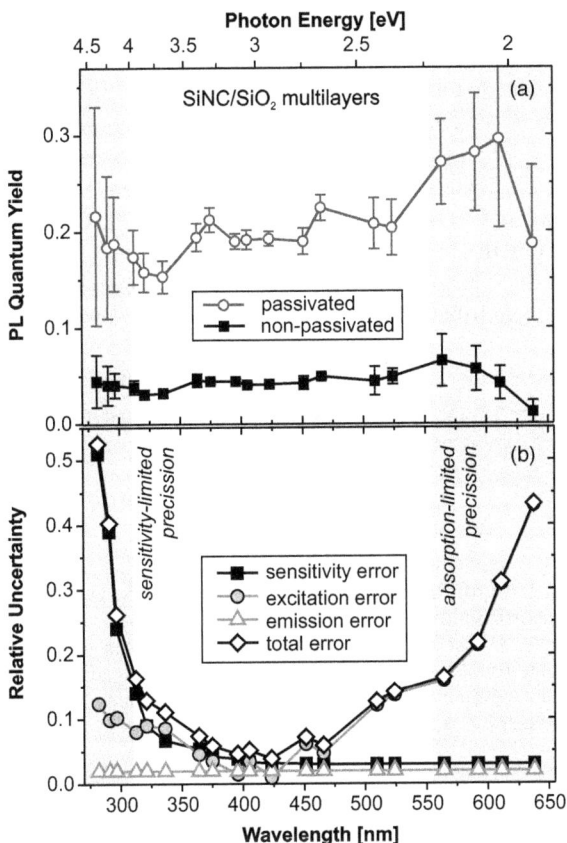

Figure 7. (a) PL QY excitation spectra for a pair of SiNC/SiO$_2$ multilayer samples: the non-passivated sample (black rectangles) and the sample passivated by annealing in hydrogen (open circles). (b) Error analysis of the above PL QY experiments: the total uncertainty (open diamonds) is combined (Equations (20) and (21)) from the excitation (grey circles), emission (open triangles) and sensitivity (black squares) errors. The sensitivity error is dominant at the short-wavelength (UV) side, while the absorption error dominates the long-wavelength side (due to very low absorption of these wavelengths by the sample).

components in Equation (21) are plotted in the lower panel (Figure 7(b)). The lowest contribution comes from the emission signal, while the excitation and correction uncertainty dominates the long- and short-wavelength edge, respectively.

The main purpose of the above error analysis (simplified in some points) was to identify the dominant source of uncertainty and possibly take measures to reduce it. In the presented example, we could improve precision of PL QY determination by increasing absorbance of samples (making thicker active layers) and by improving sensitivity of the experimental apparatus in the UV spectral range.

6. Conclusions

We have described the apparatus and the experimental procedure for determination of absolute QYs of luminescence in a broad spectral range using excitation by LEDs (with excellent temporal stability of emission) and application of an IS. In contrast to commercial spectrometers oriented to the determination of PL QY in organic fluorophores, our set-up is designed for QY measurements of solid-state nanomaterials with relatively low absorption and/or PL QY.

The radiometry theory of ISs was exploited to propose the simplest possible methodology of PL QY determination which involves measurements in only two configurations: (i) a TS under the direct excitation by an LED and (ii) a RS (or none sample – an empty sphere) excited by an LED. Such a procedure is considerably simpler than that one commonly used to determine PL QY in cuvettes with fluorophores when a sample and reference are measured both under the direct and indirect excitation (shifted out of the exciting beam). Moreover, it is proved that a variable emission pattern of different LEDs has no fatal influence on the PL QY determination with an IS.

Then we formulated practical guidelines for the PL QY experiment and calculations as well as an example experiment on Si nanocrystalline layers. The error analysis of our method is provided identifying the dominant contribution either from the absorption signal or the sensitivity limits of the apparatus.

Finally, we have to stress that the presented comprehensive description of PL QY techniques based on the IS theory has quite general applicability not restricted to the described apparatus and easily adaptable to other experimental configurations.

Acknowledgements

The author thanks Prof. I. Pelant (Institute of Physics, Czech Academy of Sciences, Prague) for critical reading of the manuscript and Prof. M. Zacharias' group from IMTEK, University of Freiburg, Germany, for providing excellent nanocrystalline silicon multilayer samples. He also expresses his appreciation to former and current students, A. Fucikova, A. Raichlova and M. Greben for using the PL QY set-up and so enabling to uncover some problems of the technique. The research leading to these results has received funding from the European Community's Seventh Framework Programme (FP7/2007 – 2013) under grant agreement no. 245977 (project NASCEnT).

Notes

1. The direct coupling of rays from either an excitation source or an emitting sample into an output channel without several reflections from IS walls must be avoided. Otherwise, the randomizing (integrating) role of an IS will be violated and strong signal artefacts generated. The narrow acceptance cone of an output channel itself can possibly avoid the direct signal coupling without using baffles.
2. Note, photons can be wavelength-shifted also by non-elastic scattering (mainly Raman scattering). We suppose that it can be either neglected or corrected by using the RS. Non-linear optical effects are excluded as very low excitation power densities are used.
3. In the case of a sample placed on an IS wall, incident photons passing through a sample are reflected back and pass for the second time (if not absorbed). Neglecting reflections on sample interfaces as well as losses during the back-reflection from an IS wall, the double passage is equivalent to the single passage through a sample with the double thickness. For convenience, the absorbance A used from now on will have meaning of the absorption probability during a double passage.

References

[1] Melhuish. Nomenclature, symbols, units and their usage in spectrochemical analysis: IV. Molecular luminescence spectroscopy. Pure Appl Chem. 1984;56:231–245.

[2] Valenta J, Fucikova A, Vácha F, Adamec F, Humpolíčková J, Hof M, Pelant I, Kůsová K, Dohnalová K, Linnros J. Light-emission performance of silicon nanocrystals deduced from the single quantum dot spectroscopy. Adv Funct Mater. 2008;18:2666–2672.

[3] Demas JN, Crosby GA. The measurements of photoluminescence quantum yields. J Phys Chem. 1971;75:991–1024.

[4] Rurack K. Fluorescence quantum yields: methods of determination and standards standardization and quality assurance in fluorescence measurements I. Techniques. In: Resch-Genger U, editor. Springer Series on Fluorescence, vol. 5. Berlin, Heidelberg: Springer; 2008. p. 101–145.

[5] Würth C, Lochmann C, Spieles M, Pauli J, Hoffmann K, Schüttrigkeit T, Franzl T, Resch-Genger U. Evaluation of a commercial integrating sphere setup for the determination of absolute photoluminescence quantum yields of dilute dye solutions. Appl Spectrosc. 2010;8:733–741.

[6] Resch-Genger U, Rurack K. Determination of the photoluminescence quantum yield of dillute dye solutions (IUPAC technical report). Pure Appl Chem. 2013;85:2005–2026.
[7] Würth C, Grabolle M, Pauli J, Spieles M, Resch-Genger U. Relative and absolute determination of fluorescence quantum yields of transparent samples. Nat Protoc. 2013;8:1535–1550.
[8] Lakowicz JR. Principles of fluorescence spectroscopy. 3rd ed. New York: Springer Science + Bussiness Media LLC; 2006.
[9] Timmerman D, Valenta J, Dohnalová K, de Boer WDAM, Gregorkiewicz T. Step-like enhancement of luminescence quantum yield of silicon nanocrystals. Nat Nanotechnol. 2011;6:710–713.
[10] Feng J, Nasiatka J, Wong J, Chen X, Hidalgo S, Vecchione T, Zhu H, Javier Palomares F, Padmore HA. A stigmatic ultraviolet-visible monochromator for use with a high brightness laser driven plasma light source. Rev Sci Instrum. 2013;84:085114.
[11] Pimputkar S, Speck JS, DenBaars SP, Nakamura S. Prospects for LED lighting. Nat Photon. 2010;3:180–182.
[12] Pelant I, Valenta J. Luminescence spectroscopy of semiconductors. Oxford: Oxford University Press; 2012.
[13] Schubert EF. Light-emitting diodes. 2nd ed. Cambridge: Cambridge University Press; 2006.
[14] de Mello JC, Wittmann HF, Friend RH. An improved experimental determination of external photoluminescence quantum efficiency. Adv Mater. 1997;9:230–232.
[15] McCluney WR. Introduction to radiometry and photometry. Boston, MA: Artech House; 1994.
[16] Goebel DG. Generalized integrating-sphere theory. Appl Opt. 1967;6:125–128.
[17] LabSphere Inc. Technical guide: integrating sphere theory and applications. LabSphere Inc. 2008. Available from: http://www.labsphere.com
[18] James JF. Spectrograph design fundamentals. Cambridge: Cambridge University Press; 2007.
[19] Resch-Genger U, DeRose PC. Characterization of photoluminescence measuring systems (IUPAC technical reports). Pure Appl Chem. 2012;84:1815–1835.
[20] Ahn TS, Al-Kaysi RO, Müller AM, Wentz KM, Bardeen CJ. Self-absorption correction for solid-state photoluminescence quantum yields obtained from integrating sphere measurements. Rev Sci Instrum. 2007;78:086105.
[21] Würth C, Grabolle M, Pauli J, Spieles M, Resch-Genger U. Comparison of methods and achievable uncertainties for the relative and absolute measurement of photoluminescence quantum yields. Anal Chem. 2011;83:3431–3439.
[22] Hartel AM, Hiller D, Gutsch S, Löper P, Estradé S, Peiró F, Garrido B, Zacharias M. Formation of size-controlled silicon nanocrystals in plasma enhanced chemical vapor deposition grown SiO_xN_y/SiO_2 superlattices. Thin Solid Films. 2011;520:121–127.
[23] Hughes IG, Hase TPA. Measurements and their uncertainties: a practical guide to modern error analysis. Oxford: Oxford University Press; 2010.

19

Sensitivity optimization for the index sensors based on waveguide metallic photonic crystals through angle-resolved tuning

Jian Zhang[a,b] and Xinping Zhang[a,b]*

[a]Institute of Information Photonics Technology, Beijing University of Technology, Beijing, People's Republic of China; [b]College of Applied Sciences, Beijing University of Technology, Beijing, People's Republic of China

An optical sensor device based on waveguide metallic photonic crystals for the detection of the environmental refractive-index change is investigated. The sensitivity of the device is found to be improved through optimizing the angle of incidence of the light. This is based on the overlap of the waveguide resonance mode with the most sensitive spectral position of the spectroscopic response of the localized surface plasmon. Experimental investigations were performed on the glucose/water solution with different concentrations. The measurement data verified the mechanisms for the optimization of the sensor sensitivity by adjusting the incident angle of the light, which show excellent linearity and high reproducibility.

Keywords: waveguide metallic photonic crystals; localized surface plasmon; waveguide resonance mode; angle-resolved tuning; refractive-index sensor

1. Introduction

Metallic photonic crystals (MPCs) are widely used in optical switching,[1] optical filters,[2,3] and sensors.[4–6] A number of sensors based on localized surface plasmon resonance (LSPR) or particle plasmon resonance (PPR) have been demonstrated.[7,8] Through coupling with the waveguide resonance mode, the sensitivity of the sensors based on PPR can be enhanced significantly, where a sensor device has been demonstrated for the detection of specific bioreactions.[9] This kind of sensor defined by the so-called waveguide MPCs is featured with simplicity in the realization of the nanostructure sensor chip, small sensing volumes, easier control and modification of parameters of the nanostructures.

In this work, we demonstrate that the sensitivity of this kind of sensor device can be enhanced further by optimizing the angle of incidence. Liquid samples of glucose/water solutions with different concentrations are used as the detection targets.

2. Fabrication of the waveguided MPC structures for index sensors

2.1. Fabrication and characterization of MPCs

Solution-processible method [10,11] based on interference lithography and colloidal gold nanoparticles is employed for the fabrication of MPCs. This shows obvious advantages of

---

*Corresponding author. Email: zhangxinping@bjut.edu.cn

simplicity, high efficiency, low cost and large-area homogeneity, as compared with the conventional methods. In this paper, we employ a two-stage annealing process. In the first stage, an annealing temperature of $350°C$ is employed and the photoresist grating functions as the template for controlling the shape and volume of gold in the formation of the precursor MPC of the one-dimensional array of gold nanostructures. In the second stage, the sample is annealed at $450°C$, the photoresist is removed through complete evaporation without destroying the periodical arrangement of the gold nanostructures.

The master grating made of photoresist (S1805 from Rohm and Haas) was fabricated by interference lithography, where a 325 nm laser from Kimmon Electric Co. was split into two beams with a separation angle of about $48°$ before the two arms were overlapped onto the photoresist film. Then, a colloidal solution of gold nanoparticles was spin-coated onto the master grating before the annealing processes. The two-stage annealing process was performed in an muffle furnace.

The MPC chip composed of periodically arranged gold nanostructures on a waveguide of indium tin oxide (ITO) was the central part of the sensor. Figure 1 shows the scanning electron microscopic (SEM) and the atomic force microscopic (AFM) images of the gold nanostructures.

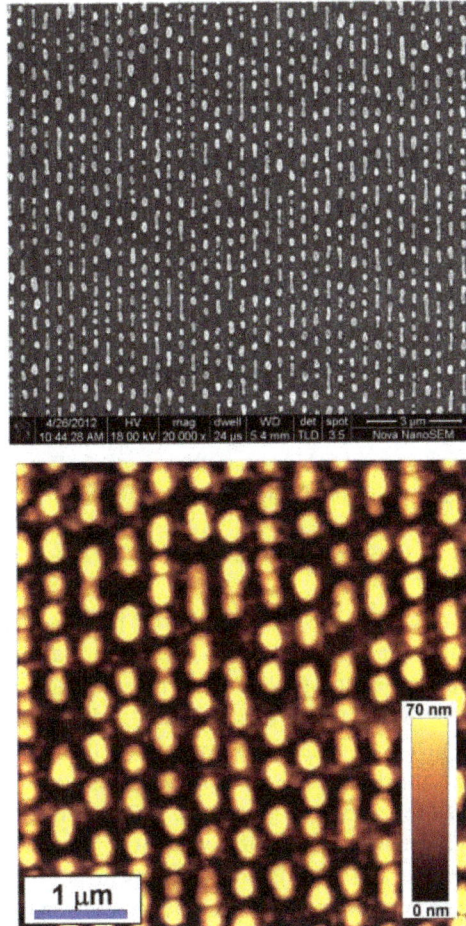

Figure 1.　SEM and AFM height images of waveguided MPC structures.

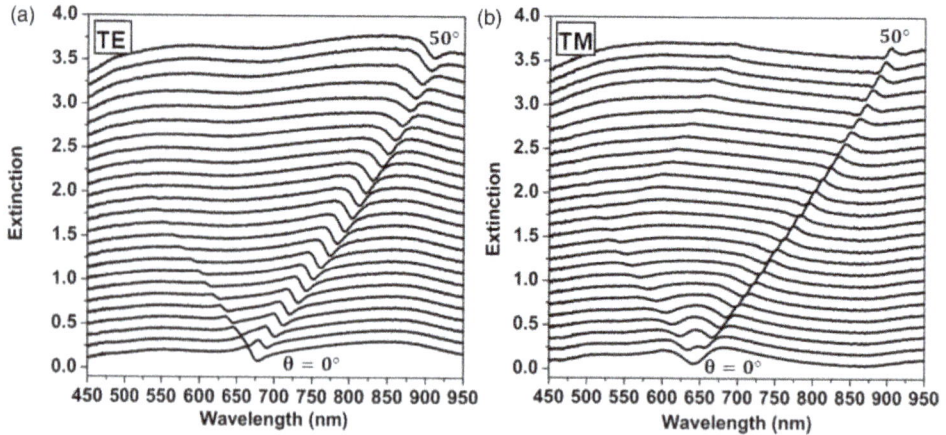

Figure 2.   Optical extinction spectra for TE and TM polarizations in air, respectively.

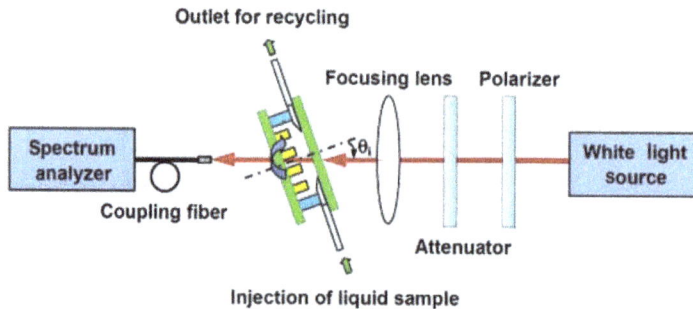

Figure 3.   Schematic drawing of the experimental setup.

The grating structures have a period of about 400 nm and a modulation depth of about 70 nm. The waveguide was provided by a 200-nm ITO layer coated on the glass substrate with a thickness of 1 mm and an effective area of $10 \times 10 \, mm^2$. Optical extinction spectra of the structures shown in Figure 2 were measured in air at an incident angle increased from 0° to 50° in steps of 2° for transverse electric (TE) (a) and transverse magnetic (TM) (b) polarizations.

2.2.   The design of the sensor system

The basic principles of the sensor device based on MPCs have been described in detail in [9]. Figure 3 shows the schematic of the sensor device. The device was mounted on a rotation stage so that white light could be incident on the MPCs at tunable angles. The incident white light passed through a polarizer, an attenuator, a focusing lens and the window of chamber holding the MPC chip before reaching a fiber-coupled spectrometer (USB4000 from Ocean Optics). The spectrometer was used to measure the transmission or extinction spectra through the sensor device. The liquid sample, which is the solution of glucose in water, is injected from the inlet into the chamber and drained through the outlet back to the container of the liquid sample. The circulation channel connecting the chamber and the container is constructed by soft silicon rubber tubes. The spectroscopic response of the waveguided MPC device was tunable by changing the angle of incidence, as the waveguide resonance mode was strongly dependent on the angle of incidence. The sensor signal could be maximized by optimizing the incident angle of the white light.

3.   Sensitivity optimization of the sensor based on waveguided MPCs

3.1.   *Definition of the sensor signal spectrum*

Figure 4 illustrates the definition of the sensor signal using a measured extinction spectrum. The signal has been measured by taking two transmission spectra $I_S(\lambda)$ and $I_0(\lambda)$ before performing the following calculation:

$$E_S(\lambda) = -\log_{10}\left[\frac{I_S(\lambda)}{I_0(\lambda)}\right],$$

where $I_S(\lambda)$ is the transmission spectra of the circulation of glucose solutions with different concentrations and $I_0(\lambda)$ is the blank spectrum of pure water. The amplitude of the sensor signal is defined by the peak-to-valley difference of the extinction spectrum. Thus, the sensitivity of the sensor device is expressed as $S = \Delta A/\Delta n$, where Δn is the practical change in the environmental refractive index.

3.2.   *Measurements on the sensor signal at different concentrations of the glucose/water solution for different angles of incidence*

The optical extinction spectroscopic properties were measured with the concentration of the glucose/water solutions increased from 1% to 10%. For each concentration, the optical extinction spectrum was measured at an incident angle increased from 0° to 28°. Figure 5(a)–(g) corresponds to incident angles of 0°, 5°, 10°, 15°, 20°, 25°, and 28°. Between each two adjacent changes of the solution concentration, pure water was circulated for 10 min to clean the channels and the MPC chip.

As shown in Figure 5(a)–(g), with increase in the angle of incidence, both the spectral position and the amplitude of the sensor signal change dramatically with change in the angle of incidence. The sensor signal resulting from the longer wavelength branch of the coupled mode between the waveguide and plasmonic resonance modes [9] shifts from about 670 to about 820 nm as the incident angle is increased from 0° to 28°. A maximum amplitude of the sensor signal can be observed in Figure 5(f) for an incident angle of 25°. This indicates a possible optimization of the sensitivity of the sensor device through change in the angle of incidence. It should be noted that the scaling of the vertical axis was not kept unchanged for the measured spectra in Figure 5. According to the measurement using an Abbe refractometer, the refractive index increases from

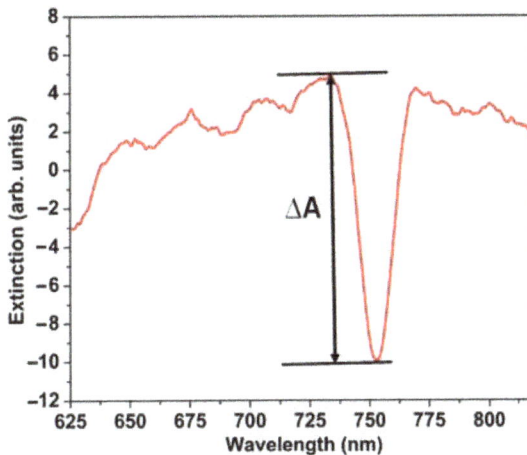

Figure 4.   Schematic illustration of the definition of the sensor signal.

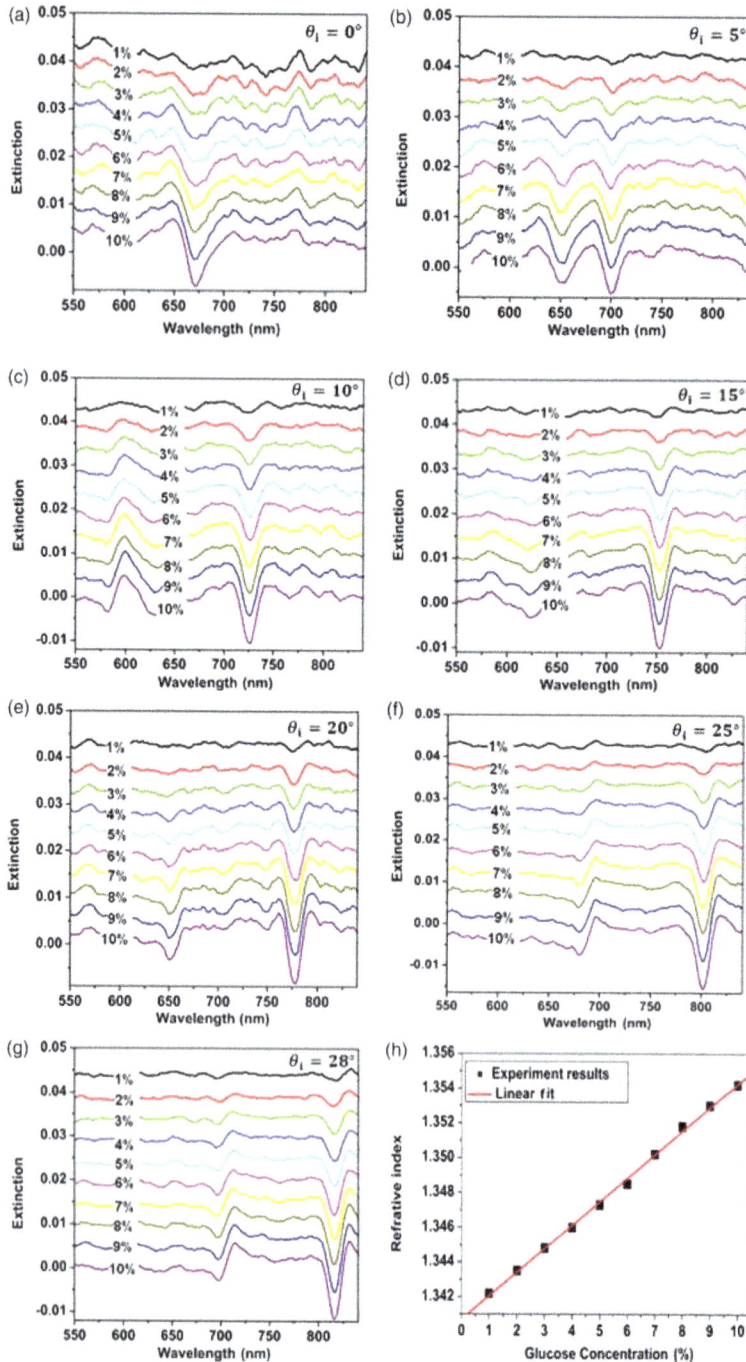

Figure 5. (a)–(g) Optical extinction spectra of glucose/water solutions with the concentration increased from 1% to 10% (W/W) using pure water as the blank sample at different incident angles: (a) 0°, (b) 5°, (c) 10°, (d) 15°, (e) 20°, (f) 25°, (g) 28°. (h) The refractive index of the glucose/water solution measured by an Abbe refractometer with the concentration increased from 1% to 10%.

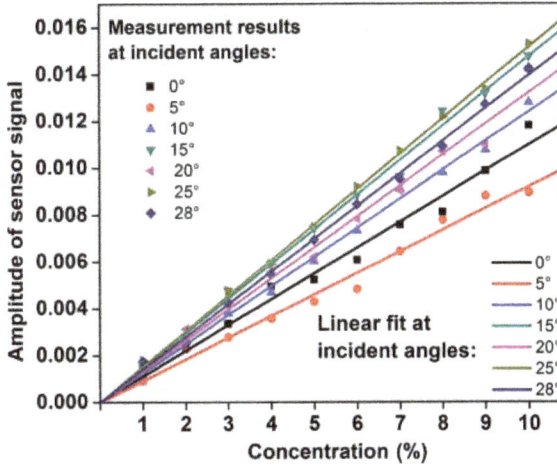

Figure 6.   Measurements on the amplitude of the sensor signal as a function of the concentration of the sample solution at different angles of incidence. The solid lines show the linear fit to each group of the measurements.

1.330 to 1.334 as the concentration of the glucose/water solution is increased from 1% to 10%. Figure 5(h) shows a good linear relationship between concentration and refractive index of the corresponding glucose solution.

To demonstrate the enhancement of the sensitivity of the sensor device, we make a comparison between the variations of the amplitude of the sensor signal as a function of the solution concentration at different angles of incidence, as shown in Figure 6. The measurement data are given by dots with different shapes and the lines with different colors are the corresponding linear fits. It should be noted that the data in Figure 6 have been shifted vertically so that all of them pass through the point of (0, 0) of the coordinates considering that the intensity of the light source changes with time, which may lead to the vertical shift of the amplitude data. Clearly, the smallest slope of both the measured data and the fitting line is observed for an incident angle of 5°, whereas at an incident angle of 25°, the measurement data have the largest slope and the slope drops again when the incident angle is increased further to 28°. Actually, the slope of the plots in Figure 5 corresponds exactly to the sensitivity of the sensor by $\Delta A / \Delta n$. Thus, the optimized value of the incident angle is 25° for this sensor device. The sensitivity of the device is measured to be about 4.6 per refractive-index units (RIU); however, this value reduces to about 3.3/RIU at an incident angle of 5°, indicating an enhancement factor of about 1.4 at an 25° incident angle.

All of the measurements in Figure 6 were performed on the TM polarization [11] at room temperature, where the definitions of the polarization direction and the angle of incidence are given in [11].

4.   Conclusions

We demonstrate an index sensor using waveguided MPCs consisting of periodically arranged gold nanostructures. The coupling between the LSPR and waveguide resonance mode of the waveguide grating structures is the basic principle and the basic mechanism for enhancing the sensitivity of the sensor device. Systematic measurements on spectroscopic response of the sensor device to the change of the concentrations of the glucose/water solutions found that the sensitivity of the sensor may be improved by changing the angle of incidence. The experimental data show a highest sensitivity at an incident angle of 25° with an enhancement factor of about 1.4 with respect to that at an incident angle of 5°.

Acknowledgements

The authors acknowledge the National Key Basic Research Program of China (2013CB922404) and the National Natural Science Foundation of China (11074018, 11274031) for the support.

References

[1] Nau D, Bertram RP, Buse K, Zentgraf T, Kuhl J, Tikhodeev SG, Gippius NA, Giessen H. Optical switching in metallic photonic crystal slabs with photoaddressable polymers. Appl Phys B. 2006;82:543–547.

[2] Sharon A, Rosenblatt D, Friesem AA. Resonant grating waveguide structures for visible and near-infrared radiation. J Opt Soc Am A. 1997;14:2985–2993.

[3] Zhang XP, Liu HM, Tian JR, Song YR, Wang L, Song JY, Zhang GZ. Optical polarizers based on gold nanowires fabricated using colloidal gold nanoparticles. Nanotechnology. 2008;19:1–6;285202.

[4] Nau D, Seidel A, Orzekowsky RB, Deb SH, Giessen H. Hydrogen sensor based on metallic photonic crystal slabs. Opt Lett. 2010;35:3150–3152.

[5] Zhang XP, Dou F, Liu HM. Molecular concentration sensor based on the diffraction resonance mode of gold nanowire gratings. Nanotechnology. 2010;21:1–7;335501.

[6] Malyarchuk V, Hua F, Mack NH, Velasquez VT, White JO, Nuzzo RG, Rogers JA. High performance plasmonic crystal sensor formed by soft nanoimprint lithography. Opt Express. 2005;13:5669–5675.

[7] Roh S, Chung T, Lee B. Overview of the characteristics of micro-and nano-structured surface plasmon resonance sensors. Sensors. 2011;11:1565–1588.

[8] Zhang S, Bao K, Halas NJ, Xu H, Nordlander P. Substrate-induced fano resonances of a plasmonic nanocube: a route to increased-sensitivity localized surface plasmon resonance sensors revealed. Nano Lett. 2011;11:1657–1663.

[9] Zhang XP, Ma XM, Dou F, Zhao PX, Liu HM. A biosensor based on metallic photonic crystals for the detection of specific bioreactions. Adv Funct Mater. 2011;21:4219–4227.

[10] Zhang XP, Sun BQ, Friend RH, Guo HC, Nau D, Giessen H. Metallic photonic crystals based on solution-processible gold nanoparticles. Nano Lett. 2006;6:651–655.

[11] Zhang XP, Sun BQ, Guo HC, Tetreault N, Giessen H, Friend RH. Large-area two-dimensional photonic crystals of metallic nanocylinders based on colloidal gold nanoparticles. Appl Phys Lett. 2007;90:1–3;133114.

20

Dependence of activation energy and lattice strain on TiO$_2$ nanoparticles?

Hossain Milani Moghaddam[a,b,c]* and Shahruz Nasirian[a,d]

[a]*Department of Physics, University of Mazandaran, Babolsar, Iran;* [b]*Nano and Biotechnology Research Group, University of Mazandaran, Babolsar, Iran;* [c]*Nanotechnology group, Hariri Scientific Foundation, Babol, Iran;* [d]*Department of Basic Sciences, University of Science and Technology of Mazandaran, Babol, Iran*

Titanium dioxide nanopowders were synthesised by the sol–gel method using TiCl$_4$ as a precursor under argon gas atmosphere. Effect of alteration of lattice strain with a change of synthesis parameters on the size of nanocrystallites and activation energy of phase transformation from anatase to rutile was investigated. Activation energy is the minimum energy required for overcoming the energy barrier for phase transformation between the two phases. The growth of nanocrystallite and a decrease in the positive lattice strain occurred when the calcinations increased from 300°C to 550°C. After the calcination increased at 600°C, the sign of the lattice strain was altered from positive to negative, the growth of crystallite continued and phase transformation occurred. In spite of the fact that the anatase crystallites grew with the increase of the gelatinisation time, the activation energy decreased. In addition, the rapid decrease of the lattice strain in the as-prepared sample in lower mixing time causes an increase of the activation energy, and the mass fraction of the rutile phase occurred.

Keywords: TiO$_2$ nanoparticles; sol–gel method; lattice strain; phase transformation; activation energy

1. Introduction

Titanium dioxide has been receiving substantial attention in the recent years due to its wide range of commercial applications, such as a photocatalysts, in white pigments, in gas sensors, in hygienic and medical instruments, in ceramics, and in vehicles external surfaces [1–6]. The applications of TiO$_2$ strongly depend on the high homogeneity and definite phase composition, particle size, high surface area and porosity. Moreover, in the evaluation of thermal stability and mechanical properties of nanocrystalline TiO$_2$, grain size, surface free energy, activation energy of phase transformation and initial strain are very important factors [6–16].

Titania naturally occurs in three polymorphs: anatase, rutile and brookite. Both anatase and brookite are metastable phases and rutile is a stable one [6,17–22]. Each phase exhibits different physical and chemical properties, which are determined by their structures. Moreover, the theoretical study and experimental results have shown that the photocatalytic and photovoltaic properties of TiO$_2$ nanoparticles with two mixed polymorphic phases (anatase and rutile) are better than pure anatase TiO$_2$ nanoparticles [21–23]. One of the methods used to produce titania

*Corresponding author. Email: milani@umz.ac.ir; hossainmilani@yahoo.com

nanopowders having two different polymorphic phases (anatase and rutile) is the phase transformation and for this purpose one needs to crystallise the as-prepared titanium hydroxide at high temperatures ($\geq 500°C$) [6–7, 24]. The actual transformation behaviour for the preparation of biphase TiO_2 depends on initial particle size, impurity content, starting phase and calcination temperature. Furthermore, the preparation of mixed phase titanium oxide with lower activation energy will be useful both in saving energy and in getting better properties. Recently, the phase transformation is described by means of different methods. For example, Cooper et al. [24] and Mohammadi et al. [25] reported a good explanation for a phase transformation in TiO_2 nanopowders via the Gibbs activation energy. Moreover, Li et al. [26] reported that the activation energy of the phase transformation from anatase to rutile and the transition onset point were decreased and the mass fraction of rutile was increased with the decrease of initial size of anatase TiO_2 nanoparticles. The phase transformation from anatase to rutile in TiO_2 nanopowders is also proportional to the alternation of the nature of lattice forces and the lattice strain. Inagaki et al. [27] have stated that the lattice strain of TiO_2 nanomaterials is obtained using Williamson–Hall method and fitting of the powder diffraction data. They have also reported that the better fit of experimental data points confirms the uniformity of the lattice strain. In addition, they reported that the lattice strain and the content of anatase phase decreased and the crystalline size increased with increasing calcination temperature and the phase transformation from anatase to rutile occurred.

In spite of the fact that many studies have been done on the lattice strain and activation energy of the phase transformation from anatase to rutile structure, there is no study, known to us, which attempts to explain the relationship between the activation energy of the phase transformation of titanium oxide and the growth rate of rutile structure by the alteration of the lattice strain.

In this article, TiO_2 nanoparticles have been synthesised by the sol–gel route using $TiCl_4$ as a precursor. The lattice strain and the size of the crystallites were then extracted using Williamson–Hall model. Moreover, we have computed the activation energy of the phase transformation from anatase to rutile by Arrhenius formula. We have also studied the effect of the change of the calcination and/or the mixing time on the lattice strain, as well as tried to determine the relationship between the alteration of the lattice strain and the activation energy of the phase transformation from anatase to rutile. The roles of both the effects on the growth of nanocrystallites were also investigated. It is demonstrated that the study of the alternation of the lattice strain is useful in finding the initial temperature and the rate of the phase transformation from anatase to rutile, accompanying the explanation for the decrease of the activation energy.

2. Experimental technique

2.1. Sample preparation

TiO_2 nanoparticles were synthesised via the sol–gel method. In a typical synthesis, 2 mL of titanium tetrachloride (99.5% Merck) was slowly added dropwise into 20 mL of ethanol (99.8% Merck) under stirring at a temperature of about 22°C and under argon gas atmosphere. A transparent yellowish solution was formed after adding all the $TiCl_4$. The pH of the transparent yellowish solution was nearly 1.0–1.5. After stirring the solution for several hours under air atmosphere with 88% humidity, a colourless sol was formed, which was then aged for 3 h. We then prepared the gel-solution in this way using ultrasonic waves for 30 min (exposed with ultrasonic waves at a frequency of 40 kHz and 60 Watt power).

The resulting solution was vaporised at 80°C until a dry-gel was obtained. The dry-gel precursor was calcined at different temperatures (at ramping rate of 5°C/ min) for 1 h to form TiO_2 nanopowders.

2.2. *Characterisation*

Characterisation was accomplished using X-ray powder diffraction (XRD). XRD was carried out using a GBC X-ray diffractometer (XRD; GBC MMA) with copper radiation (Cu Kα, $\lambda = 1.54$ Å) through a graphite monochromator and step-scanning measurements in a 2θ range from 10°C to 80°C, with a step of 0.05 2θ and a counting time of 3 s per step with a working voltage of 30 kV. The morphology, particle size and size-distribution of the TiO_2 powders were characterised by transmission electron microscopy (TEM). TEM study was carried out on a Philips CM-120 electron microscopy instrument. The samples for TEM were prepared by dispersing the final powders in ethanol; the dispersion was then dropped on carbon–copper grids.

3. Results and discussion

The XRD patterns of the as-prepared samples calcined at 300–750°C and gelatinised for various times are shown in Figures 1 and 2. The XRD patterns of the TiO_2 powder calcined up to about 500°C indicate the formation of pure anatase TiO_2 phase (anatase: JCPDS file no. 21-1272). No peaks of rutile or brookite phase were detected. It is well known that Scherrer's formula provides only the lower bound to the crystallite size [28]. Here, the crystallite size value is calculated using Scherrer's formula (Table 1) to compare with the values obtained from Williamson–Hall model and TEM micrographs results. The Scherrer's formula is given as follows [19,28],

$$D_S = \frac{K\lambda}{\beta_{hkl} \cos \theta_{hkl}} \tag{1}$$

where D_S is the crystallite size, K the shape factor ($K = 0.9$), $\lambda = 1.54$ Å the wavelength of the X-ray radiation, θ_{hkl} the Bragg diffraction angle and β_{hkl} the broadening of the *hkl* diffraction peak measured at half of its maximum intensity (in radian).

The volume weighted crystallite size, D_{W-H}, and the lattice strain, ε, are estimated by analysing the XRD peak broadening, using the Williamson–Hall method [27–30],

$$\beta_{hkl} \cos \theta_{hkl} = \frac{K\lambda}{D_{W-H}} + 4\varepsilon \sin \theta_{hkl} \tag{2}$$

where K is the shape factor ($K = 0.9$), $\lambda = 1.54$ Å the wavelength of the X-ray radiation, θ_{hkl} the Bragg diffraction angle and β_{hkl} the full width at half maximum intensity observed (in radian). It was calculated using Gaussian fitting of the XRD peaks. Plots were drawn by taking $4 \sin \theta_{hkl}$ along X-axis and $\beta_{hkl} \cos \theta_{hkl}$ along Y-axis, as shown in Figures 3 and 4. The strain present in the material and the crystallite size were, respectively, extracted from the slope and the intercept of the linear fit made to the plot (Table 1). It seems that the conventional Williamson–Hall model is an appropriate approach for the estimation of the lattice strain due to less scatter of data points from the linear fit. The better fit of the experimental data points confirms the uniformity of the lattice

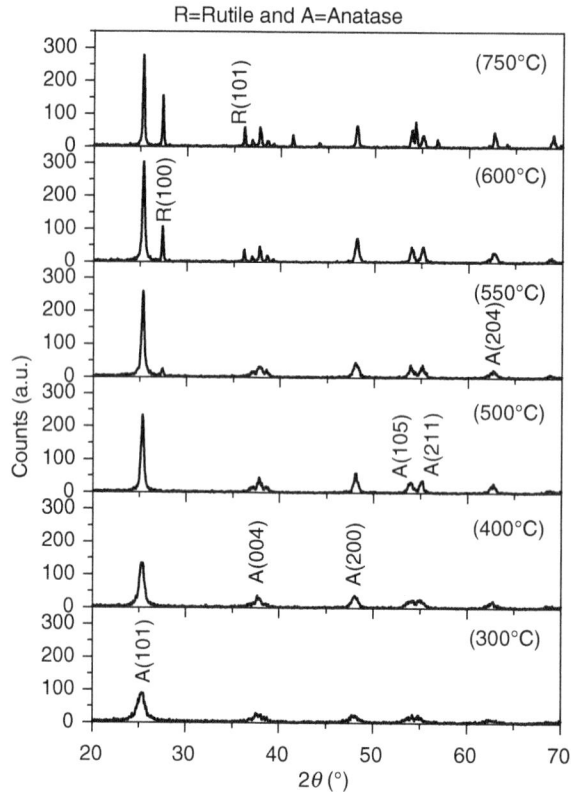

Figure 1.   X-ray diffraction pattern of TiO_2 nanopowders that calcined at different temperatures for 1 h. The sol–gel solution gelatinised for 24 h.

strain. Figures 5 and 6 show the lattice strain versus the calcination temperature and the crystallite size obtained by the Williamson–Hall model. According to Figures 5 and 6, the lattice strain was positive when the samples calcined at 300°C. Moreover, in a calcination step at temperatures ranging from 300°C to 550°C, the size of the nanocrystalline increased and the lattice strain decreased as its value remained positive. However, when the calcination increased from 550°C to 600°C, the sign of the lattice strain was altered from positive to negative and the size of the anatase crystallite continued growth. The lattice strain is positive (negative) when crystal lattice is under the influence of tensile (compressive) forces [31,32]. Apparently, this is because of the alteration in pure resultant forces from the surface to the volume of the crystallites and a change of the octahedral blocks on TiO_2 crystal lattice [8,33–36]. Therefore, the change of the force in the region after 550°C was caused by the distortion of each (all) octahedral(s) and the alteration of the lattice strain in the crystalline lattice; a phase transformation was also demonstrated to occur. The lattice strain was negative in the region after 600°C and there was no change in the sign of the lattice strain. Moreover, the driving force for grain growth results from the reductions of free energy of the system by decreasing the total grain boundary energy [16]. Detailed studies of the activation energy of the phase transformation and the relationship between it and the lattice strain are an urgent require meet.

    The mass fraction of rutile (X_R) in the crystal lattice can be calculated based on the relationship between the integrated intensities of anatase (1 0 1) and rutile (1 1 0) peaks by the following

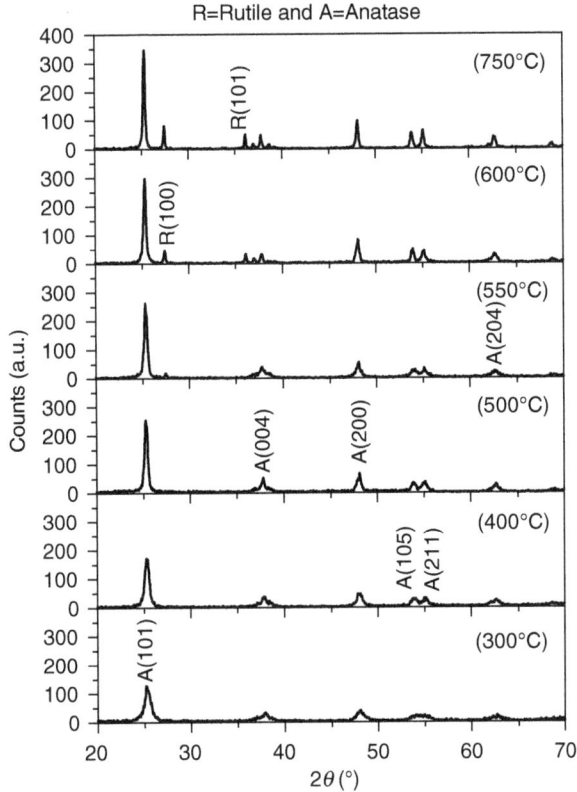

Figure 2. X-ray diffraction pattern of TiO$_2$ nanopowders that calcined at different temperatures for 1 h. The sol–gel solution gelatinised for 120 h.

formula [25],

$$X_{R} = \left[1 + 0.79 \left(\frac{I_{A}}{I_{R}} \right) \right]^{-1} \qquad (3)$$

where I_{A} the integrated peak intensities of the anatase, I_{R} the integrated peak intensities of the rutile and 0.79 the scattering coefficient. In addition, the XRD patterns showed that the mass fractions of anatase phase were more than rutile phase even up to the calcination at 750°C. The mass fractions of the rutile phase were 0.18 (0.31) and 0.17 (0.25) when the samples were mixed for 24 and 120 h and calcined at 600°C (750°C), respectively. The growth of rutile phase was also better in the gelatinising process for 24 h because of the smaller crystallite size of anatase TiO$_2$. This is in agreement with the reported results by Li et al. [26] and Zhang and Banfield [33], who had concluded that anatase crystallite size was the main factor lowering anatase-to-rutile transformation temperature. Furthermore, this is in agreement with the reported results by Inagaki et al. [27], who had suggested that the phase transformation from anatase to rutile structure is mainly governed by the lattice strain and increase in the crystallite size. A decrease in the lattice strain mainly occurs due to the partial transformation of anatase to rutile.

Table 1.   The lattice strain of the as-prepared samples calcined at 300–750°C and gelatinised for various times.

	Gelatinising time (h)							
	24				120			
	Anatase (1 0 1) Crystalline size (nm)				Anatase (1 0 1) Crystalline size (nm)			
Calcination temperature (°C)	W–H method	Scherrer formula	Lattice strain	X_R (%)	W–H method	Scherrer formula	Lattice strain	X_R (%)
300	10.54	7.23	0.00402	0	17.05	11	0.00219	0
400	19.52	12.15	0.00303	0	22.91	14.8	0.00148	0
500	33.16	20.1	0.0015	0	36.57	24.3	0.00095	0
550	39.6	24.2	0.00011	9.2	43.3	26.6	0.00015	7.3
600	44.71	28.06	−0.00094	18.1	47.96	32.55	−0.00075	17.3
750	61.06	37.85	−0.0005	31	51.91	36.2	−0.00091	25.2

Notes: Moreover, the crystallite size values (nm) obtained by Scherrer's equation and Williamson–Hall (W–H) method at different experimental conditions.

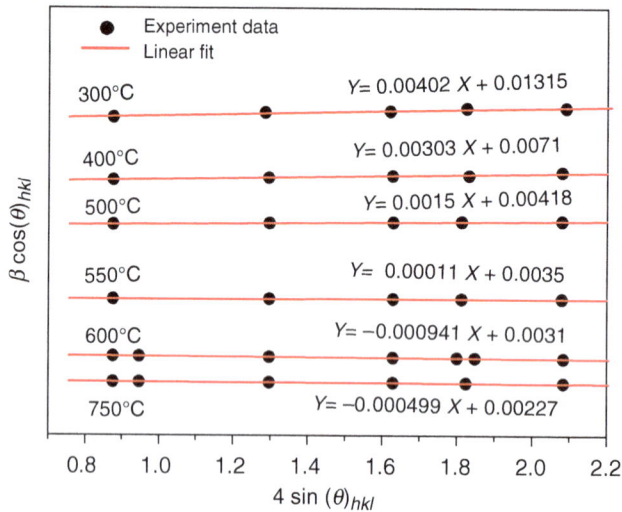

Figure 3.   Williamson–Hall plots. The samples gelatinised for 24 h then calcined at different temperatures for 1 h.

We have calculated the activation energy of the phase transformation from anatase to rutile using Avrami and Arrhenius equations. It has been shown that kinetics of the phase transformation is followed by Avrami formula as follows [34].

$$X_R = 1 - \exp(-k \cdot t) \tag{4}$$

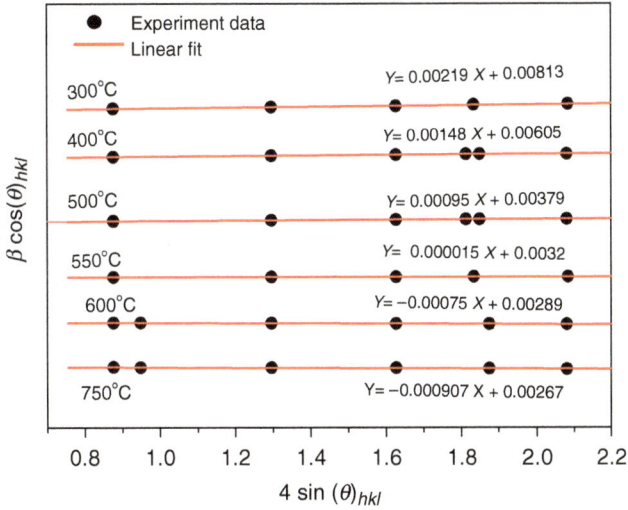

Figure 4.   Williamson–Hall plots. The samples gelatinised for 120 h, then calcined at various temperatures for 1 h.

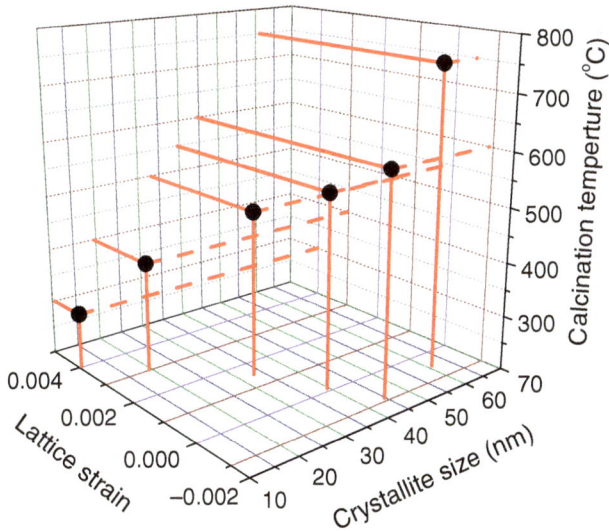

Figure 5.   (Colour online) Plot of lattice strain vs. crystallite size and calcination temperature. The sol–gel solution mixed for 24 h.

where t is the time and k the kinetic constant. The kinetic constant can be calculated for each transformation fraction. The activation energy of the phase transformation can be obtained according to the Arrhenius formula [26,34,35].

$$\ln k = \left(\frac{-E_{\mathrm{a}}}{R \cdot T}\right) + \ln k_{\mathrm{o}} \qquad (5)$$

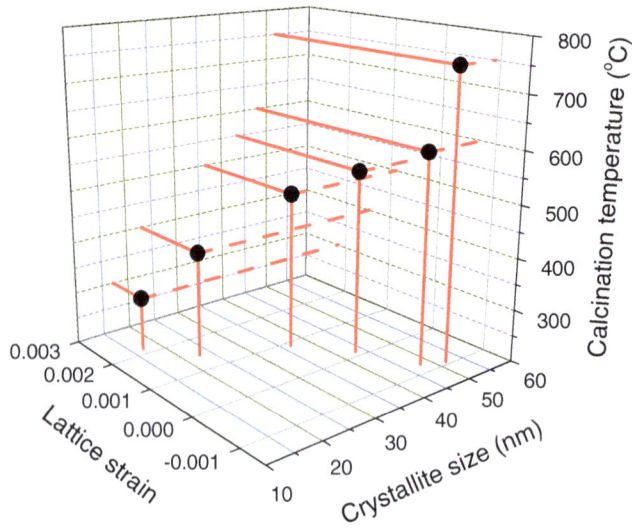

Figure 6.   (Colour online) Plot of lattice strain vs. crystallite size and calcination temperature. The sol–gel solution mixed for 120 h.

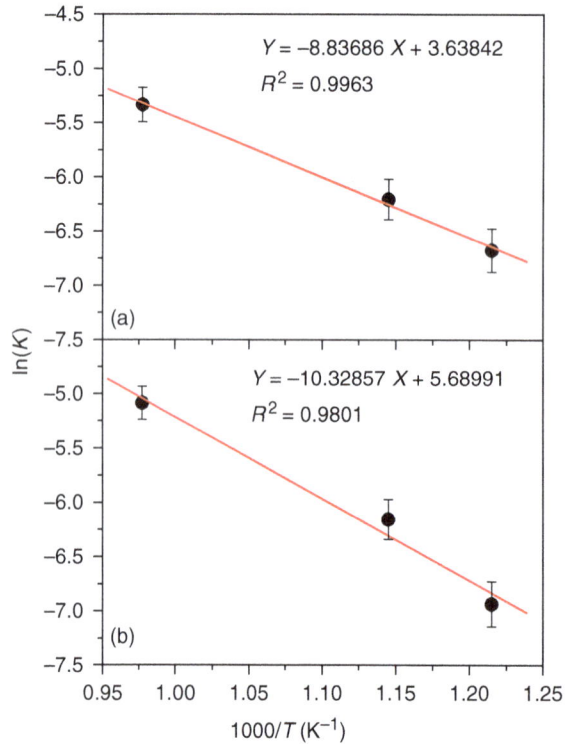

Figure 7.   Arrhenius plots for samples obtained at different times as: (a) 24 and (b) 120 h.

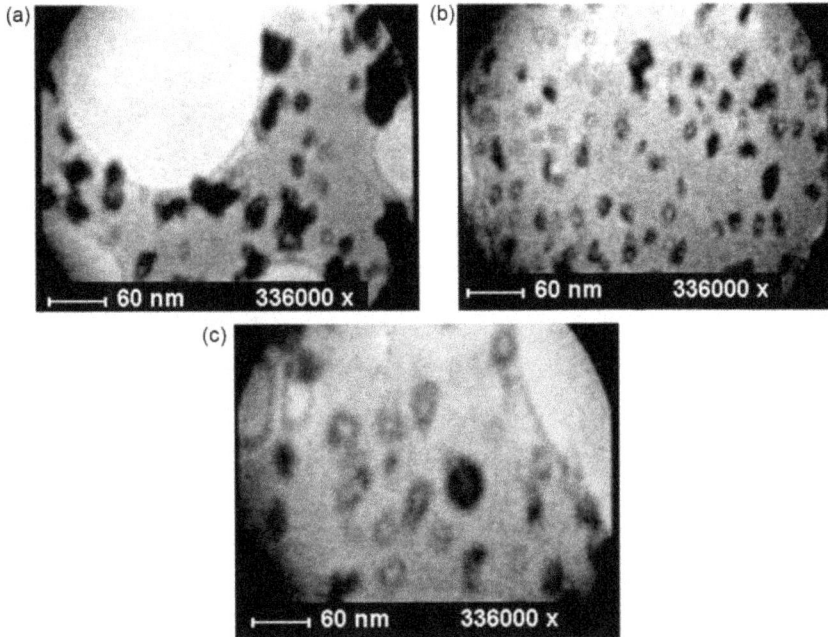

Figure 8. TEM images of TiO$_2$ powders calcined at (a) 400°C, (b) 500°C and (c) 750°C for 1 h. The gelatinising time was 120 h.

where k_o is a material properties constant, E_a the activation energy, T the temperature in Kelvin and R the universal gas constant (8.314 J/mol K).

From the Arrhenius plots of various kinetic constants for the as-prepared samples obtained at different experimental conditions (Figure 7) and with the calculation of slope, the activation energies were obtained as 85.87 and 73.47 kJ/mol, when the samples were mixed for 24 and 120 h and calcined at different temperatures, respectively. The activation energy of the phase transformation also decreased with the increase of the gelatinisation time. In spite of the fact that the crystallite size value decreased with the decrease of the gelatinisation time, the activation energy of the phase transformation increased. Moreover, the transition onset point decreased with the decrease of the size of the initial anatase crystallite; however, the activation energy increased. This is not in relative agreement with the reported results by Li et al. [26] who have concluded that the size of the initial anatase nanoparticle was the main factor in determining the activation energy of the phase transformation from anatase to rutile. Finally, the decrease in the lattice strain and the increase in crystallite size of initial anatase phase induced the decrease in the activation energy of the phase transformation from anatase to rutile. However, the faster decrease of the lattice strain in the as-prepared sample during the lower gelatinisation time with the increase of the calcination temperature from 300°C to the transition onset point gave rise to the larger activation energy. As a result, the size of the crystallites and the activation energy of phase transformation depended on the value of the lattice strain and its circumstances.

Figure 8(a)–(c) shows the TEM image of the TiO$_2$ nanoparticles which gelatinised for 120 h and calcined at 400°C, 500°C and 750°C, respectively. The as-prepared powder in the lower calcination consisted of the particles with poor agglomeration and was completely crystalline. TEM images indicated that when the calcination increased from 400°C to 500°C, the homogeneity of the particles was improved and the mean grain sizes were 15 ± 5 nm and 20 ± 5 nm, respectively. Moreover, the crystallite size value increased with the increase of calcination to 750°C and the average crystallite size was 49 ± 5 nm. Approximately 200 particles were counted for each sample and the size standard deviations were 2.7, 2.1 and 1.7 for the calcination temperatures 400°C, 500°C and 750°C, respectively. Then, there is a good agreement between the crystallite size values obtained by Williamson–Hall method (Table 1) and those obtained by TEM images. Furthermore, the results of the TEM images show significant difference from the obtained crystallite size values obtained by Scherrer formula.

4. Conclusions

The results show that an increase in the calcination or/and the gelatinising time causes an increase of the average size of the anatase TiO$_2$ nanoparticles although the increase in calcination causes a decrease of the lattice strain. In other words, crystallinity of anatase phase was improved. Moreover, there was a strong relationship between an increase of the particle size and a decrease of the lattice strain. The sign of the lattice strain was changed from positive to negative in the region after 550°C in various mixing times and phase transformation from anatase-to-rutile structure occurred.

In spite of the fact that the initial anstase crystallite size increased with an increase of the gelatinisation time, the activation energy of the phase transformation from anatase-to-rutile structure decreased. In addition, the rapid decrease of the lattice strain in the as-prepared sample in the lower gelatinisation time causes an increase of the activation energy of the phase transformation from anatase to rutile and the mass fraction of the rutile structure.

Acknowledgements

This study was financially supported by the Hariri Scientific Foundation in Babol.

References

[1] A. Fujishima and K. Honda, *Electrochemical photolysis of water at a semiconductor electrode*, Nature 238 (1972), pp. 37–38.
[2] W. Wang, B. Gu, L. Liang, W.A. Hamilton, and D.J. Wesolowski, *Synthesis of rutile (a-Tio₂) nanocrystals with controlled size and shape by low-temperature hydrolysis: Effects of solvent composition*, J. Phys. Chem. B 108 (2004), pp. 14789–14792.
[3] T. Trung and C.S. Ha, *One-component solution system to prepare nanometric anatase TiO₂*, Mater. Sci. Eng., C 24 (2004), pp. 19–22.
[4] B. O'Regan and M. Grätzel, *A low-cost, high-efficiency solar cell based on dye-sensitized colloidal TiO₂ films*, Nature 353 (1991), pp. 737–740.
[5] S. Tojo, T. Tachikawa, M. Fujitsuka, and T. Majima, *Oxidation processes of aromatic sulfides by hydroxyl radicals in colloidal solution of TiO₂ during pulse radiolysis*, Chem. Phys. Lett. 384 (2004), pp. 312–316.
[6] L. Castaneda and M. Terrones, *Synthesis and structural characterization of novel flower-like titanium dioxide nanostructures*, Physica B 390 (2007), pp. 143–146.

[7] C. Xiaobo and S.S. Mao, *Titanium dioxide nanomaterials: Synthesis, properties, modifications and application*, Chem. Rev. 107 (2007), pp. 2891–2906.

[8] S. Mahshid, M. Askari, and M. Sasani Ghamsari, *Synthesis of TiO₂ nanoparticles by hydrolysis and peptization of titanium isopropoxide solution*, J. Mater. Process. Technol. 189 (2007), pp. 296–300.

[9] M. Kaneko and I. Okura, *Photocatalysis Science and Technology*, Kodansha/Springer Press, Berlin/Heidelberg/New York, 2002, pp. 57–260.

[10] E.M. El-Maghraby, *Effect of Sn ratio on the photocatalytic degradation of methylene blue and soot of ink by TiO₂ − SnO₂ nanostructured thin films*, Physica B 405 (2010), pp. 2385–2389.

[11] W. Guo, Z. Lin, X. Wang, and G. Song, *Sonachemical synthesis of nanocrystalline TiO₂ by hydrolysis of titanium alkoxides*, Microelectron. Eng. 66 (2003), pp. 95–101.

[12] V. Ramaswamy, N.B. Jagtap, S. Vijayanand, D.S. Bhange, and P.S. Awati, *Photocatalytic decomposition of methylene blue on nanocrystalline titania prepared by different methods*, Mater. Res. Bull. 42 (2008), pp. 1145–1152.

[13] M. Pal, J. Garcia Serrano, P. Santiago, and U. Pal, *Size-controlled synthesis of spherical TiO₂ nanoparticles: Morphologycrystallization, and phase transition*, J. Phys. Chem. C 111 (2007), pp. 96–102.

[14] K.M. Lee, V. Suryanarayanan, and K.C. Ho, *The influence of surface morphology of TiO₂ coating on the performance of dye-sensitized solar cells*, Sol. Energy Mater. Sol. Cells 90 (2006), pp. 2398–2404.

[15] J. Brinker and G.W. SchererSol-gel Science 1st ed., Academic Press, San Diego, CA, 1990, pp. 50–720.

[16] H. Zhang, B. Chen, and J.F. Banfield, *The size dependence of the surface free energy of titania nanocrystals*, Phys. Chem. Chem. Phys. 11 (2009), pp. 2553–2558.

[17] Y. Wang, H. Cheng, Y. Hao, J. Ma, W. Li, and S. Cai, *Characterization and photo-electric behaviors of Fe(III) doped TiO₂ nanoparticles*, J. Mater. Sci. 34 (1999), pp. 3721–3729.

[18] H. Cheng, J. Ma, Z. Zhao, and L. Qi, *Hydrothermal preparation of uniform nanosize rutile and anatase particles*, Chem. Mater. 7 (1995), pp. 663–671.

[19] Y. Wang, Y. Hao, H. Cheng, J. Ma, B. Xu, W. Li, and S. Cai, *The photoelectrochemistry of transition metal-iondoped TiO₂ nanocrystalline electrode and higher solar cell conversion efficiency based on Zn+2-doped TiO₂ electrode*, J. Mater. Sci. 34 (1999), pp. 2773–2779.

[20] O. Carp, C.L. Huisman, and A. Reller, *Photoinduced reactivity of titanium dioxide*, Prog. Solid State Chem. 32 (2004), pp. 33–177.

[21] I. Zumeta, R. Espinosa, J.A. Ayllon, X. Doménech, R. Rodríguez-Clemente, and E. Vigil, *Comparative study of nanocrystalline TiO₂ photoelectrodes based on characteristics of nanopowder used*, Sol. Energy Mater. Sol. Cells 76 (2003), pp. 15–24.

[22] Q. Shen, K. Katayama, T. Sawada, M. Yamaguchi, Y. Kumagai, and T. Toyoda, *Photoexcited hole dynamics in TiO₂ nanocrystalline films characterized using a lens-free heterodyne detection transient grating technique*, Chem. Phys. Lett. 419 (2006), pp. 464–468.

[23] J.H. Jho, D.H. Kim, S.J. Kim, and K.S. Lee, *Synthesis and photocatalytic property of a mixture of anatase and rutile TiO₂ doped with Fe by mechanical alloying process*, J. Alloys Compd. 459 (2008), pp. 386–389.

[24] I.L. Cooper, T.A. Egerton, and F. Qiu, *A model of growth of titanium dioxide crystals with simultaneous transformation from anatase to rutile*, J. Eur. Ceram. Soc. 29 (2009), pp. 637–646.

[25] M.R. Mohammadi, M.C. Cordero-Cabrera, M. Ghorbani, and D.J. Fray, *Synthesis of high surface area nanocrystalline anatase-TiO₂ powders derived from particulate sol-gel route by tailoring processing parameters*, J. Sol–Gel Sci. Technol. 40 (2006), pp. 15–23 and therein.

[26] W. Li, C. Ni, H. Lin, C.P. Huang, and S. Ismat Shah, *Size dependence of thermal stability of TiO₂ nanoparticles*, J. Appl. Phys. 96 (2004), pp. 6663–6668.

[27] M. Inagaki, R. Nonaka, B. Tryba, and A.W. Morawski, *Dependence of photocatalytic activity of anatase powders on their crystallinity*, Chemosphere 64 (2006), pp. 437–445.

[28] K. Venkateswarlu, A. Chandra Bose, and N. Rameshbabu, *X-ray peca broadening studies of nanocrystalline hydroxyapatite by Williamson-Hall analysis*, Physica B 405 (2010), pp. 4256–4261.

[29] E. Manova, P. Aranda, M. Angeles Martin-Luengo, S. Letaief, and E. Ruiz-Hitzky, *New titania-clay nanostructured porous materials*, Microporous Mesoporous Mater. 131 (2010), pp. 252–260.

[30] G. Colon, M.C. Hidalogo, J.A. Navio, A. Kubacka, and M. Fernandez-Garcia, *Influence of sulfur on the structural, surface properties and photocatalytic activity of sulfated TiO₂*, Appl. Catal., B 90 (2009), pp. 633–641.

[31] M. Croft, Z. Zhong, N. Jisrawi, I. Zakharchenko, R.L. Holtz, J. Skaritka, T. Fast, K. Sadananda, M. Lakshmipathy, and T. Tsakalakos, *Strain profiling of fatigue crack overload effects using energy dispersive X-ray diffraction*, Int. J. Fatigue 27 (2005), pp. 1408–1419.

[32] P. Ghosh, J. Oliva, E. De la Rosa, K. Kanta Haldar, D. Solis, and A. Patra, *Enhancement of upconversion emission of LaPO₄:Er@Yb core-shell nanoparticles/nanorods*, J. Phys. Chem. C 112 (2008), pp. 9650–9658.

[33] H. Zhang and J.F. Banfield, *Understanding polymorphic phase transformation behavior during growth of nanocrystalline aggregates: Insights from Tio₂*, J. Phys. Chem. B 104 (2000), pp. 3481–3487.

[34] J. Yang, B.J. McCoy, and G. Madras, *Distribution kinetics of polymer crystallization and the Avrami equation*, J. Chem. Phys. 122 (2005), pp. 64901–64911.

[35] M.R. Mohammadi, D.J. Fray, and A. Mohammadi, *Sol-gel nanostructured titanium dioxide: Controlling the crystal structure, crystallite size, phase transformation, packing and ordering*, Microporous Mesoporous Mater. 112 (2008), pp. 392–402.
[36] H. Zhang, B. Chen, and J.F. Banfield, *The size dependence of the surface free energy of titania nanocrystals*, Phys. Chem. Chem. Phys. 11 (2009), pp. 2553–2558.

21

Hydrothermal synthesis of CeO$_2$ nanorods using a strong base–weak acid salt as the precipitant

Xiaobin Yin, Youjin Zhang*, Zhiyong Fang, Zhenyu Xu and Wei Zhu

Department of Chemistry, University of Science and Technology of China, Hefei 230026, P.R. China

Single-crystalline CeO$_2$ nanorods of the size of about 30 nm diameter and 300 nm length were synthesised by using a strong base–weak acid salt as the precipitant, by a simple hydrothermal method without any surfactant. The structure and properties of the CeO$_2$ nanorods were characterised by X-ray diffraction, X-ray photoelectron spectroscopy, high-resolution transmission electron microscopy, selected-area electron diffraction, UV–visible absorption and scanning electron microscopy field emission. Reaction parameters, such as the reaction temperature, the reaction time, the molar ratio of the reactants and the pH value, were investigated by a series of control experiments. The effect of the strong base–weak acid salt in the synthesis of the CeO$_2$ nanorods was discussed.

Keywords: CeO$_2$ nanorods; hydrothermal method; strong base–weak acid salt

1. Introduction

In the past decades, the synthesis, characterisation and application of nanomaterials have been studied widely. Among these, rare earth nanomaterials have drawn much attention, due to its unique properties arising from the availability of the 4f shell. Cerium oxide is one of the most reactive rare earth metal oxides due to its particular functions in optical, electrical and magnetic fields, and it has been widely utilised in UV blockers, solid state fuel cells, gas sensors and catalysts [1–4].

It is well known that the properties of a material are strongly related to its composition, crystal type, shape and size. CeO$_2$ with morphologies such as nanoparticles, nanospheres, nanocubes, nanorods, nanowires and nanotubes has been prepared by many research groups [5–10]. One-dimensional nanomaterials offer great potential as building blocks for applications in nanoscale electronics, photonics and some other areas [11]. As an important class of one-dimensional nanostructure, nanorods have attracted enormous technological and scientific interest [12–14].

Many routes have been employed to synthesise one-dimensional CeO$_2$ nanostructure materials, such as sol–gel process, chemical vapour deposition, electrochemical deposition technique, microemulsion method, template method, thermal decomposition and hydrothermal synthesis [15–21]. Among these methods, hydrothermal synthesis is well known as one-step low-temperature synthesis, powder reactivity and shape control, and it is regarded as one of the most effective and economical routes [22].

In traditional hydrothermal synthesis of CeO$_2$ nanostructures, strong base such as NaOH or KOH is used as precipitants [23,24]. In the reaction system, the OH$^-$ releases so rapidly that

---

*Corresponding author. Email: zyj@ustc.edu.cn

the reaction is not easy to be controlled. In addition, although the use of surfactants has been testified to be a helpful approach in the synthesis of CeO_2 recently [25,26], the removal of surfactants is still a problem because of the strong force between surfactants and final crystal surfaces [27].

In this study, large-scale single-crystalline CeO_2 nanorods were synthesised at $150°C$ for $13\,h$ through a simple hydrothermal method without any surfactant. $Na_2S \cdot 9H_2O$ was used as a strong base–weak acid salt precipitant to make the approach controllable. To the best of our knowledge, no previous studies focusing on the use of strong base–weak acid salt of $Na_2S \cdot 9H_2O$ for the preparation of CeO_2 nanorods have been reported. The effect of the conditions on the product morphology was investigated by a series of experiments and the structure and properties of the CeO_2 nanorods were characterised. Moreover, to study the effect of $Na_2S \cdot 9H_2O$ in the synthesis of the CeO_2 nanorods, a comparative experiment using NaOH as the replacement of $Na_2S \cdot 9H_2O$ was carried out.

2. Experimental section

All chemicals were of analytical grade and used as purchased without further purification, except for $Na_2S \cdot 9H_2O$, which was stored at freezing point. In a typical procedure, $3\,mmol$ $Ce(NO_3)_3 \cdot 6H_2O$ and $9\,mmol$ $Na_2S \cdot 9H_2O$ were dissolved in $20\,mL$ of distilled water with stirring. Then the mixture was transferred into a 60-mL Teflon liner and some distilled water was added up to 80% of the total volume. It was put into a stainless autoclave and maintained at $150°C$ for $13\,h$ and then let to cool to ambient temperature, naturally. The precipitates were filtrated and washed several times with distilled water and then dried at $60°C$ for $6\,h$. The product was finally obtained.

X-ray diffraction (XRD) pattern was examined by a Japan Rigaku D/max-rA X-ray diffractometer equipped with graphite-monochromatised high-intensity Cu-Kα radiation ($\lambda = 0.15478\,nm$). The scanning rate was $0.05° s^{-1}$ in the 2θ range from $10°$ to $70°$. X-ray photoelectron spectroscopy (XPS) analysis was performed on an ESCALAB MK II X-ray photoelectron spectrometer, using Mg-Kα radiation as the excitation source. High-resolution transmission electron microscopy (HRTEM) images and selected-area electron diffraction (SAED) pattern were taken on a JEOL 2010 HRTEM performed at $200\,kV$. UV–visible absorption (UV–vis) spectrum was measured by a UV–visible spectrophotometer (Shimadzu UV-2401PC) using ethanol as the dispersant. Field-emission scanning electron microscopy (FESEM) images were obtained on a JEOL-6300F field-emission scanning electron microscope with an accelerating voltage of $15\,kV$.

3. Results and discussion

3.1. *The structure of the as-synthesised product*

The phase purity and crystal structure of all samples were examined by XRD. Figure 1 shows the XRD patterns of the as-obtained CeO_2 samples, which can be indexed to a face-centred cubic pure phase (JCPDS card no. 34-0394, $a = 0.5411\,nm$).

Figure 2(a) shows the transmission electron microscopy (TEM) image of the CeO_2 nanorods. The SAED pattern (inset) of the as-obtained sample indicates the single crystalline nature of the CeO_2 nanorod. Figure 2(b) shows the HRTEM image of the product which was taken from the body of one CeO_2 nanorod. It displays clearly the $(1\,1\,1)$, $(2\,0\,0)$ and $(2\,2\,0)$ lattice fringes with the interplanar spaces of 0.32, 0.28 and $0.19\,nm$, respectively. The result reveals that the CeO_2

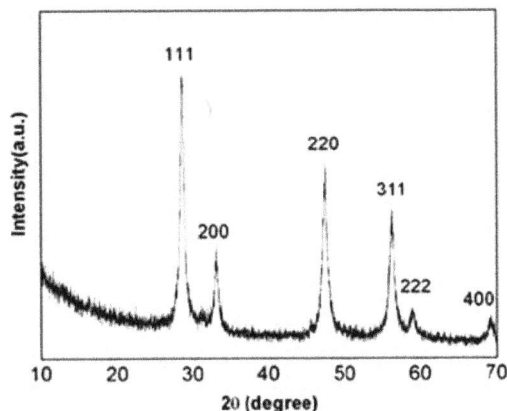

Figure 1. XRD pattern of the as-obtained sample.

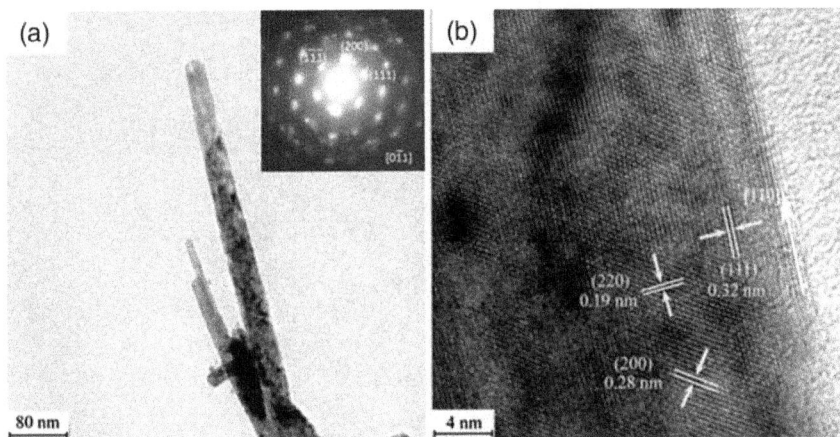

Figure 2. (a) TEM image and SAED pattern (inset) of the product; (b) HRTEM image.

nanorod shows a one-dimensional growth structure with a preferred growth direction along the [1 1 0] [26].

3.2. *The morphologies of the as-synthesised products*

The effect of the reaction conditions was investigated in this study. Figure 3 shows the FESEM images of the samples obtained at different temperatures for 13 h. Figure 3(a) displays that the product morphology at 75°C is irregular which has been testified to be impure phase by XRD. As the temperature is at 100°C, a number of the CeO_2 nanorods are found, but their size is not well-proportioned (Figure 3b). When the temperature reaches 150°C, a large number of homogeneous CeO_2 nanorods are obtained (Figure 3c). If the temperature increases continuously, it can be seen from the Figure 3(d) and (e) that the CeO_2 nanorods are destroyed gradually.

The effect of the reaction time and the molar ratio of the starting reagents were also investigated. Time-independent experience indicates that 10 h is proper to obtain regular CeO_2 nanorods (Figure 4c). If the reaction time is less or more than 10 h, the products are shorter and thicker

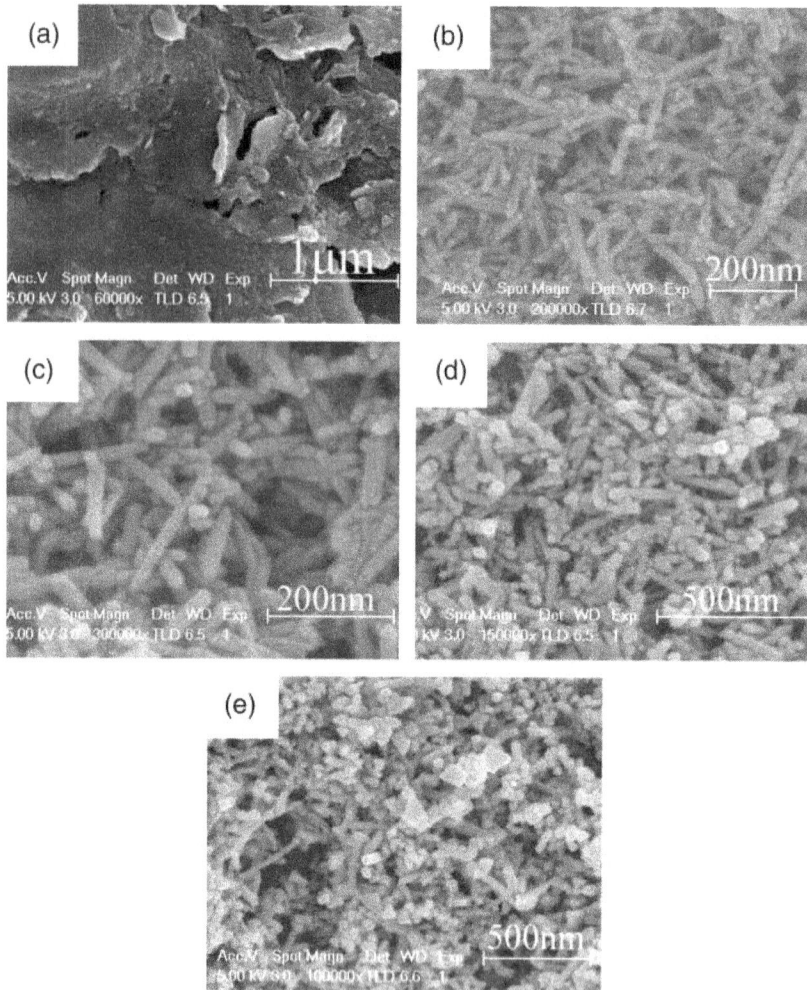

Figure 3.   FESEM images of the samples obtained at different reaction temperatures for 13 h: (a) 75°C; (b) 100°C; (c) 150°C; (d) 200°C; and (e) 225°C.

(Figure 4a, b and d). The molar ratio of $Ce(NO_3)_3 \cdot 6H_2O$ to $Na_2S \cdot 9H_2O$ in this reaction shows almost no influence on the product morphology (Figure 4e–h).

The pH value of the mixture solution was found to be an important factor in the synthesis of the pure phase CeO_2 nanorods. To study the pH influence, HNO_3 or $NH_3 \cdot H_2O$ was used to adjust the pH value of the solution. When the pH value of the solution is below 10, no pure phase product is testified. Only if the pH value is above 10, can the product of the pure phase CeO_2 nanorods be acquired.

3.3.   X-ray photoelectron spectroscopy

The product of the CeO_2 nanorods obtained at 150°C for 13 h with the molar ratio 1:3 was characterised by XPS spectra for the estimate of its composition and purity (Figure 5). The survey spectrum of the product suggests that there is no other metal element except for Cerium on the surface of the sample. From the Ce, 3d core level peak (insetted in Figure 5), it is clear that the

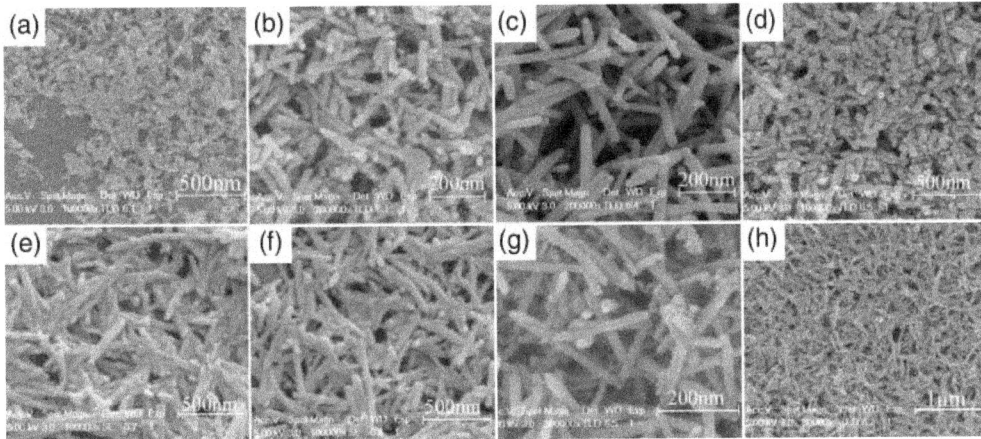

Figure 4.   FESEM images of the samples obtained at 150°C: different reaction times with the molar ratio 1:3: (a) 4 h; (b) 7 h; (c) 10 h; (d) 13 h: different molar ratios for 10 h: (e) 1:1; (f) 1:2; (g) 1:3 and (h) 1:4.

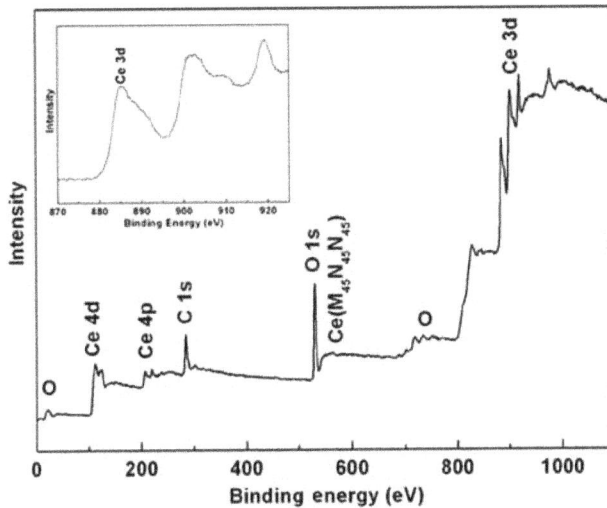

Figure 5.   XPS survey spectra of the CeO_2 nanorods and the Ce 3d core level spectrum (inset).

cerium exists in the Ce (IV) oxidation state (885 eV) without any impurity of the Ce (III) oxidation state [28].

3.4.   *UV–vis absorption*

The UV–vis absorption property of the CeO_2 nanorods was investigated. As shown in Figure 6, the spectrum illustrates a strong absorption band at 336 nm in the UV range. The charge-transfer transitions from O 2p to Ce 4f bands are considered to be the causation of this absorption property [23].

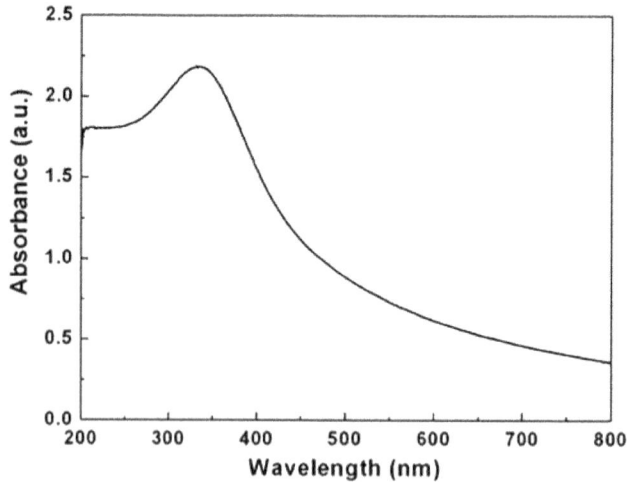

Figure 6.   UV-vis absorption spectrum of the as-obtained product.

3.5.   *The effect of Na₂S in the synthesised process*

The possible reaction mechanism of this approach may be as follows: The Na_2S hydrolysis gives rise to OH^- ions. Hydrate Ce^{3+} ions are oxidised by the O_2 existing in the aqueous solution and form complex $[Ce(H_2O)_x(OH^-)_y]^{(4-y)+}$, which can be regarded as the precursor of the cerium oxide [29]. Then, it reacts with the OH^- and CeO_2 is finally formed [30].

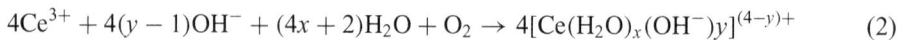

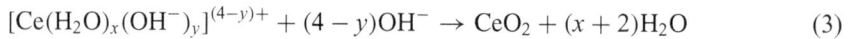

$$S^{2-} + 2H_2O \leftrightarrow 2OH^- + H_2S \tag{1}$$

$$4Ce^{3+} + 4(y-1)OH^- + (4x+2)H_2O + O_2 \rightarrow 4[Ce(H_2O)_x(OH^-)y]^{(4-y)+} \tag{2}$$

$$[Ce(H_2O)_x(OH^-)_y]^{(4-y)+} + (4-y)OH^- \rightarrow CeO_2 + (x+2)H_2O \tag{3}$$

To further investigate the role of the $Na_2S \cdot 9H_2O$ in the hydrothermal system, comparative experiment was carried out under the same hydrothermal conditions. When $Na_2S \cdot 9H_2O$ was replaced by NaOH, the product CeO_2 was irregular nanoparticles.

It is clear that the strong base–weak acid salt $Na_2S \cdot 9H_2O$ plays a significant role in the preparation of CeO_2 nanorods. The hydrolysis of the S^{2-} not only gives rise to OH^-, but also generates H_2S. On the one hand, the formation of the CeO_2 nanorods consumes OH^-, which makes the Equation (1) move to the right. On the other hand, as the system is airtight, the pressure of H_2S gets so high that it dissolves again, which makes the Equation (1) move to the left. As a result, the pH value of the reaction system is maintained in a proper level throughout the whole reaction process, which is propitious to CeO_2 nanorod growth. But when the strong base NaOH is used as the precipitant, the OH^- releases so rapidly that the CeO_2 growing speed is not easy to be controlled, which is unfavourable to the formation of the CeO_2 nanorods.

4.   Conclusion

In summary, large-scale single-crystalline CeO_2 nanorods with the size of about 30 nm diameter and 300 nm length have been successfully prepared through a simple hydrothermal method without any surfactant. In this approach, $Na_2S \cdot 9H_2O$ was utilised as the precipitant and it made the process controllable. The temperature and the pH value of the reaction system were confirmed to be the crucial factors for the formation of CeO_2 nanorods. Moreover, the effect of $Na_2S \cdot 9H_2O$ in the synthesis of the CeO_2 nanorods is discussed deeply. Hydrothermal method using strong base–weak acid salt as precipitant might offer an excellent approach to design other similar nanomaterials.

References

[1] R.X. Li, S. Yabe, M. Yamashita, S. Momose, S. Yoshida, S. Yin, and T. Sato, *Synthesis and UV-shielding properties of ZnO- and CaO-doped CeO2 via soft solution chemical process*, Solid State Ion 151 (2002), pp. 235–241.

[2] H.T. Chen, Y.M. Choi, M.L. Liu, and M.C. Lin, *A first-principles analysis for sulfur tolerance of CeO2 in solid oxide fuel cells*, J. Phys. Chem. C 111 (2007), pp. 11117–11122.

[3] L. Liao, H.X. Mai, Q. Yuan, H.B. Lu, J.C. Li, C. Liu, C.H. Yan, Z.X. Shen, and T. Yu, *Single CeO2 nanowire gas sensor supported with Pt nanocrystals: Gas sensitivity, surface bond states, and chemical mechanism*, J. Phys. Chem. C 112 (2008), pp. 9061–9065.

[4] K. Minami, T. Masui, N. Imanaka, L. Dai, and B. Pacaud, *Redox behavior of CeO2 − ZrO2− Bi2O3 and CeO2 − ZrO2 − Y2O3 solid solutions at moderate temperatures*, J. Alloys Compd. 408–412 (2006), pp. 1132–1135.

[5] Z.L. Wang and X.D. Feng, *Polyhedral shapes of CeO2 nanoparticles*, J. Phys. Chem. B 107 (2003), pp. 13563–13566.

[6] Y.J. Zhang, Q.X. Hu, Z.Y. Fang, T. Cheng, K.D. Han, and X.Z. Yang, *Self-assemblage of single/multiwall hollow CeO2 microspheres through hydrothermal method*, Chem. Lett. 35 (2006), pp. 944–945.

[7] Z.Q. Yang, K.B. Zhou, X.W. Liu, Q. Tian, D.Y. Lu, and S. Yang, *Single-crystalline ceria nanocubes: Size-controlled synthesis, characterization and redox property*, Nanotechnology 18 (2007), pp. 1–4.

[8] C.W. Sun, H. Li, H.R. Zhang, Z.X. Wang, and L.Q. Chen, *Controlled synthesis of CeO2 nanorods by a solvothermal method*, Nanotechnology 16 (2005), pp. 1454–1463.

[9] S. Kumar and S.K. Chakarvarti, *Large-scale synthesis of uniform nickel nanowires and their characterisation*, J. Exp. Nanosci. 5 (2010), pp. 126–133.

[10] W.Q. Han, L.J. Wu, and Y.M. Zhu, *Formation and oxidation state of CeO2−x nanotubes*, J. Am. Chem. Soc. 127 (2005), pp. 12814–12815.

[11] Y. Huang, X.F. Duan, Q.Q. Wei, and C.M. Lieber, *Directed assembly of one-dimensional nanostructures into functional networks*, Science 291 (2001), pp. 630–633.

[12] P.X. Huang, F. Wu, B.L. Zhu, X.P. Gao, H.Y. Zhu, T.Y. Yan, W.P. Huang, S.H. Wu, and D.Y. Song, *CeO2 nanorods and gold nanocrystals supported on CeO2 nanorods as catalyst*, J. Phys. Chem. B 109 (2005), pp. 19169–19174.

[13] G.R. Patzke, A. Michailovski, F. Krumeich, R. Nesper, J.D. Grunwaldt, and A. Baiker, *One-step synthesis of submicrometer fibers of MoO3*, Chem. Mater. 16 (2004), pp. 1126–1134.

[14] D.E. Zhang, X.M. Ni, H.G. Zheng, X.J. Zhang, and J.M. Song, *Fabrication of rod-like CeO2: Characterization, optical and electrochemical properties*, Solid State Sci. 8 (2006), pp. 1290–1293.

[15] G.S. Wu, T. Xie, X.Y. Yuan, B.C. Cheng, and L.D. Zhang, *An improved sol–gel template synthetic route to large-scale CeO2 nanowires*, Mater. Res. Bull. 39 (2004), pp. 1023–1028.

[16] D. Barreca, A. Gasparotto, C. Maccato, C. Maragno, and E. Tondello, *Toward the innovative synthesis of columnar CeO2 nanostructures*, Langmuir 22 (2006), pp. 8639–8641.

[17] R. Inguanta, S. Piazza, and C. Sunseri, *Template electrosynthesis of CeO2 nanotubes*, Nanotechnology 18 (2007), pp. 1–6.

[18] Y.J. He, B.L. Yang, and G.X. Cheng, *On the oxidative coupling of methane with carbon dioxide over CeO2/ZnO nanocatalysts*, Catal. Today 98 (2004), pp. 595–600.

[19] K.L. Yu, G.L. Ruan, Y.H. Ben, and J.J. Zou, *Convenient synthesis of CeO2 nanotubes*, Mater. Sci. Eng. B 139 (2007), pp. 197–200.

[20] S.F. Wang, F. Gu, C.Z. Li, and H.M. Cao, *Shape-controlled synthesis of CeOHCO3 and CeO2 microstructures*, J. Cryst. Growth 307 (2007), pp. 386–394.

[21] L. Yan, R.B. Yu, J. Chen, and X.R. Xing, *Template-free hydrothermal synthesis of CeO2 nano-octahedrons and nanorods: Investigation of the morphology evolution*, Cryst. Growth Des. 8 (2008), pp. 1474–1477.

[22] H.X. Mai, L.D. Sun, Y.W. Zhang, R. Si, W. Feng, H.P. Zhang, H.C. Liu, and C.H. Yan, *Shape-selective synthesis and oxygen storage behavior of ceria nanopolyhedra, nanorods, and nanocubes*, J. Phys. Chem. B 109 (2005), pp. 24380–24385.

[23] Y.W. Zhang, R. Si, C.S. Liao, and C.H. Yan, *Facile alcohothermal synthesis, size-dependent ultraviolet absorption, and enhanced CO conversion activity of ceria nanocrystals*, J. Phys. Chem. B 107 (2003), pp. 10159–10167.

[24] F.H. Scholes, A.E. Hughes, S.G. Hardin, P. Lynch, and P.R. Miller, *Influence of hydrogen peroxide in the preparation of nanocrystalline ceria*, Chem. Mater. 19 (2007), pp. 2321–2328.

[25] C. S. Pan, D. S. Zhang, and L.Y. Shi, *CTAB assisted hydrothermal synthesis, controlled conversion and CO oxi-dation properties of CeO₂ nanoplates, nanotubes, and nanorods*, J. Solid State Chem. 181 (2008), pp. 1298–1306.

[26] A. Vantomme, Z.Y. Yuan, G.H. Du, and B.L. Su, *Surfactant-assisted large-scale preparation of crystalline CeO₂ nanorods*, Langmuir 21 (2005), pp. 1132–1135.

[27] C.S. Pan, D.S. Zhang, L.Y. Shi, and J.H. Fang, *Template-free synthesis, controlled conversion, and CO oxida-tion properties of CeO₂ nanorods, nanotubes, nanowires, and nanocubes*, Eur. J. Inorg. Chem. (2008), pp. 2429–2436.

[28] D.S. Zhang, H.X. Fu, L.Y. Shi, C.S. Pan, Q. Li, Y.L. Chu, and W.J. Yu, *Synthesis of CeO₂ nanorods via ultrasonication assisted by polyethylene glycol*, Inorg. Chem. 46 (2007), pp. 2446–2451.

[29] X.H. Liao, J.M. Zhu, J.J. Zhu, J.Z. Xu, and H.Y. Chen, *Preparation of monodispersed nanocrystalline CeO₂ powders by microwave irradiation*, Chem. Commun. 10 (2001), pp. 937–938.

[30] K.B. Zhou, Z.Q. Yang, and S. Yang, *Highly reducible CeO₂ nanotubes*, Chem. Mater. 19 (2007), pp. 1215–1217.

22

Fabrication of vertical graphitic nano-partitions for anti-reflection function of silicon solar cells

Jeff Tsung-Hui Tsai[a]*, Wen-Ching Shih[b], Jian-Min Jeng[b], Yen-Tang Chiao[b], Chin-Tze Hwang[b] and Jyi-Tsong Lo[b]

[a]*Institute of Optoelectronic Sciences, National Taiwan Ocean University, Taiwan;*
[b]*Graduate Institute of Electro-Optical Engineering, Tatung University, Taipei 10452, Taiwan, ROC*

Vertical standing, thin graphitic nano-partitions were fabricated on quartz and silicon substrates using a reactive plasma sputtering system. These nano-sized partitions act as light reflectors, enhancing the solar absorption of omni-directional incident light. The structure of nano-partitions doubles the output power of the solar cell when illuminated from a 30° to 150° angle of incidence. Such nano-partitions also enhance solar efficiency when reflected light is used to produce more photon absorption. Measurement of reflectance has shown a minimum of 0.49% at a wavelength of 580 nm when nano-partitions are used. The average reflectance of the entire visible regime is 0.85%, which is equivalent to the conventional multi-layer anti-reflection coatings. We also found that this nano-partition system has superior anti-reflection efficiency in a UV regime. This 3D anti-reflection structure provides a promising route to the fabrication of high-efficiency solar cells.

Keywords: sputtering; amorphous carbon; graphite; optical properties characterisation

1. Introduction

Of all the renewable energy sources available, solar power is one of the most environmentally friendly. Electricity produced from photovoltaic cells does not result in air or water pollution. As solar cells are designed to convert available light into electrical energy, the more sunlight absorbed by active photovoltaic materials, the more electrical power will be generated. Conventional approaches to enhancing solar cell efficiency by applying an anti-reflection coating (ARC) to the silicon surface increase the absorption the sunlight. The most common design of such ARCs uses a 90° angle of incidence (AOI). This is the direction of illumination that is perpendicular to the surface of the solar cell [1–3]. Another approach that has been widely investigated for enhancement of solar cell efficiency is to reduce light reflection by texturing the silicon surface [4–6]. Recently, chemically etched porous silicon has been used as an ARC for high-efficiency solar cells [7]. Creating a mixture of air and silicon produces a porous structure that reduces

---

*Corresponding author. Email: thtsai@ttu.edu.tw

the macroscopic refractive index of the layer, and subsequently, reduces the reflection that is due to the index mismatch at the interface. This relies on a nano-structure to create a reduction in the graded refractive index (GRIN) to minimise reflection. The effect of this type of structure is known as the Moths' Eye Principle [8].

Nevertheless, owing to the variations in the illuminated area when the AOI changes, solar cell power output will dramatically decrease when the AOI declines. When setting up a stationary solar panel, it must be tilted up at a certain angle, depending on the latitude of the location, in order to maximise the power output. The electrical power output of a stationary solar panel is also dependent on the position of the sun. Most stationary solar panels generate 84% less power at dawn and dusk compared to the power output at midday. Therefore, an ARC system will also improve the outdoor solar cell efficiency if it can increase the reflection of low sunlight onto the solar cell surface.

In this article, we present a method for producing an effective ARC that can maintain the efficiency of a solar cell from high to low AOI. Using a reactive sputtering system, we generated dense nano-sized vertical standing graphitic films as the reflector on a silicon surface. This sub-wavelength nanostructure gives the solar cell surface an anti-reflection function which has large transmission bands with respect to both wavelength and AOI.

2. Experimental

The vertical standing graphitic nano-partition films were sputter-deposited using a planar conventional balanced dc magnetron unit, with a target (100 mm in diameter and 12 mm thick) composed of graphite disk. To produce films with different porosities, the amount of inlet gas and substrate temperatures were varied in different experiments. The target-to-substrate distance was 75 mm. The reacting sputtering gases were argon, hydrogen, and methane, at a constant pressure of 7.5 mTorr, with the optimum condition of gas ratio at $10:6:5$ sccm. The quartz and p- or n-type ($\rho = 1$–$100\,\Omega$-cm) single crystal wafers of (100)-oriented silicon were used as substrates. Prior to film deposition, the target was pre-sputtered on the shield in order to remove contaminations from the target surface and stabilise the magnetron discharge parameters. The magnetron was operated in an optimum regime of about 75 to 100 W power, and this allowed a fairly good deposition rate (0.05 nm/sec) while minimising non-controllable heating of the growing film by plasma radiation. Under the above deposition conditions, the substrate temperature during film condensation was maintained at 375–425°C. No biasing voltage was applied to the substrate during deposition. As a result, uniform arrays of vertical standing graphitic partitions were deposited on silicon wafers. The vertical films were deposited with a thickness of about 30 nm and a height of 500–800 nm, as measured by the scanning electron microscopy (SEM) method shown in Figure 1(a). The porosity of vertical standing graphitic partitions, from dense to sparse, can be controlled by the deposition temperature, as shown in Figure 1(b) and (c). From the top view micrographic, a higher process temperature gives less open space on the silicon surface from this 3-D structure.

The specimens for the structure study were prepared as follows. The as-deposited graphitic films were removed from the silicon substrate by immersion in DI water with an ultrasound bath for 2 h. They were then transferred both onto a copper grid of the kind used for electron microscopy, and onto a silicon wafer. The copper grid was sent for structure study in a transmission electron microscope (TEM) JEM-100CX II (JEOL),

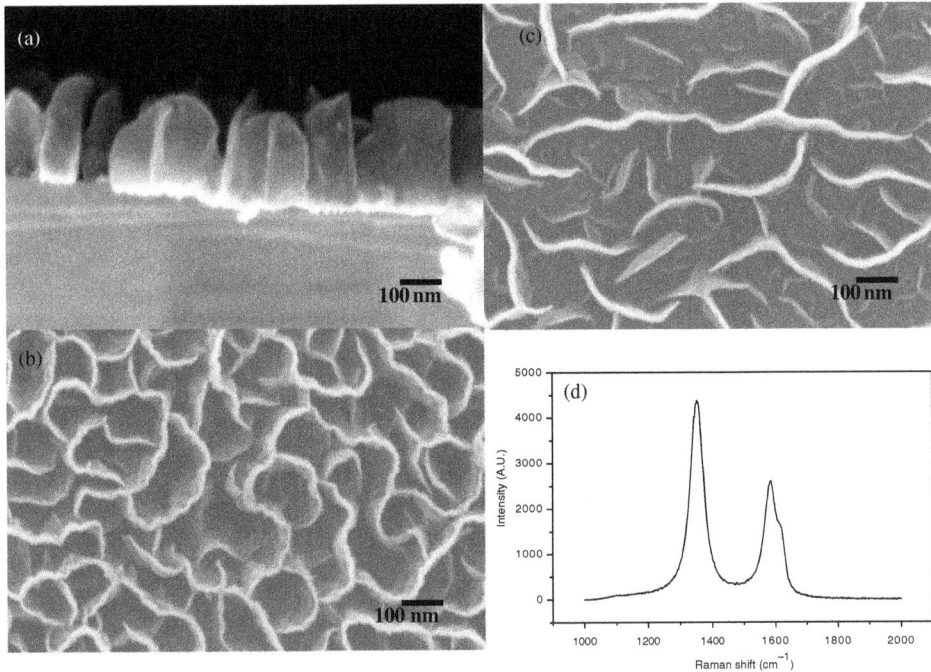

Figure 1. SEM micrographs of the vertical standing graphitic nano-partitions. (a) A cross-sectional view showing vertical standing films. (b) and (c) Top view of nano-partitions, where the deposition temperatures are 375°C and 450°C, respectively. (d) Raman spectrum of the graphitic nano-partitions.

operating at 100 kV. The carbon structure of the silicon wafer with graphitic films was then determined based on its Raman spectra. Based on the TEM investigation, these carbon films appear mostly amorphous with graphitic clusters embedded. From the Raman spectra, the graphitic structure of the carbon film can be observed by determining the intensity ratio from the graphitic peak (I_G) to the disorder peak (I_D), as shown in Figure 1(d).

3. Results and discussion

The optical properties of the vertical nano-partition and the total anti-reflection performance were measured. The hemispherical reflectance of the vertical nano-partition using an integrating sphere is shown in Figure 2(a) from the sparse nano-partitions sample. The reflectance from planar silicon and vertical nano-partition surfaces were compared using unpolarised light. The anti-reflection performance of silicon in the UV to visible wavelength range was relatively poor in the control sample, which had a very high reflectance of 34.5%. When integrated nano-partitions were on the silicon surface, as in the experimental sample, the reflectance was dramatically reduced. The results indicated that such ARCs have an average reflectance of 1.19% over a 200–800 nm range. However, more importantly, we demonstrated an average hemispherical reflectance of 0.7% in the UV to visible range.

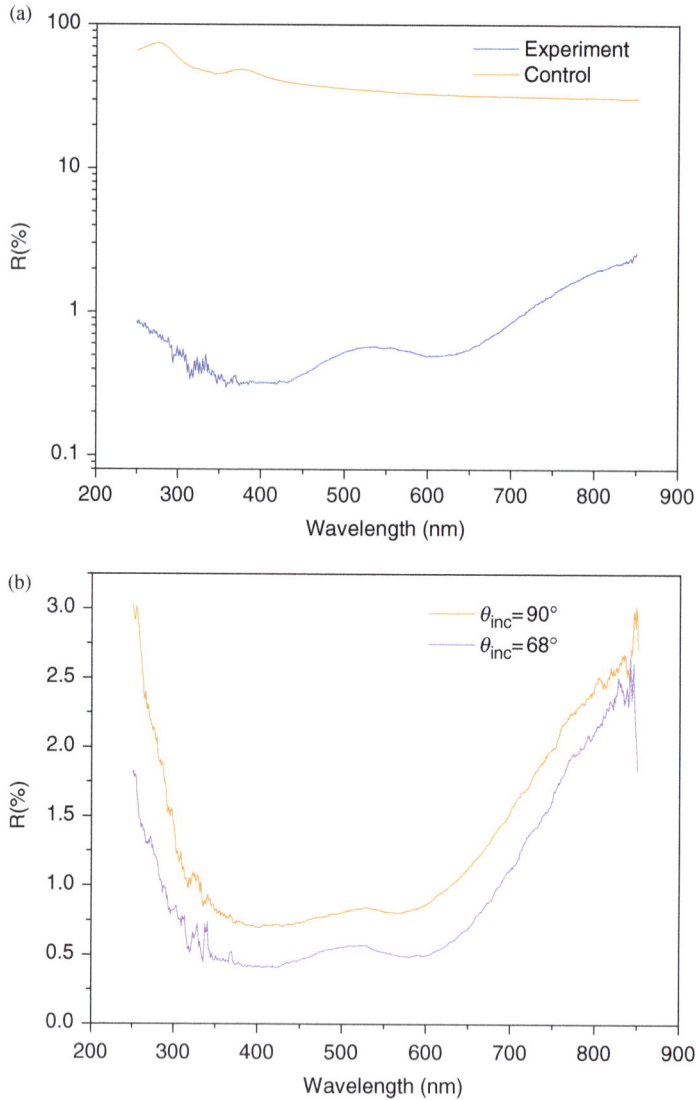

Figure 2. (a) Comparison between a bare silicon surface and a nano-partition ARC; the reflectance reduces by nearly two orders. (b) Reflectance measurement of the vertical nano-partitions for 90° and 68° AOI.

In order to perform as a good anti-reflector, the structure must show low reflectance over a wide range of AOI values. Generally, nano-partitions are slightly sensitive to the AOI in the UV–VIS region. When we decreased the AOI to 68°, the hemispherical reflectance decreased in the centre of the visible range at 580 nm, as shown in Figure 2(b).

Photovoltaic measurements were performed in a structure in which a commercial thin film solar cell covered a quartz window topped with vertical standing nano-partitions.

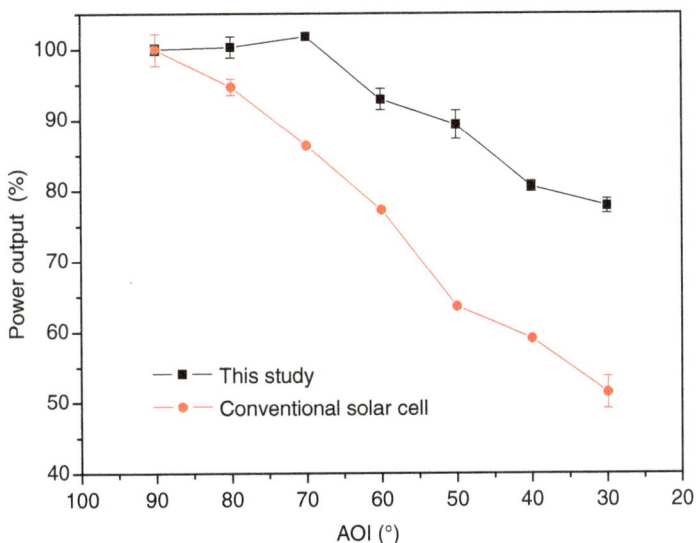

Figure 3. The nano-partition ARC window on a commercial thin film silicon solar cell maintains its power output under angled illumination.

We studied the solar cell power output with a short circuit current (I_{sc}) with an open circuit voltage (V_{oc}) by illuminating it with simulated sunlight of AM1.5. Compared to the control sample, which was a thin film solar cell only, the power output from the ARC window-assisted solar cell demonstrated less sensitivity to the AOI. Figure 3 shows the solar cell output power versus the AOI. The power output decreased with the AOI as the photon flux decreased with illuminated angles by the factor of $\cos \theta$. Hence, the power output decreased by 50% in the control sample when the AOI was 30°. Conversely, the vertically aligned nano-partition-coated solar cell maintained the power output from 60° to 120° AOI, and power only decayed 20% when the AOI was 30° (Figure 3). Compared to the conventional solar cell with ARC, the solar cell with nano-partition ARC doubles the total output power when integrated with an AOI from 0° to 180°. This result indicates that this approach to the preparation of vertical nano-partition ARCs for solar cells enhances solar efficiency and may be suitable for a dawn-to-dusk stationary solar panel which generates electrical power throughout the day.

4. Conclusion

In summary, we have proposed an effective approach to improving solar cell efficiency when sunlight illuminates the panel surface with a low AOI. By depositing vertical graphitic nano-partitions, the functional ARC provides a strong anti-reflection ability. From our preliminary investigation, this approach improves solar efficiency by up to 30% for angled illuminations.

References

[1] P. Doshi, G.E. Jellison, and A. Rohatgi, *Characterization and optimization of absorbing plasma-enhanced chemical vapor deposited antireflection coatings for silicon photovoltaics, Appl,* Opt. 36 (1997), pp. 7826–7837.

[2] H. Takato, M. Yamanaka, Y. Hayashi, R. Shimokawa, I. Hide, S. Gohda, F. Nagamine, and H. Tsuboi, *Effects of Optical Confinement in Textured Antireflection Coating using ZnO Films for Solar Cells,* Jpn. J. Appl Phys 31 (1992), pp. 1665–1667.

[3] K. Ramanathan, M.A. Contreras, C.L. Perkins, S. Asher, F.S. Hasoon, J. Keane, D. Young, M. Romero, W. Metzger, R. Noufi, J. Ward, and A. Duda, *Properties of 19.2% efficiency $ZnO/CdS/CuInGaSe_2$ thin-film solar cells,* Prog. Photovoltaics Res. Appl. 11 (2003), pp. 225–230.

[4] S. Koynov, M.S. Brandt, and M. Stutzmann, *Black nonreflecting silicon surfaces for solar cells,* Appl. Phys. Lett. 88 (2006), 203107.

[5] P. Lalanne and G.M. Morris, *Antireflection behavior of silicon subwavelength periodic structures for visible light,* Nanotechnology 8 (1997), 53.

[6] C.H. Sun, B. Jiang, and P. Jiang, *Broadband moth-eye antireflection coatings on silicon,* Appl. Phys. Lett. 92 (2008), 061112.

[7] L. Stalmans, J. Poortmans, H. Bender, M. Caymax, K. Said, E. Vazsonyi, J. Nijs, and R. Mertens, *Porous silicon in crystalline silicon solar cells: A review and the effect on the internal quantum efficiency,* Prog. Photovolt. Res. Appl. 6 (1998), pp. 233–246.

[8] P. Vukusic and J.R. Sambles, *Photonic structures in biology,* Nature 424 (2003), pp. 852–855.

Permissions

All chapters in this book were first published in Nanoscience Methods, by Taylor & Francis Online; hereby published with permission under the Creative Commons Attribution License or equivalent. Every chapter published in this book has been scrutinized by our experts. Their significance has been extensively debated. The topics covered herein carry significant findings which will fuel the growth of the discipline. They may even be implemented as practical applications or may be referred to as a beginning point for another development.

The contributors of this book come from diverse backgrounds, making this book a truly international effort. This book will bring forth new frontiers with its revolutionizing research information and detailed analysis of the nascent developments around the world.

We would like to thank all the contributing authors for lending their expertise to make the book truly unique. They have played a crucial role in the development of this book. Without their invaluable contributions this book wouldn't have been possible. They have made vital efforts to compile up to date information on the varied aspects of this subject to make this book a valuable addition to the collection of many professionals and students.

This book was conceptualized with the vision of imparting up-to-date information and advanced data in this field. To ensure the same, a matchless editorial board was set up. Every individual on the board went through rigorous rounds of assessment to prove their worth. After which they invested a large part of their time researching and compiling the most relevant data for our readers.

The editorial board has been involved in producing this book since its inception. They have spent rigorous hours researching and exploring the diverse topics which have resulted in the successful publishing of this book. They have passed on their knowledge of decades through this book. To expedite this challenging task, the publisher supported the team at every step. A small team of assistant editors was also appointed to further simplify the editing procedure and attain best results for the readers.

Apart from the editorial board, the designing team has also invested a significant amount of their time in understanding the subject and creating the most relevant covers. They scrutinized every image to scout for the most suitable representation of the subject and create an appropriate cover for the book.

The publishing team has been an ardent support to the editorial, designing and production team. Their endless efforts to recruit the best for this project, has resulted in the accomplishment of this book. They are a veteran in the field of academics and their pool of knowledge is as vast as their experience in printing. Their expertise and guidance has proved useful at every step. Their uncompromising quality standards have made this book an exceptional effort. Their encouragement from time to time has been an inspiration for everyone.

The publisher and the editorial board hope that this book will prove to be a valuable piece of knowledge for researchers, students, practitioners and scholars across the globe.

List of Contributors

D. Sathish Chander
Department of Physics, Indian Institute of Technology Kanpur, Kanpur 208016, India
Department of Mechanical Engineering, Indian Institute of Technology Kanpur, Kanpur 208016, India

J. Ramkumar
Department of Mechanical Engineering, Indian Institute of Technology Kanpur, Kanpur 208016, India

S. Dhamodaran
Department of Physics, Indian Institute of Technology Kanpur, Kanpur 208016, India

J. Brett Kimbrell, Walter J. Steward, Farooq A. Khan, Anne C. Gaquere-Parker and Douglas A. Stuart
Department of Chemistry, University of West Georgia, Carrollton, GA 30118, USA

Chritopher M. Crittenden
Department of Chemistry, University of Texas at Austin, Austin, TX 78712, USA

Thanh Son Le
Institute of Environmental Technology, Vietnam Academy of Science and Technology, 18 Hoang Quoc Viet Road, Cau Giay District, Hanoi, Vietnam

Marc Cretin, Patrice Huguet and Philippe Sistatb
Institut Européen des Membranes, ENSCM-CNRS-Université Montpellier 2, CC047, Place Eugène Bataillon, 34095 Montpellier, France

Frederic Pichot
Centrale de Technologies en Micro et Nanoélectronique, Université Montpellier 2, Place Eugène Bataillon, 34095 Montpellier, France

Laura K. Braydich-Stolle, Alicia B. Castle, Elizabeth I. Maurer and Saber M. Hussain
Applied Biotechnology Branch, Human Effectiveness Directorate, Air Force Research Laboratory, Wright-Patterson AFB, OH, USA

Xiaoqing Zhang, Xianping Fan, Xvsheng Qiao and Qun Luo
State Key Laboratory of Silicon Materials, Department of Materials Science and Engineering, Zhejiang University, Hangzhou 310027, P.R. China

Haifeng Guo
Institute for Advanced Ceramics, School of Materials Science and Engineering, Harbin Institute of Technology, Harbin 150080, China
PetroChina Pipeline R&D Center, Langfang, Hebei 065000, China

Feng Ye
Institute for Advanced Ceramics, School of Materials Science and Engineering, Harbin Institute of Technology, Harbin 150080, China

A.B. Dongil
Instituto de Catálisis y Petroleoquímica, CSIC, c/Marie Curie No. 2, Cantoblanco, 28049 Madrid, Spain

B. Bachiller-Baeza and I. Rodríguez-Ramos
Instituto de Catálisis y Petroleoquímica, CSIC, c/Marie Curie No. 2, Cantoblanco, 28049 Madrid, Spain
Grupo de Diseño y Aplicación de Catalizadores Heterogéneos, Unidad Asociada UNED-CSIC (ICP), Spain

A. Guerrero-Ruiz
Dpto. Química Inorgánica y Técnica, Fac. de Ciencias, UNED, C/ Senda del Rey no 9, 28040, Madrid, Spain

Ludwig Vinches, Nicolas Testori, Gérald Perron and Stéphane Hallé
École de technologie supérieure, 1100 Notre-Dame Ouest, Montréal, QC, Canada H3C 1K3

Patricia Dolez
École de technologie supérieure, 1100 Notre-Dame Ouest, Montréal, QC, Canada H3C 1K
CTT Group, 3000 rue Boullé, Saint-Hyacinthe, QC, Canada J2S 1H9

Kevin J. Wilkinson
Department de chimie, Université de Montréal, C.P. 6128, succ. Centre-ville, Montréal, QC, Canada H3C 3J7

S.N. Soni, J. Bajpai and A.K. Bajpai
Department of Chemistry, Bose Memorial Research Laboratory, Government Autonomous Science College, Jabalpur – 482 001, Madhya Pradesh, India

Devasish Chowdhury
Material Science Division, Polymer Unit, Institute of Advanced Study in Science and Technology, Paschim Boragaon, Garchuk, Guwahati 781035, Assam, India

Samrat Roy Choudhury, Saheli Pradhan and Arunava Goswami
Agricultural and Ecological Research Unit, Biological Sciences Division, Indian Statistical Institute, 203 Barrackpore Trunk Road, Kolkata 700108, India

Z. Jubri
Department of Engineering Sciences and Mathematics, College of Engineering, Universiti Tenaga Nasional, 43000 Kajang, Selangor, Malaysia

M.Z. Hussein, A. Yahayab and Z. Zainal
Department of Chemistry, Universiti Putra Malaysia, 43400 Serdang, Selangor, Malaysia

S.R. Bonde, D.P. Rathod, A.P. Ingle, R.B. Ade, A.K. Gade and M.K. Rai
Department of Biotechnology, SGB Amravati University, Amravati 444602, Maharashtra, India

Xingbao Wang
Xinhai Senior High School, Lianyungang, Jiangsu 222006, China

Weipeng Guan
Zhejiang Great Southeast Plastic Group Corp., Zhuji, Zhejiang 311809, China

Patricia M. Perillo and Daniel F. Rodriguez
Comisión Nacional de Energía Atómica, CAC, Grupo MEMS, Av. Gral. Paz 1499 (1650) Bs. As., Argentina

S. Chakrabarty and K. Chatterjee
Department of Physics and TechnoPhysics, Vidyasagar University, Midnapore 721102, West Bengal, India

Irfan Ahmad Mir and D. Kumar
Department of Applied Chemistry & Polymer Technology, Delhi Technological University (Delhi College of Engineering), Shahbad Daulatpur, Bawana Road, Delhi-110042, India

Jan Valenta
Department of Chemical Physics & Optics, Faculty of Mathematics & Physics, Charles University, Ke Karlovu 3, Prague 2 CZ-121 16, Czech Republic

Jian Zhang and Xinping Zhang
Institute of Information Photonics Technology, Beijing University of Technology, Beijing, People's Republic of China
College of Applied Sciences, Beijing University of Technology, Beijing, People's Republic of China

Hossain Milani Moghaddam
Department of Physics, University of Mazandaran, Babolsar, Iran
Nano and Biotechnology Research Group, University of Mazandaran, Babolsar, Iran
Nanotechnology group, Hariri Scientific Foundation, Babol, Iran

Shahruz Nasirian
Department of Physics, University of Mazandaran, Babolsar, Iran

Department of Basic Sciences, University of Science and Technology of Mazandaran, Babol, Iran

Xiaobin Yin, Youjin Zhang, Zhiyong Fang, Zhenyu Xu and Wei Zhu
Department of Chemistry, University of Science and Technology of China, Hefei 230026, P.R. China

Jeff Tsung-Hui Tsai
Institute of Optoelectronic Sciences, National Taiwan Ocean University, Taiwan

Wen-Ching Shih, Jian-Min Jeng, Yen-Tang Chiao, Chin-Tze Hwang and Jyi-Tsong Lo
Graduate Institute of Electro-Optical Engineering, Tatung University, Taipei 10452, Taiwan, ROC

Index

www.ingramcontent.com/pod-product-compliance
Lightning Source LLC
Chambersburg PA
CBHW061939190326
41458CB00009B/2776

* 9 7 8 1 6 3 2 3 8 5 5 7 4 *